CAMBRIDGE MONOGRAPHS ON MATHEMATICAL PHYSICS

General editors: P.V. Landshoff, D.R. Nelson, D.W. Sciama, S. Weinberg

STATISTICAL FIELD THEORY

Volume 1: From Brownian motion to renormalization
and lattice gauge theory

STATISTICAL FIELD THEORY

Volume 1
From Brownian motion to renormalization and lattice gauge theory

CLAUDE ITZYKSON
JEAN-MICHEL DROUFFE

Commissariat à l'Energie Atomique
Centre d'Etudes Nucléaires de Saclay
Service de Physique Théorique

CAMBRIDGE
UNIVERSITY PRESS

CAMBRIDGE UNIVERSITY PRESS

Cambridge, New York, Melbourne, Madrid, Cape Town, Singapore,
São Paulo, Delhi, Dubai, Tokyo, Mexico City

Cambridge University Press
The Edinburgh Building, Cambridge CB2 8RU, UK

Published in the United States of America by
Cambridge University Press, New York

www.cambridge.org
Information on this title: www.cambridge.org/9780521408059

First published 1989
First paperback edition 1991
Reprinted 1992, 1994, 1997

A catalogue record for this publication is available from the British Library

Library of Congress Cataloguing-in-Publication Data is available

ISBN 978-0-521-34058-8 Hardback
ISBN 978-0-521-40805-9 Paperback

Contents

Contents of Volume 2

Preface

Some ten years ago, when completing with J.-B. Zuber a previous text on *Quantum Field Theory*, the senior author was painfully aware that little mention was made that methods in statistical physics and Euclidean field theory were coming closer and closer, with common tools based on the use of path integrals and the renormalization group giving insights on global structures. It was partly to fill this gap that the present book was undertaken. Alas, over the five years that it took to come to life, both subjects have undergone a new evolution. Disordered media, growth patterns, complex dynamical systems or spin glasses are among the new important topics in statistical mechanics, while superstring theory has turned to the study of extended systems, Kaluza–Klein theories in higher dimensions, anticommuting coordinates ... in an attempt to formulate a unified model including all known interactions. New and sophisticated techniques have invaded statistical physics, ranging from algebraic methods in integrable systems to fractal sets or random surfaces. Powerful computers or special devices provide "experimental" means for a new brand of theoretical physicists. In quantum field theory, applications of differential topology, geometry, Riemannian manifolds, operator theory ... require a deeper background in mathematics and a knowledge of some of its most recent developments. As a result, when surveying what has been included in the present volume in an attempt to uncover the basic unity of these subjects, the authors have the same unsatisfactory feeling of not being able to bring the reader really up to date. It is presumably the fate of such endeavours to always come short of accomplishing their purpose.

With these shortcomings fully admitted, we have tried to present to the reader an overview of the main themes which justify the title "Statistical field theory." This interpretation of Euclidean field theory offers a new language, effective computing means, as

xi

well as a natural and consistent short-distance cutoff. In other words, it allows one to give a global meaning to path integrals, to discover possible anomalies arising from integration measures, or to understand in simple terms systems with redundant variables such as gauge models. The theory of continuous phase transitions provides a bridge between probabilistic mechanics and continuous field theory, using the renormalization group to filter out relevant operators and interactions. Many authors contributed to these views, culminating in the work of K. Wilson and his collaborators and followers, which promoted the renormalization group as a universal tool to analyse the large-distance behaviour. It still retains its value, while new developments take place, particularly with conformal, or local scale invariance coming to prominence in the study of two-dimensional systems.

The content of this book is naturally divided into two parts. The first six chapters describe in succession Brownian motion, its anti-commutative counterpart in the guise of Onsager's solution to the two-dimensional Ising model, the mean-field or Landau approximation, scaling ideas exemplified by the Kosterlitz–Thouless theory for the XY-transition, the continuous renormalization group applied to the standard φ^4 theory, the simplest typical case, and lattice gauge theory as an attempt to understand quark confinement in chromodynamics.

The next five chapters (in volume 2) cover more diverse subjects. We give an introduction to strong coupling expansions and various means of analyzing them. We then briefly introduce Monte Carlo simulations with an emphasis on the applications to gauge theories. Next we turn to the significant advances in two-dimensional conformal field theory, with a lengthy presentation of the methods as well as early results. A chapter on simple disordered systems includes sample applications of fermionic techniques with no pretence at completeness. The final chapter is devoted to random geometry and an introduction to the Polyakov model of random surfaces which illustrates the relations between string theory and statistical physics.

At the price of being perhaps a bit repetitive, we have tried in the first part to introduce the subject in an elementary fashion. It is, however, assumed that the reader has some familiarity with thermodynamics as well as with quantum field theory. We often switch from one to the other interpretation, assuming that it will not be disturbing once it is realized that the exponential of the action plays the role of the Boltzmann–Gibbs statistical weight. The last chapters cover subjects still in fast evolution.

Many important subjects could unfortunately not be covered. In random order they include dynamical critical phenomena, renormalization of σ-models or non-Abelian gauge fields except for a mention of lowest order results, topological aspects, classical solutions, instantons, monopoles, anomalies (except for the conformal one). Integrable systems are missing apart from the two-dimensional Ising model. Quantum gravity *à la Regge* is only mentioned. The list could, of course, be made much longer. Our involvement in some of the topics has certainly produced obvious biases and overemphases in certain instances. We have tried, as much as possible, to correct for these defects as well as for the numerous omissions by including at the end of each chapter a section entitled "Notes." Here we quote our sources, original articles, reviews, books and complementary material. These notes are purposely scattered through the volume, as we are sure that our quotations are very incomplete. A fair bibliography in such a large domain is beyond human capacities. Should any one feel that his or her work has not been reported or not properly mentioned, he or she is certainly right and we present our most sincere apologies. On the other hand we did not hesitate to use and sometimes follow very closely some articles or reviews which served our purpose. For instance chapter 5 is built around the definitive contributions of E. Brézin, J.-C. Le Guillou, J. Zinn-Justin and G. Parisi. Except for some further elaboration by the authors themselves, it was futile to try to improve on their work. Further examples are mentioned in the notes. It is the very nature of a survey such as this one to be inspired largely by other people's works. We hope that we did not distort or caricature them.

A book might give the illusion, especially to students, that some knowledge has become definitive and that the authors understand every part of it. This is a completely false view. No one can really fully master even his own subject, and this is luckily a source of progress. It is in the process of learning, of objecting, of finding misprints and errors, in rediscovering for oneself, that one gets the real benefits. It is very likely that, in spite of our care, many errors have crept in here and there. We welcome gladly comments and criticisms.

It was very hard to keep uniform notation throughout the text, even sometimes in the same chapter. This is a standard difficulty, especially when traditional notation in a given domain comes into conflict with those used in another one, and a compromise is necessary. We hope that this will not be a source of confusion for the reader.

We have added appendices which generally gather material in very concise form. They should be supplemented by further reading. For instance appendix C of chapter 9 is obviously insufficient to describe finite and infinite Lie algebras and their representations. This appendix is, rather, meant to induce the interested reader to study the subject further. This is also the nature of several sections where the degree of mathematical sophistication seems to increase beyond the standard background, reflecting recent trends. It was felt difficult to omit these developments but also impossible to give a proper complete introduction.

Included in small type here and there are comments, exercises and short complements ... It was felt inappropriate to develop a scholarly set of problems. In this respect the whole text can be read as a problem book.

One of the "heroes" of the whole subject of statistical physics, in one guise or another, is still to this day our old friend the Ising model. We keep a few bottles of good old French wine for the lucky person who solves it in three dimensions. It would seem appropriate to create in the theoretical physics community a prize for its solution, analogous to the one founded at the beginning of the century for the proof of Fermat's theorem. Both subjects have a similar flavour, being elementary to formulate. While it is to be presumed that the answer itself is to a large extent inessential, they motivated creative efforts (and still do) which go largely beyond the goal of solving the problem itself.

Among the many books which either overlap or amply complement the present one, the foremost are of course those in the series edited by C. Domb and M.S. Green and now J. Lebowitz, entitled *Phase transitions and critical phenomena* and published through the years by Academic Press (New York). We freely refer to this series in the notes. Let us also quote here a few others, again without pretence at exhaustivity. On the statistical side, K. Huang, *Statistical mechanics*, Wiley, New York (1963); H.E. Stanley, *In-*

troduction to phase transitions and critical phenomena, Oxford University Press (1971); S.K. Ma, *Modern theory of critical phenomena*, Benjamin, New York (1976) and *Statistical mechanics*, World Scientific, Singapore (1985); D.J. Amit, *Field theory, the renormalization group and critical phenomena*, 2nd edition, World Scientific, Singapore (1984).

Books on the path integral approach to field theory are by now numerous. Among them, the classical one is R.P. Feynman and A.R. Hibbs, *Quantum mechanics and path integrals*, McGraw Hill, New York (1965). Further aspects are covered in C. Itzykson and J.-B. Zuber, *Quantum field theory*, McGraw Hill, New York (1980); P. Ramond, *Field theory, a modern primer*, Benjamin/Cummings, Reading, Mass. (1981); J. Glimm and A. Jaffe, *Quantum physics*, Springer, New York (1981). To fill some gaps on other developments in field theory, see S. Coleman, *Aspects of symmetries*, Cambridge University Press (1985); S. Treiman, R. Jackiw, B. Zumino, E. Witten *Current algebra and anomalies*, World Scientific, Singapore (1985), and to learn about integrable systems, R. Baxter *Exactly solved models in statistical mechanics*, Academic Press, New York (1982), and M. Gaudin *La fonction d'onde de Bethe*, Masson, Paris (1983). Of course, many more books are mentioned in the notes. We are also aware that several important texts are either in preparation or will appear in the near future.

Our knowledge of English remains to this day very primitive and we apologize for our cumbersome use of a foreign language. This lack of fluency has prevented us of any attempt at humour which would have been sometimes more than welcome.

We would have never undertaken writing, were it not for the teaching opportunities that we were given by several universities and schools. One of the authors (C.I.) is grateful to his colleagues from the "Troisième cycle de Suisse Romande" in Lausanne and Geneva, from the "Département de Physique de l'Université de Louvain La Neuve" and from the "Troisième cycle de physique théorique" in Marseille for giving him the possibility to teach what became parts of this text, as well as to the staff of these institutions for providing secretarial help in preparing a French unpublished manuscript. The other author (J.M.D.) acknowledges similar opportunities afforded by the "Troisième cycle de physique théorique" in Paris.

The final and certainly most pleasant duty is, of course, to thank all those, friends, colleagues, collaborators, students and secretaries who have helped us through the years. A complete list should include all the members of the Saclay "Service de physique théorique", together with its numerous visitors and the members of the many departments, institutions and meetings which offered us generous hospitality and stimulation.

Particular thanks go to our very long time friends and colleagues R. Balian, M. Bander, M. Bauer, D. Bessis, E. Brézin, A. Cappelli, A. Coste, F. David, J. des Cloizeaux, C. De Dominicis, E. Gardner, M. Gaudin, B. Derrida, J.-M. Luck, A. Morel, P. Moussa, H. Orland, G. Parisi, Y. Pomeau, R. Lacaze, H. Saleur, R. Stora, J. Zinn-Justin, and J.-B. Zuber for friendly collaborations, endless discussions and generous advice. The final form of the manuscript owes a great deal to Dany Bunel. Let her receive here our warmest thanks for her tireless help. We are also very grateful to M. Porneuf and to the documentation staff, M. Féron, J. Delouvrier and F. Chétivaux.

Last but not least, we thank the Commissariat à l'Energie Atomique, the Institut de Recherche Fondamentale and the Service de Physique Théorique for their support.

Saclay, 1988

1

FROM BROWNIAN MOTION
TO EUCLIDEAN FIELDS

It may seem surprising to start our study with a description of
Brownian motion. However, this offers an interesting introduction
to the concept of Euclidean quantum field, and an intuitive understanding of the role of dimensionality. The effective (or Hausdorff)
dimension two of Brownian curves is particularly significant. It
means that two such curves fail to intersect, hence to interact, in
dimension higher than four. This is illustrated in the first section
of this chapter, which also discusses the transition from a discrete
to a continuous walk. A similar analysis for interacting fields, pi-
oneered by K. Symanzik, is presented in the second section. It
is related to strong coupling, or high temperature, expansions, to
be studied later, in particular in chapter 6 of this volume and
chapter 7 of volume 2. The introduction of n-component fields
provides the means to incorporate "self-avoiding" walks in the
limit $n \rightarrow 0$. We conclude this chapter with an analysis of elementary one-dimensional systems. This enables us to introduce
the useful concept of transfer matrix.

1.1   Brownian motion

1.1.1   Random walks

We begin with a discussion of random walks on a regular,
infinite lattice in d-dimensional Euclidean space. Each site has
q neighbours, where q is called the *coordination number* of the
lattice. At regular time intervals, separated by an amount $\Delta t = 1$,
a walker jumps from one site towards a neighbouring one, chosen
at random. The probability of landing on any particular adjacent
site is $1/q$. Successive jumps are considered to be statistically
independent events. We choose a (hyper)-cubic lattice generated

1

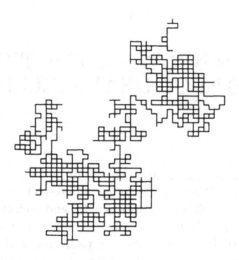

Fig. 1   A two-dimensional random walk of 2000 steps.

by d orthonormal vectors $\mathbf{e}_{(1)}$, ..., $\mathbf{e}_{(d)}$ such that $\mathbf{e}_{(\mu)} \cdot \mathbf{e}_{(\nu)} = \delta_{\mu\nu}$. Sites are located at $\mathbf{x} = x^\mu \mathbf{e}_{(\mu)}$, where the coordinates x^μ are integers. For this lattice, $q = 2d$. Figure 1 shows such a two-dimensional random walk of 2000 steps. Note the presence of large patches where the walker visited all sites.

Let us determine the conditional probability $P(\mathbf{x}_1, t_1; \mathbf{x}_0, t_0)$ for the walker to be on the site $\mathbf{x}_1$ at time t_1, knowing that his initial position was $\mathbf{x}_0$ at time t_0. For $t_1 = t_0$, we have

$$P(\mathbf{x}_1, t_0; \mathbf{x}_0, t_0) = \delta_{\mathbf{x}_1, \mathbf{x}_0} \tag{1}$$

The notation $\delta_{\mathbf{x}_1, \mathbf{x}_0}$ is shorthand for the product $\prod_{\mu=1}^d \delta_{x_1^\mu x_2^\mu}$. The probability P is defined for $t_1 \geq t_0$ and depends only on the differences $t_1 - t_0$, $\mathbf{x}_1 - \mathbf{x}_0$, because of the translational invariance in both time and space. At fixed t_1, the normalization condition reads

$$\sum_{\mathbf{x}_1} P(\mathbf{x}_1, t_1; \mathbf{x}_0, t_0) = 1 \tag{2}$$

and each probability P is either positive or zero.

The discrete formulation depends strongly on the choice of lattice. For instance, if $t_1 - t_0$ is even (odd), then so is the sum of coordinates of $\mathbf{x}_1 - \mathbf{x}_0$. However, we are only interested in those

$t = 0$	$t = 1$	$t = 2$	$t = 3$
			1
			3 . 3
		1	3 . 9 . 3
	1	2 . 2	
1	1 . 1	1 . 4 . 1	1 . 9 . 9 . 1
	1	2 . 2	
		1	3 . 9 . 3
			3 . 3
			1
1	4	16	64

Fig. 2 Relative probabilities for a two-dimensional random walk on a square lattice. The normalization factor is 4^{-t}.

asymptotic properties which are independent of the particular lattice structure chosen.

A recurrence relation between the probabilities P at successive times t and $t+1$ follows from the fact that the walker can reach the point $\mathbf{x}$ only if, one unit of time before, he was on a neighbouring site $\mathbf{x}' = \mathbf{x} \pm \mathbf{e}_{(\mu)}$

$$P(\mathbf{x}, t + 1; \mathbf{x}_0, t_0) = \frac{1}{2d} \sum_{\substack{\mathbf{x}' \\ \text{neighbour of } x}} P(\mathbf{x}', t; \mathbf{x}_0, t_0) \qquad (3)$$

This relation generalizes Pascal's construction for the binomial coefficients, which corresponds to $d = 1$. Now P is completely determined by equations (1) to (3). Figure 2 illustrates the two-dimensional case.

Equation (3) can be rewritten, using a discretized version, Δ_r, of the Laplacian operator Δ,

$$\Delta_r f(\mathbf{x}) = \frac{1}{2d} \sum_{\mu=1}^{d} [f(\mathbf{x} + \mathbf{e}_{(\mu)}) + f(\mathbf{x} - \mathbf{e}_{(\mu)}) - 2f(\mathbf{x})] \qquad (4)$$

as

$$P(\mathbf{x}, t + 1; \mathbf{x}_0, t_0) - P(\mathbf{x}, t; \mathbf{x}_0, t_0) = \Delta_r P(\mathbf{x}, t; \mathbf{x}_0, t_0) \qquad (5)$$

This is a finite difference approximation to the diffusion equation in continuous space

$$\left(\frac{\partial}{\partial t} - \Delta \right) P = 0 \qquad (6)$$

A Fourier transform allows one to solve equations (3) or (5). Writing

$$P(\mathbf{x}, t; \mathbf{x}_0, t_0) = \int_{-\pi}^{\pi} \frac{d^d\mathbf{k}}{(2\pi)^d} e^{i\mathbf{k}\cdot\mathbf{x}} \tilde{P}(\mathbf{k}, t) \tag{7}$$

we deduce

$$\tilde{P}(\mathbf{k}, t+1) = \frac{1}{d} \sum_{\mu=1}^{d} \cos k_\mu \tilde{P}(\mathbf{k}, t), \qquad \text{with} \quad \tilde{P}(\mathbf{k}, t_0) = e^{-i\mathbf{k}\cdot\mathbf{x}_0}$$

as follows from the initial condition at $t = t_0$. Hence the solution is

$$P(\mathbf{x}_1, t_1; \mathbf{x}_0, t_0) = \int_{-\pi}^{\pi} \frac{d^d\mathbf{k}}{(2\pi)^d} e^{i\mathbf{k}\cdot(\mathbf{x}_1-\mathbf{x}_0)} \left(\frac{1}{d} \sum_\mu \cos k_\mu \right)^{t_1-t_0} \tag{8}$$

We want to obtain the asymptotic properties of this solution at large distance with respect to the lattice spacing and for long times. It is convenient to rescale distances and time using a lattice spacing a rather than 1, and a time interval τ rather than 1. Now $\mathbf{e}_{(\mu)} \cdot \mathbf{e}_{(\nu)} = a^2\delta_{\mu\nu}$. Performing the substitutions $t \to t/\tau$, $\mathbf{x} \to \mathbf{x}/a$ and $\mathbf{k} \to a\mathbf{k}$, equation (8) becomes

$$P(\mathbf{x}_1 - \mathbf{x}_0, t_1 - t_0) = a^d \int_{-\pi/a}^{\pi/a} \frac{d^d\mathbf{k}}{(2\pi)^d} e^{i\mathbf{k}\cdot(\mathbf{x}_1-\mathbf{x}_0)} \left(\frac{1}{d} \sum_\mu \cos ak_\mu \right)^{(t_1-t_0)/\tau} \tag{9}$$

We now take the limit as a and τ go to zero, keeping the distances and time intervals fixed. We consider a volume Δx around $\mathbf{x}$ which is large with respect to the elementary lattice volume a^d, but which is also sufficiently small to ensure that P remains nearly constant within Δx; this last requirement is fulfilled if $(t_1 - t_0)/\tau$ is also large. This permits a probability density $p = P/a^d$ to be defined as

$$p(\mathbf{x}_1 - \mathbf{x}_0, t_1 - t_0)\Delta x = \sum_{\mathbf{x}_1' \in \mathbf{x}_1 + \Delta x} P(\mathbf{x}_1' - \mathbf{x}_0; t_1 - t_0)$$

$$\approx \frac{\Delta x}{a^d} P(\mathbf{x}_1 - \mathbf{x}_0, t_1 - t_0) \tag{10}$$

and thus

$$p(\mathbf{x}_1 - \mathbf{x}_0, t_1 - t_0) = \lim_{a,\tau \to 0} \int_{-\pi/a}^{\pi/a} \frac{d^d k}{(2\pi)^d} e^{i\mathbf{k}\cdot(\mathbf{x}_1 - \mathbf{x}_0)} \left(\frac{1}{d} \sum_\mu \cos a k_\mu \right)^{(t_1 - t_0)/\tau} \tag{11}$$

This limit is nontrivial only when a and τ vanish in such a way that the ratio τ/a^2 is kept fixed, as shown by the expansion of the cosine

$$\left(\frac{1}{d} \sum_\mu \cos a k_\mu \right)^{(t_1 - t_0)/\tau} = \left(1 - \frac{a^2}{2d} \mathbf{k}^2 + \cdots \right)^{(t_1 - t_0)/\tau} \to e^{-(t_1 - t_0)\mathbf{k}^2} \tag{12}$$

in which the time scale has been fixed using

$$\tau = \frac{1}{2d} a^2 \tag{13}$$

Hence

$$p(\mathbf{x}_1 - \mathbf{x}_0, t_1 - t_0) = \int_{-\infty}^{+\infty} \frac{d^d k}{(2\pi)^d} \exp[-(t_1 - t_0)\mathbf{k}^2 + i(\mathbf{x}_1 - \mathbf{x}_0) \cdot \mathbf{k}]$$

$$= \frac{1}{[4\pi(t_1 - t_0)]^{d/2}} \exp\left(-\frac{(\mathbf{x}_1 - \mathbf{x}_0)^2}{4(t_1 - t_0)} \right) \tag{14}$$

This is the well-known kernel of the diffusion equation in continuous space. It is positive, symmetric, and satisfies

$$\int d^d \mathbf{x}\, p(\mathbf{x}, t; \mathbf{x}_0, t_0) = 1 \tag{15a}$$

$$\lim_{t_1 \to t_0} p(\mathbf{x}_1, t_1; \mathbf{x}_0, t_0) = \delta^d(\mathbf{x}_1 - \mathbf{x}_0) \tag{15b}$$

$$\left(\frac{\partial}{\partial t} - \Delta \right) p(\mathbf{x}, t; \mathbf{x}_0, t_0) = 0 \tag{15c}$$

$$t_2 > t_1 > t_0 \Rightarrow$$

$$\int d^d \mathbf{x}_1\, p(\mathbf{x}_2, t_2; \mathbf{x}_1, t_1) p(\mathbf{x}_1, t_1; \mathbf{x}_0, t_0) = p(\mathbf{x}_2, t_2; \mathbf{x}_0, t_0) \tag{15d}$$

Condition (15a) is the conservation law for the probabilities (this is the continuous counterpart of equation (2)), while (15b) describes the initial conditions. Equation (15c) is the diffusion equation. Finally, (15d) expresses the obvious fact that the walker was certainly somewhere at an intermediate time t_1. This last relation, characteristic of a Markov process, is compatible with the convolution properties of Gaussian integrals and will be used later to construct a path integral representation which is the continuous counterpart of a similar property on the lattice.

The continuous diffusion law (14) is isotropic and translationally invariant, whereas the discrete version only presents the cubic symmetries. In the limiting procedure, we fixed the physical time intervals $t_1 - t_0$ and distances $\mathbf{x}_1 - \mathbf{x}_0$ and considered the case where the spatial and time steps a and τ vanished, with the ratio τ/a^2 kept fixed. To be quite rigorous, $t_1 - t_0$ and the components of $\mathbf{x}_1 - \mathbf{x}_0$ should be multiples of τ and a respectively. Of course, we could have proceeded differently, taking $\mathbf{x}_1 - \mathbf{x}_0$ and $t_1 - t_0$ large with respect to the spacings. The view adopted above, namely to let a and τ go to zero, is comparable to the ultraviolet limit in field theory, where a cutoff factor (here $1/a$, a natural scale for momenta $\mathbf{k}$) tends to infinity, whereas physical (measurable) quantities are kept fixed. We also used here a cutoff τ^{-1} for frequencies, with $\tau^{-1} \sim a^{-2}$. This will be later interpreted by considering that the typical Brownian curve has a "dimension" 2; this is called the *Hausdorff dimension*. When the curve is followed at a constant speed a/τ, the typical distance behaves as $|\mathbf{x}_1 - \mathbf{x}_0| \sim (t_1 - t_0)^{\frac{1}{2}}$. In other words, if a Brownian curve is contained almost certainly in a volume Ω and if its length $t_1 - t_0$ is multiplied by λ, the volume subsequently needed to enclose it will be $\lambda^{d/2}\Omega$. Defining an exponent ν characteristic of the end-to-start distance by

$$|\mathbf{x}_1 - \mathbf{x}_0| \sim (t_1 - t_0)^{\nu} \tag{16}$$

the value of ν for Brownian motion is therefore

$$\nu = \frac{1}{2} \tag{17}$$

This is a first example of a so-called *critical exponent*. Much more will be said on this subject in the following chapters.

The result (16)-(17) is explained in an elementary way by the central limit theorem of probability theory. Consecutive steps are uncorrelated random variables, and the mean quadratic sum of distances behaves as the number of terms, i.e. the number of steps.

Verify that a random walk on a two-dimensional triangular lattice yields the same continuous limit (14) as on a square lattice.

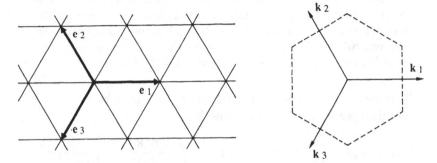

Fig. 3   The triangular lattice and its first Brillouin zone.

On a triangular lattice, each site has $q = 6$ neighbours. It is convenient, as shown in figure 3, to use a redundant basis of 3 vectors $\mathbf{e}_{(i)}$ such that $\sum \mathbf{e}_{(i)} = 0$, $\mathbf{e}_{(i)}^2 = 1$, $\mathbf{e}_{(i)} \cdot \mathbf{e}_{(j)} = -\frac{1}{2}$ $(i \neq j)$. The conditional probability that the walker is on site $\mathbf{x}$ at time $t+1$ is expressed as

$$P(\mathbf{x}, t+1) = \frac{1}{6} \sum_{i=1}^{3} P(\mathbf{x} + \mathbf{e}_{(i)}, t) + P(\mathbf{x} - \mathbf{e}_{(i)}, t)$$

Lattice sites are labelled by three integers $\{n_i\}$, with $\mathbf{x} = \sum n_i \mathbf{e}_{(i)}$. A constant can be added to the n_i's without changing $\mathbf{x}$. Similarly, the conjugate quantity $\mathbf{k} = \sum k_i \mathbf{e}_{(i)}$ is characterized by the three numbers k_i, on which we impose the additional constraint $\sum k_i = 0$. Therefore,

$$\mathbf{k} \cdot \mathbf{x} = \frac{3}{2} \sum k_i n_i \quad , \quad \mathbf{k}^2 = \frac{3}{2}(k_1^2 + k_2^2 + k_3^2)$$

The integration domain for $\mathbf{k}$ is the hexagon $|k_i| < \pi$, the area of which is $2\pi^2 \sqrt{3}$. Consequently,

$$P(\mathbf{x}, t) = \int_{\text{hexagon}} \frac{d^2 \mathbf{k}}{2\pi^2 \sqrt{3}} e^{i\mathbf{k} \cdot \mathbf{x}} \left[\frac{1}{3} \left(\cos \frac{3}{2} k_1 + \cos \frac{3}{2} k_2 + \cos \frac{3}{2} k_3 \right) \right]^t$$

Changing the scale according to $\mathbf{x} \to \mathbf{x}/a$, $\mathbf{k} \to \mathbf{k}a$, $t \to t/\tau$, $t = a^2/8$, we obtain in the limit $a \to 0$ the same result as before

$$p(\mathbf{x}, t) = \int \frac{d^2 \mathbf{k}}{(2\pi)^2} e^{-t\mathbf{k}^2 + i\mathbf{k} \cdot \mathbf{x}} = \frac{1}{4\pi t} e^{-\mathbf{x}^2/4t}$$

using the fact that there are $2\Delta x/a^2 \sqrt{3}$ lattice sites in an area Δx. This confirms the universality of the isotropic limit.

In arbitrary dimension the previous construction generalizes into the projection of a $(d+1)$-dimensional hypercubic lattice on the hyperplane $x_1 + \cdots + x_{d+1} = 0$. Its dual is the cross section by this hyperplane. When $d = 3$, this gives the body centered (b.c.c.) cubic lattice dual to the face centered cubic (f.c.c.) lattice.

The mean time spent on a given site (which may be either finite or infinite) leads to the Green function

$$G(\mathbf{x}_1 - \mathbf{x}_0) = \sum_{t_1=t_0}^{\infty} P(\mathbf{x}_1, t_1; \mathbf{x}_0, t_0) = \int_{-\pi}^{\pi} \frac{d^d\mathbf{k}}{(2\pi)^d} \frac{e^{i\mathbf{k}\cdot(\mathbf{x}_1-\mathbf{x}_0)}}{1 - \dfrac{1}{d}\sum_{\mu} \cos k_{\mu}}$$

(18)

which obeys

$$G(\mathbf{x}_1 - \mathbf{x}_0) = \delta_{\mathbf{x}_1,\mathbf{x}_0} + \frac{1}{2d}\sum_{\mu} G(\mathbf{x}_1 + \mathbf{e}_{(\mu)} - \mathbf{x}_0) + G(\mathbf{x}_1 - \mathbf{e}_{(\mu)} - \mathbf{x}_0)$$

(19)

or, equivalently,

$$\Delta_r G(\mathbf{x}_1 - \mathbf{x}_0) = -\delta_{\mathbf{x}_1,\mathbf{x}_0}$$

(20)

This relation justifies the name of *Green function* for the operator $-\Delta_r$; G is symmetric, positive, and well defined for $d > 2$. When $d = 1$ or 2, the mean time spent on a given site is infinite, as can be seen from the infrared divergence of the integral (18). The integrand indeed behaves as $d^d\mathbf{k}/\mathbf{k}^2$ when $\mathbf{k}$ is small. However, one can introduce a subtracted quantity

$$G^s(\mathbf{x}_1 - \mathbf{x}_0) = G(\mathbf{x}_1 - \mathbf{x}_0) - G(0)$$

$$= \sum_{t_1=t_0}^{\infty} P(\mathbf{x}_1, t_1; \mathbf{x}_0, t_0) - P(\mathbf{x}_0, t_1; \mathbf{x}_0, t_0)$$

$$= \int \frac{d^d\mathbf{k}}{(2\pi)^d} \frac{\cos \mathbf{k}\cdot(\mathbf{x}_1 - \mathbf{x}_0) - 1}{1 - \dfrac{1}{d}\sum_{\mu}\cos k_{\mu}}$$

(21)

which is now well-defined whatever the value of d. It still satisfies equation (20), since we have only formally subtracted a (possibly infinite) constant. However, the positivity property is lost, as shown by the one-dimensional result

$$d = 1 \qquad\qquad G^s(\mathbf{x}_1 - \mathbf{x}_0) = -|\mathbf{x}_1 - \mathbf{x}_0|$$

(22)

We define the generating function for the probabilities

$$G(\mathbf{x}_1 - \mathbf{x}_0, \lambda) = \lambda \sum_{t_1 = t_0}^{\infty} \lambda^{t_1 - t_0} P(\mathbf{x}_1, t_1; \mathbf{x}_0, t_0) \qquad (23)$$

which coincides for $\lambda = 1$ with the Green function previously considered. When $|\lambda| < 1$, the sum is always convergent and has an integral representation

$$G(\mathbf{x}_1 - \mathbf{x}_0, \lambda) = \int \frac{d^d \mathbf{k}}{(2\pi)^d} \frac{e^{i\mathbf{k} \cdot (\mathbf{x}_1 - \mathbf{x}_0)}}{\lambda^{-1} - \frac{1}{d} \sum_{\mu} \cos k_{\mu}} \qquad (24)$$

It satisfies the relation

$$\left[-\Delta_r + (\lambda^{-1} - 1) \right] G(\mathbf{x}, \lambda) = \delta_{\mathbf{x}, \mathbf{0}} \qquad (25)$$

This function is the analog of the massive propagator in particle physics. Introducing again the lattice spacing $\mathbf{x} \to \mathbf{x}/a$, $\mathbf{k} \to \mathbf{k}a$, the continuous limit is obtained for $1/\lambda \equiv 1 + m^2 a^2 / 2d$, in the limit $a \to 0$

$$g(\mathbf{x}, m^2) = \lim_{a \to 0} \frac{1}{2da^{d-2}} G\left(\frac{\mathbf{x}}{a}, \lambda \right) = \int_{-\infty}^{+\infty} \frac{d^d \mathbf{k}}{(2\pi)^d} \frac{e^{i\mathbf{k} \cdot \mathbf{x}}}{\mathbf{k}^2 + m^2} \qquad (26)$$

1.1.2 The sum over paths

The properties of Brownian motion can be understood in terms of sums over paths. This seems quite natural in the present context, and will later be the origin of fruitful generalizations. This interpretation follows directly from relation (3): a path of length $t + 1$ is obtained by adding one step to a path of length t. Iterating this relation leads to

$$P(\mathbf{x}_1, t_1; \mathbf{x}_0, t_0) = \frac{\text{Number of paths joining } \mathbf{x}_0 \text{ to } \mathbf{x}_1 \text{ with } t_1 - t_0 \text{ steps}}{\text{Number of paths originating from } \mathbf{x}_0 \text{ with } t_1 - t_0 \text{ steps}}$$

$$(27)$$

The denominator is equal to $(2d)^{t_1 - t_0} = q^{t_1 - t_0}$. Another derivation follows from formula (8), when one replaces cosines by exponentials, expands the integrand and then integrates over $\mathbf{k}$ term by term. This equality is interpreted by saying that all paths covered during the time $t_1 - t_0$ are equally probable. Formula

(27) can also be rewritten as

$$P(\mathbf{x}_1, t_1; \mathbf{x}_0, t_0) = \frac{1}{(2d)^{t_1-t_0}} \sum_{\substack{x(t) \\ t=1,\ldots,t_1-t_0-1}} \prod_{t=0}^{t_1-t_0-1} \delta_{1, \sum_{\mu=1}^{d} |x^\mu(t+1) - x^\mu(t)|}$$

(28)

which states explicitly that elementary steps have unit length, and $\mathbf{x}(0) = \mathbf{x}_0$, $\mathbf{x}(t_1 - t_0) = \mathbf{x}_1$. However, this expression does not yield easily the continuous limit. Let us use a more compact notation. Let ω be a path joining $\mathbf{x}_0$ to $\mathbf{x}_1$. By definition, this is a set of lattice links such that its boundary $\partial\omega$, defined as the set of sites shared by an odd number of links of ω, is $\{\mathbf{x}_0, \mathbf{x}_1\}$. To be precise, the definition of ω implies a prescription on how to describe it as a function of time. Equation (27) is thus rewritten as

$$P(\mathbf{x}_1, t_1; \mathbf{x}_0, t_0) = \sum_{\substack{\omega, \partial\omega = \{\mathbf{x}_0, \mathbf{x}_1\} \\ |\omega| = t_1 - t_0}} \left(\frac{1}{2d}\right)^{|\omega|}$$

(29)

where $|\omega|$ is the number of links of ω. These expressions are easily generalized to the Green function, and, from (23) with $\lambda < 1$,

$$G(\mathbf{x}_1 - \mathbf{x}_0, \lambda) = \lambda \sum_{\omega, \partial\omega = \{\mathbf{x}_0, \mathbf{x}_1\}} \left(\frac{\lambda}{2d}\right)^{|\omega|}$$

(30)

The mean time spent at $\mathbf{x}_1$ is obtained in the limit $\lambda \to 1$.

This formula would suggest that its continuous counterpart implies a sum over weighted paths, where the weight should be an exponential of the form $\exp(-A|\omega|)$. With an abusive extension of the language used in quantum mechanics, the argument of the exponential (up to its sign) is called the *action*. In equilibrium statistical mechanics, it would be identified (up to a factor $\beta = 1/kT$) with the energy of a classical configuration (here a path going from $\mathbf{x}_0$ to $\mathbf{x}_1$). In any case, the correct procedure to reach the continuous asymptotic limit is to group together a very large number of infinitesimal terms when the spacing a goes to zero. This results in a very different form of the action. The preceding formulae are not really adapted to this operation. We shall rather proceed as follows.

Let us remark that (with $a^2 = 2d\tau$)

$$g(\mathbf{x}, m^2) = \lim_{a \to 0} \frac{1}{2da^{d-2}} G\left(\frac{\mathbf{x}}{a}, \lambda \equiv \left(1 + \frac{m^2 a^2}{2d}\right)^{-1}\right)$$

$$= \lim_{a \to 0} \frac{1}{2da^{d-2}} \sum_{t/\tau=0}^{\infty} \left(1 + \frac{m^2 a^2}{2d}\right)^{-t/\tau} P\left(\frac{\mathbf{x}}{a}, \frac{t}{\tau}; 0, 0\right)$$

$$= \lim_{a \to 0} \frac{a^2}{2d} \sum_{t/\tau=0}^{\infty} \exp\left(-\frac{m^2 a^2 t}{2d\tau}\right) p(\mathbf{x}, t)$$

$$= \int_0^{\infty} dt\, e^{-m^2 t} p(\mathbf{x}, t)$$

$$(31)$$

One can also obtain this result directly by combining equations (26) and (14). We recall the convolution property (15*d*)

$$p(\mathbf{x}_1, t_1; \mathbf{x}_0, t_0) = \int d^d\mathbf{x}\, p(\mathbf{x}_1, t_1; \mathbf{x}, t) p(\mathbf{x}, t; \mathbf{x}_0, t_0) \qquad t_0 < t < t_1$$

$$(32)$$

Dividing $t_1 - t_0$ into an arbitrary number of intervals and using the expression (14), we write

$$p(\mathbf{x}_f, t_f; \mathbf{x}_i, t_i) = \int \prod_{j=1}^{n-1} \frac{d^d\mathbf{x}_j}{\left[4\pi(t_{j+1} - t_j)\right]^{d/2}} \frac{1}{\left[4\pi(t_1 - t_i)\right]^{d/2}}$$

$$\times \exp\left[-\frac{1}{4} \sum_{j=0}^{n-1} \frac{(\mathbf{x}_{j+1} - \mathbf{x}_j)^2}{t_{j+1} - t_j}\right]$$

$$(33)$$

(with $\mathbf{x}_0 = \mathbf{x}_i$, $\mathbf{x}_n = \mathbf{x}_f$). This gives a precise meaning to the symbolic formula

$$p(\mathbf{x}_f, t_f; \mathbf{x}_i, t_i) = \int_{\substack{\mathbf{x}(t_i)=\mathbf{x}_i \\ \mathbf{x}(t_f)=\mathbf{x}_f}} \mathcal{D}\mathbf{x}(t) \exp\left[-\frac{1}{4} \int_{t_i}^{t_f} dt\, \dot{\mathbf{x}}^2\right]$$

$$(34)$$

where $\mathcal{D}\mathbf{x}(t)$ implies integrals on the positions at intermediate times, normalized as indicated in (33). Equation (31) leads to a similar relation for $g(\mathbf{x}, m^2)$

$$g(\mathbf{x}, m^2) = \int_0^{\infty} dt \int_{\substack{\mathbf{x}(0)=0 \\ \mathbf{x}(t)=\mathbf{x}}} \mathcal{D}\mathbf{x}(t') \exp\left[-\int_0^t dt'(m^2 + \frac{1}{4}\dot{\mathbf{x}}(t')^2)\right]$$

$$(35)$$

These path integrals in the continuum – which, despite their symbolic appearance, have a rigorous meaning indicated by formula (33) – seem quite different from their discrete counterparts. The action is now $\int_0^t dt'(m^2 + \frac{1}{4}\dot{\mathbf{x}}(t')^2)$, with a distance scale defined by the inverse of m. This form comes from the short-distance behaviour of the continuous limit $p(\mathbf{x}, t) \sim \exp(-\mathbf{x}^2/4t)$, and not from its discrete partner P. Indeed, a small increment in the continuum is the sum of an infinite number of lattice steps. It should be pointed out that, in the continuous notation, the length of a path would be $\int_0^t dt' \sqrt{\dot{\mathbf{x}}(t')^2}$. Finally, the discrete expressions were independent from a given parametrization of the path. This is not the case in the action involved in (35).

1.1.3   The dimension two of Brownian curves

In three-dimensional space, surfaces generically intersect along curves. In four dimensions, they meet only at isolated points. They do not intersect for higher dimensions, apart from exceptional cases. From this point of view, we will see in this section that the characteristics of Brownian motion look more similar to those of a surface than to those of a curve. These properties will be the origin of an intuitive understanding of some fundamental features of critical phenomena.

To be more specific, we consider again a random walk on a cubic lattice, with the probability of being on site $\mathbf{x}$ at time t given by equation (8). What is the probability for the walker to have visited the site $\mathbf{x}$ during the time interval t, assuming he was at the origin at time $t = 0$? If $\mathbf{x} = \mathbf{0}$, this is conventionnally the probability of returning to the origin, excluding the starting time. Note that the answer is *not* the sum over t of $P(\mathbf{x}, t)$; as we already pointed out, this sum is not a probability, but the mean time the walker spends at site $\mathbf{x}$. However, there are close relations between these two quantities.

Let $P_i(\mathbf{x}, t)$ be the probability of being on site $\mathbf{x}$ at time t for the ith time. As we deal with exclusive events, it is clear that

$$P(\mathbf{x}, t) = \delta_{t,0}\delta_{\mathbf{x},\mathbf{0}} + \sum_{i=1}^{\infty} P_i(\mathbf{x}, t) \qquad (36)$$

Only a finite number of terms contribute to this sum since $P_i(\mathbf{x}, t)$ vanishes whenever $i > t$. We deduce a recurrence relation from the fact that the $(i+1)$th visit occurs just after the ith one. Using the spatial homogeneity of the motion, this leads to

$$P_{i+1}(\mathbf{x}, t) = \sum_{t_1+t_2=t} P_i(\mathbf{x}, t_1) P_1(\mathbf{0}, t_2) \qquad (37)$$

Summing over all possible values $i = 1, 2, ...$, one finds

$$P(\mathbf{x}, t) - P_1(\mathbf{x}, t) - \delta_{\mathbf{x},0}\delta_{t,0} = \sum_{t_1+t_2=t} P(\mathbf{x}, t_1) P_1(\mathbf{0}, t_2) - \delta_{\mathbf{x},0} P_1(\mathbf{x}, t)$$

$$(38)$$

The required quantity, i.e. the probability $\Pi(\mathbf{x})$ of visiting the point $\mathbf{x}$ at least once, is the sum over all times of the (mutually exclusive) probabilities of visiting this point for the first time. It is convenient to introduce the generating function

$$\Pi(\mathbf{x}, \lambda) = \sum_{t=0}^{\infty} \lambda^t P_1(\mathbf{x}, t) \qquad (|\lambda| \le 1) \qquad (39)$$

so that

$$\Pi(\mathbf{x}) = \Pi(\mathbf{x}, \lambda = 1) \qquad (40)$$

Taking into account the definition of the propagator (33), relation (38) allows us to write

$$\lambda^{-1} G(\mathbf{x}, \lambda) = \delta_{\mathbf{x},0} + (1 - \delta_{\mathbf{x},0})\Pi(\mathbf{x}, \lambda) + \lambda^{-1} G(\mathbf{x}, \lambda)\Pi(\mathbf{0}, \lambda) \quad (41)$$

The additional factor λ in the definition of G is the origin of the combination $\lambda^{-1}G$. Considering the two cases $\mathbf{x} = \mathbf{0}$ and $\mathbf{x} \ne \mathbf{0}$ in (41) separately, we obtain a new remarkable interpretation of the propagator in terms of the probability Π

$$\Pi(\mathbf{0}, \lambda) = 1 - \frac{\lambda}{G(\mathbf{0}, \lambda)} \qquad (42)$$

$$\Pi(\mathbf{x}, \lambda) = \frac{G(\mathbf{x}, \lambda)}{G(\mathbf{0}, \lambda)} \qquad \text{for } \mathbf{x} \ne \mathbf{0} \qquad (43)$$

If $d \le 2$, the mean time spent on a given site is infinite, hence $G(\mathbf{x}, \lambda) = \infty$, $\Pi(\mathbf{x}, \lambda) = 1$ in the limit $\lambda \to 1$ for all $\mathbf{x}$. As was previously stated, the walker visits any point with unit probability (Pólya, 1921), according to the well-known adage *all roads lead to Rome*. When $d > 2$, the probability to return to the origin

$\Pi(\mathbf{0}) = 1 - G(\mathbf{0}, 1)^{-1}$ is positive, less than unity and decreases as $1/d$ for large d. Its expansion in $1/d$ follows from equation (18)

$$G(\mathbf{0},1) = 1 + \frac{1}{2d} + 3\left(\frac{1}{2d}\right)^2 + 12\left(\frac{1}{2d}\right)^3 + 60\left(\frac{1}{2d}\right)^4 +$$
$$+ 355\left(\frac{1}{2d}\right)^5 + 2380\left(\frac{1}{2d}\right)^6 + 17430\left(\frac{1}{2d}\right)^7 + \qquad (44)$$
$$+ 134190\left(\frac{1}{2d}\right)^8 + \mathcal{O}\left(\left(\frac{1}{2d}\right)^9\right)$$

This states that, for d large, the time spent at the origin tends towards the time interval between two jumps. Correspondingly, the probability of returning to the origin is very small

$$\Pi(\mathbf{0}) = \left(\frac{1}{2d}\right) + 2\left(\frac{1}{2d}\right)^2 + 7\left(\frac{1}{2d}\right)^3 + 35\left(\frac{1}{2d}\right)^4 + 215\left(\frac{1}{2d}\right)^5 +$$
$$+ 1501\left(\frac{1}{2d}\right)^6 + 11354\left(\frac{1}{2d}\right)^7 + 88978\left(\frac{1}{2d}\right)^8 + \mathcal{O}\left(\left(\frac{1}{2d}\right)^9\right)$$
$$\qquad (45)$$

On the other hand, in the vicinity of $d = 2$, we have

$$G(\mathbf{0},1) = \frac{2}{\pi(d-2)} + \cdots, \qquad \Pi(\mathbf{0}) = 1 - \frac{1}{2}\pi(d-2) + \cdots \quad (46)$$

These two expressions are plotted as a function of d in figure 4.

Similarly, for a point distinct from the origin, we have $\Pi(\mathbf{x}) = 1$ whenever $d \leq 2$. If $d > 2$, this probability is less than unity and decreases with the distance to the origin. Asymptotically,

$$\Pi(\mathbf{x}) \underset{|\mathbf{x}|\to\infty}{\longrightarrow} (1 - \Pi(\mathbf{0}))\frac{2d}{(d-2)S_d} \times \frac{1}{|\mathbf{x}|^{d-2}} \qquad (47)$$

where S_d is the area of the unit sphere in d-dimensional space

$$S_d = \frac{2\pi^{d/2}}{\Gamma(d/2)} \qquad (48)$$

Notice also the one-dimensional relations

$$G(\mathbf{0},\lambda) = \int_{-\pi}^{\pi} \frac{dq}{2\pi} \frac{1}{1 - \lambda\cos q} = \frac{1}{\sqrt{1-\lambda^2}}$$
$$\Pi(\mathbf{0},\lambda) = 1 - \sqrt{1-\lambda^2} \qquad (49)$$

on which it is explicitly seen that $G(\mathbf{0},1) = +\infty$ and $\Pi(\mathbf{0}) = 1$.

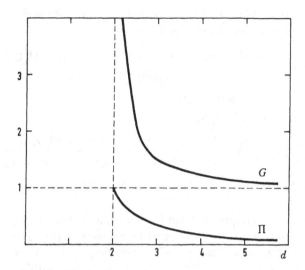

Fig. 4 $G(\mathbf{0}, 1)$ and $\Pi(\mathbf{0})$ as functions of d.

The mean time elapsed before the first visit to the site $\mathbf{x}$ (which is, of course, the time elapsed before the first return to the origin for $\mathbf{x} = \mathbf{0}$) can also be computed (see (39)) as

$$t(\mathbf{x}) = \frac{\sum\limits_{t=0}^{\infty} t P_1(\mathbf{x}, t)}{\Pi(\mathbf{x})} = \lambda \frac{\partial}{\partial \lambda} \ln \Pi(\mathbf{x}, \lambda)\bigg|_{\lambda=1} \qquad (50)$$

This time is infinite for any dimension less than or equal to four. This fact might seem rather surprising at first sight.

It is possible to give a more intuitive proof of formulae (42) and (43), using the fact that $G(\mathbf{x}, 1)$ is the mean time spent at site $\mathbf{x}$ (including the first time interval for $\mathbf{x} = \mathbf{0}$). Indeed, as $\Pi(\mathbf{0})$ is the probability of returning to the origin and $1 - \Pi(\mathbf{0})$ the probability of never returning, the average $G(\mathbf{0}, 1)$ can be decomposed into terms corresponding to 0, 1, 2,... visits with probabilities $1 - \Pi(\mathbf{0})$, $\Pi(\mathbf{0})(1 - \Pi(\mathbf{0}))$, $\Pi^2(\mathbf{0})(1 - \Pi(\mathbf{0}))$, ...; hence

$$G(\mathbf{0}, 1) = 1[1 - \Pi(\mathbf{0})] + 2\Pi(\mathbf{0})[1 - \Pi(\mathbf{0})] + 3\Pi^2(\mathbf{0})[1 - \Pi(\mathbf{0})] + \cdots$$
$$= [1 - \Pi(\mathbf{0})][1 - \Pi(\mathbf{0})]^{-2}$$
$$= [1 - \Pi(\mathbf{0})]^{-1} \qquad (51)$$

which is nothing but equation (42) for $\lambda = 1$. Similarly, at a point $\mathbf{x}$ distinct from the origin, we have

$$
\begin{aligned}
G(\mathbf{x}, 1) &= 1\Pi(\mathbf{x})[1 - \Pi(\mathbf{0})] + 2\Pi(\mathbf{x})\Pi(\mathbf{0})[1 - \Pi(\mathbf{0})] \\
&\quad + 3\Pi(\mathbf{x})\Pi^2(\mathbf{0})[1 - \Pi(\mathbf{0})] + \cdots \\
&= \Pi(\mathbf{x})[1 - \Pi(\mathbf{0})]^{-1} \\
&= \Pi(\mathbf{x})G(\mathbf{0}, 1) \tag{52}
\end{aligned}
$$

$$
\Pi(\mathbf{x}) = \frac{G(\mathbf{x}, 1)}{G(\mathbf{0}, 1)} = \frac{mean\ time\ spent\ at\ \mathbf{x}}{mean\ time\ spent\ at\ the\ origin} \tag{53}
$$

which completes this proof of equation (43).

A more refined quantity can be defined as the mean time $T(\mathbf{x}, t)$ spent at point $\mathbf{x}$ up to time t, which coincides with the mean number of visits at $\mathbf{x}$ up to time t. Of course, $\lim_{t \to \infty} T(\mathbf{x}, t) = G(\mathbf{x}, 1)$. Using a generating function for this quantity, one obtains

$$
\begin{aligned}
T(\mathbf{x}, \lambda) &= \sum_{t=0}^{\infty} \lambda^t T(\mathbf{x}, t) = \sum_{t=0}^{\infty} \lambda^t \sum_{t'=0}^{t} P(\mathbf{x}, t') \\
&= \sum_{t=0}^{\infty} \sum_{t'=0}^{\infty} \lambda^{t''} \lambda^{t'} P(\mathbf{x}, t') = \frac{1}{1-\lambda} \lambda^{-1} G(\mathbf{x}, \lambda)
\end{aligned} \tag{54}
$$

Let us now examine the intersection probabilities. A total number t of sites have been visited during the time t, some of them several times. Let $D(t)$ be the mean number of *distinct* sites visited during time t. The difference $E(t) = t - D(t)$ is a weighted average of the number of sites visited more than once, with a weight 1 if visited twice, 2 if visited three times, ... This weight is easily interpreted if the walker proceeds as Tom Thumb, leaving a stone at each visited site; the weight is the number of stones found at the last visit at a given site. Let us recall that $P_1(\mathbf{x}, t)$ is the probability for the first visit to site $\mathbf{x}$ at time t. The average number of newly visited sites at time t is hence $\sum_{\mathbf{x} \neq \mathbf{0}} P_1(\mathbf{x}, t)$, and therefore

$$
D(t) = 1 + \sum_{t'=1}^{t} \sum_{\mathbf{x} \neq \mathbf{0}} P_1(\mathbf{x}, t') \tag{55}
$$

The origin, already visited at time 0, contributes 1 in the r.h.s. The sum can be extended to $t' = 0$ since $P_1(\mathbf{x} \neq \mathbf{0}, 0) = 0$. We can compute the generating function

$$D(\lambda) = \sum_{t=0}^{\infty} \lambda^t D(t) = \sum_{t=0}^{\infty} \lambda^t + \sum_{t''=0}^{\infty} \lambda^{t''} \sum_{t'=0}^{\infty} \lambda^{t'} \sum_{\mathbf{x}\neq 0} P_1(\mathbf{x}, t')$$

$$= \frac{1}{1-\lambda} \left(1 + \sum_{\mathbf{x}\neq 0} \Pi(\mathbf{x}, \lambda) \right) = \frac{1}{1-\lambda} \left(1 + \sum_{\mathbf{x}\neq 0} \frac{G(\mathbf{x}, \lambda)}{G(\mathbf{0}, \lambda)} \right)$$

$$(56)$$

Using the relation

$$\sum_{\mathbf{x}} G(\mathbf{x}, \lambda) = \frac{\lambda}{1-\lambda}$$

one obtains

$$D(\lambda) = \frac{\lambda}{(1-\lambda)^2 G(\mathbf{0}, \lambda)}$$

$$E(\lambda) = \frac{\lambda}{(1-\lambda)^2} \left(1 - \frac{1}{G(\mathbf{0}, \lambda)} \right)$$

$$(57)$$

It is remarkable that one can find a closed expression for the generating function. The positive numbers $D(t)$ increase with t and therefore their behaviour as t goes to infinity can be deduced from the behaviour of $D(\lambda)$ in the neighborhood of $\lambda = 1$. Table I summarizes the various behaviours resulting from the expression of $G(\mathbf{0}, \lambda)$ near $\lambda = 1$, which depends on the dimension. For $d < 2$, $G(\mathbf{0}, \lambda)$ diverges with a power law. The singularity becomes logarithmic at $d = 2$. In dimension greater than two, a similar behaviour is seen for the correction to the finite limit. The table uses a constant

$$K = \Gamma\left(1 - \frac{d}{2}\right)\left(\frac{d}{2\pi}\right)^{d/2}$$

$$(58)$$

positive for $0 < d < 2$, negative for $2 < d < 4$. Only the dominant terms are displayed. While looking at this table (in which dimensions two and four appear as limiting cases), one is led to some remarks. An intuitive estimation of $D(t)$ could be provided by counting the number of sites in an effective volume $\sim t^{d/2}$ (of extension $R \sim \sqrt{t}$). This estimate is correct for $d < 2$, but fails as soon as $t^{d/2}$ increases faster than t which is the total number of visited sites. A typical logarithmic correction is found for $d = 2$; for $d > 2$, the dominant term is $tG(\mathbf{0}, 1)^{-1} = (1-\Pi(\mathbf{0}))t$,

Table I. Leading behaviour of the various quantities introduced in the text as $\lambda \to 1$ or $t \to \infty$. The d-dependent coefficient K is defined in equation (58).

	$0<d<2$	$d=2$	$2<d<4$	$d=4$
$G(0,\lambda)$	$K(1-\lambda)^{d/2-1}$	$-\dfrac{1}{\pi}\ln(1-\lambda)$	$G(0,1)+K(1-\lambda)^{d/2-1}$	$G(0,1)+\dfrac{4}{\pi^2}(1-\lambda)\ln(1-\lambda)$
$D(\lambda)$	$K^{-1}(1-\lambda)^{-d/2-1}$	$\dfrac{-\pi(1-\lambda)^{-2}}{\ln(1-\lambda)}$	$G^{-1}(0,1)(1-\lambda)^{-2}-\dfrac{K(1-\lambda)^{d/2-3}}{G(0,1)^2}$	$G^{-1}(0,1)(1-\lambda)^{-2}-\dfrac{4(1-\lambda)^{-1}}{\pi^2G^2(0,1)}\ln(1-\lambda)$
$D(t)$	$\dfrac{K^{-1}t^{d/2}}{\Gamma(1+d/2)}$	$\dfrac{\pi t}{\ln t}$	$G^{-1}(0,1)t-\dfrac{Kt^{2-d/2}}{G^2(0,1)\Gamma(3-d/2)}$	$G^{-1}(0,1)t+\dfrac{4\ln t}{\pi^2G^2(0,1)}$
$E(t)$	$t-\dfrac{K^{-1}t^{d/2}}{\Gamma(1+d/2)}$	$t\left(1-\dfrac{\pi}{\ln t}\right)$	$\Pi(0)t+\dfrac{Kt^{2-d/2}}{G^2(0,1)\Gamma(3-d/2)}$	$\Pi(0)t-\dfrac{4\ln t}{\pi^2G^2(0,1)}$
$I(t)$	$\dfrac{Kt^{2-d/2}}{\Gamma(3-d/2)}-t$	$t\left(\dfrac{\ln t}{\pi}-1\right)$	$t(G(0,1)-1)+\dfrac{Kt^{2-d/2}}{\Gamma(3-d/2)}$	$t(G(0,1)-1)-\dfrac{4\ln t}{\pi^2}$

while the fraction of sites visited several times behaves as $\Pi(\mathbf{0})t$, where $\Pi(\mathbf{0})$ decreases from 1 for $d = 2$ to 0 as d goes to infinity. The corrections to these behaviours are negligible when d is greater than four. These results are in agreement with the dimensionality two of random curves, but show that some care is required when using this concept. The behaviour $\Pi(\mathbf{0})t$ of $E(t)$ is intuitively understood by remarking that, all along the curve (hence providing the factor t), there is a probability $\Pi(\mathbf{0})$ for the path to make a loop. Since this is an overestimate of $E(t)$, the correction is negative.

As $D(t)$, $E(t)$ is of course bounded by t (recall that $G(\mathbf{0}, 1) > 1$). The self-intersections of a Brownian curve can also be discussed by weighting a site visited n times by a factor $n(n-1)/2$ (instead of $n-1$); this quantity counts the mean number of self-intersections of the curve and reads

$$I(t) = \sum_{0 \leq t_1 < t_2 \leq t} P(\mathbf{0}, t_2 - t_1) \tag{59}$$

Its generating function is

$$
\begin{aligned}
I(\lambda) &= \sum_{t=1}^{\infty} \lambda^t I(t) = \sum_{t=1}^{\infty} \lambda^t \sum_{t'=1}^{t} (t - t' + 1) P(\mathbf{0}, t') \\
&= \sum_{t_1=0}^{\infty} (t_1 + 1) \lambda^{t_1} \sum_{t_2=1}^{\infty} \lambda^{t_2} P(\mathbf{0}, t_2) \\
&= \frac{G(\mathbf{0}, \lambda) - \lambda}{\lambda (1 - \lambda)^2}
\end{aligned}
\tag{60}
$$

The dominant behaviour is also displayed in table I. Notice that $E(t)$ and $I(t)$ are similar in large dimensions (because triple and higher order intersections become very rare).

Finally, let us consider the intersection properties of two different Brownian curves in the continuous limit. $\mathcal{P}(\mathbf{x}_1, \mathbf{x}_2, \Lambda)$ will denote the probability for two curves starting from distinct points $\mathbf{x}_1 \neq \mathbf{x}_2$ to intersect on a nonempty set Λ. The length of both curves is assumed infinite. It is clear that

$$\mathcal{P}(\mathbf{x}_1, \mathbf{x}_2, \Lambda) \leq \sum_{\mathbf{z} \in \Lambda} \Pi(\mathbf{z} - \mathbf{x}_1) \Pi(\mathbf{z} - \mathbf{x}_2) \tag{61}$$

since $\Pi(\mathbf{x})$ is the probability, starting from the origin, to visit $\mathbf{x}$. This is only an inequality because the probabilities are not independent as $\mathbf{z}$ varies. Let us now perform a dilatation by a factor $\xi \to \infty$ of the positions and of the domain Λ (with the restrictions imposed by the lattice). This allows the computation of the corresponding continuous quantity $p(\mathbf{x}_1, \mathbf{x}_2, \Lambda)$

$$
p(\mathbf{x}_1, \mathbf{x}_2, \Lambda) = \lim_{\xi \to \infty} \mathcal{P}(\xi \mathbf{x}_1, \xi \mathbf{x}_2, \xi \Lambda) \leq \lim_{\xi \to \infty} \frac{\xi^d}{G(0, 1)^2}
$$

$$
\times \int_\Lambda d^d\mathbf{z} \int_{-\pi}^{\pi} \frac{d^d\mathbf{q}_1 d^d\mathbf{q}_2}{(2\pi)^{2d}} \frac{\exp\left\{i\xi[\mathbf{q}_1 \cdot (\mathbf{x}_1 - \mathbf{z}) + \mathbf{q}_2 \cdot (\mathbf{x}_2 - \mathbf{z})]\right\}}{\left(1 - \frac{1}{d}\sum_\mu \cos q_1^\mu\right)\left(1 - \frac{1}{d}\sum_\mu \cos q_2^\mu\right)}
$$

$$
\leq \lim_{\xi \to \infty} \left(\frac{2d}{G(0, 1)}\right)^2 \xi^{4-d}
$$

$$
\times \int_\Lambda d^d\mathbf{z} \int_{-\infty}^{\infty} \frac{d^d\mathbf{q}_1 d^d\mathbf{q}_2}{(2\pi)^{2d}} \frac{\exp\left\{i[\mathbf{q}_1 \cdot (\mathbf{x}_1 - \mathbf{z}) + \mathbf{q}_2 \cdot (\mathbf{x}_2 - \mathbf{z})]\right\}}{q_1^2 q_2^2}
$$

$$
\tag{62}
$$

The integral is convergent for $d > 2$; the important point is the purely dimensional factor ξ^{4-d} which vanishes for $d > 4$ in the desired limit. This shows directly that two Brownian paths have no chance to cross in an arbitrary region as soon as the dimension is greater than four. Conversely, for d strictly less than four, one can show that they certainly cross in the entire space (this property is evident if $d \leq 2$). Dimension four is the limiting case.

(1) (from G.F. Lawler, *Comm. Math. Phys.* **86**, 539 (1982)). Consider a four-dimensional lattice.

(i) If two Brownian paths of length t start from two points $\mathbf{x}_1$, $\mathbf{x}_2$, at a distance of order $\sqrt{t}$, the probability that they intersect behaves as $1/\ln t$ for large t. Hence, in the continuous limit, two paths of infinite length starting from fixed points (at a finite distance) have no chance to intersect.

(ii) If the two paths start from the same point, the corresponding probability tends to unity as their length increases (the correction probably behaves as $(\ln t)^{-\frac{1}{2}}$).

(2) The number of distinct sites visited by both paths starting from the same point, with respective lengths t_1 and t_2 is $D(t_1) + D(t_2) - D(t_1 + t_2)$. If $t_1 = \alpha t$, $t_2 = \beta t$ and $t \to \infty$, this number diverges for $d \leq 4$ (for $d = 4$, this divergence is logarithmic) and remains finite for $d > 4$.

(3) In one dimension, the probability $P_t(N)$ to visit N distinct sites during the time t $(D(t) = \sum_N N P_t(N))$ is (B. Derrida)

$$P_t(N) = q_t(N) - 2q_t(N-1) + q_t(N-2)$$

with

$$q_t(N) = \frac{2}{N+1} \sum_{\substack{1 \leq q \leq N \\ q \text{ odd}}} \left(\cos \frac{q\pi}{N+1} \right)^t \cot an \frac{q\pi}{2(N+1)}$$

As $t, N \to \infty$ with $t \gg N^2$, $P_t(N)$ is exponentially small as $(8\pi^2 t^2 / N^5) \exp(-\pi^2 t / 2N^2)$. It is expected that the d-dimensional analogous property is $P_t(N) \sim \exp(-Kt/N^{2/d})$.

(4) In three dimensions, the mean time spent at the origin $G(0, 1)$ and the escape probability $\Pi(0, 1) = 1 - G(0, 1)^{-1}$ are finite. Construct a probabilistic proof of the following results (Watson, see M.L. Glasser, I.J. Zucker, *Proc. Natl. Acad. Sci. USA*, **74**, 1800 (1977))

$$G(0, 1) = \int_{-\pi}^{\pi} \frac{d^3 k}{(2\pi)^3} \frac{1}{1 - c_1 c_2 c_3} = \frac{1}{4\pi^3} \Gamma(\tfrac{1}{4})^4 \simeq 1.3932039297$$

for the body centered cubic lattice (b.c.c.);

$$G(0, 1) = \int_{-\pi}^{\pi} \frac{d^3 k}{(2\pi)^3} \frac{1}{1 - \tfrac{1}{3}(c_1 c_2 + c_2 c_3 + c_3 c_1)}$$

$$= \frac{9\Gamma(\tfrac{1}{3})^6}{2^{14/3} \pi^4} \simeq 1.3446610732$$

for the face centered cubic lattice (f.c.c.); and

$$G(0, 1) = \int_{-\pi}^{\pi} \frac{d^3 k}{(2\pi)^3} \frac{1}{1 - \tfrac{1}{3}(c_1 + c_2 + c_3)}$$

$$= \frac{4\sqrt{6}}{\pi^2} \Gamma(\tfrac{1}{24}) \Gamma(\tfrac{5}{24}) \Gamma(\tfrac{7}{24}) \Gamma(\tfrac{11}{24}) \simeq 1.5163860591$$

for the simple cubic lattice (s.c.). In these formulae, one uses the short-hand notation $c_i = \cos k_i$.

1.2 Euclidean fields

When one performs an analytic continuation of the time to imaginary values, the Poincaré group of mechanics becomes the Euclidean group of displacements and reflections. From the superposition principle, quantum probability amplitudes appear now as sums similar to those used in statistical mechanics. Hence an evolution problem in quantum field theory is transformed into questions relative to a static thermodynamical equilibrium, where the fundamental variables are stochastic classical fields. To give a precise and rigorous meaning to the various continuous sums appearing in this formalism, the continuous space $\mathcal{R}^d$ is replaced by a lattice. This procedure leads to the problem of examining the limit of an infinitely dense lattice. Let us recall first the free field formalism. It is based on an extensive use of Gaussian integrals and we assume that the reader is familiar with the formula

$$\int \mathrm{d}X \exp\left(-\tfrac{1}{2}{}^T X A X + {}^T J X\right) = \left(\det \frac{A}{2\pi}\right)^{-1/2} \exp\left(\tfrac{1}{2}{}^T J A^{-1} J\right) \tag{63}$$

where X is a vector and the symmetric matrix A has a positive definite real part. Such integrals have already occured in the previous section.

1.2.1   Free fields

The function $G(\mathbf{x}, \lambda)$ can be interpreted as the propagator of a Euclidean free field $\varphi(\mathbf{x})$, defined on each lattice site and varying from $-\infty$ to $+\infty$. In this context, the free field notion is equivalent to a statistical Gaussian weight

$$e^{-S(\varphi)} \tag{64a}$$

where the quadratic action $S(\varphi)$ reads

$$S(\varphi) = \tfrac{1}{2} \sum_{\mathbf{x},\mathbf{x}'} \varphi(\mathbf{x})[\lambda^{-1}\delta_{\mathbf{x},\mathbf{x}'} - J_{\mathbf{x},\mathbf{x}'}]\varphi(\mathbf{x}') \tag{64b}$$

with

$$J_{\mathbf{x},\mathbf{x}'} = \begin{cases} 1/2d & \text{if } |\mathbf{x} - \mathbf{x}'| = 1 \\ 0 & \text{otherwise} \end{cases}$$

One introduces the partition function

$$Z = \int \prod_{\mathbf{x}} \frac{\mathrm{d}\varphi_{\mathbf{x}}}{\sqrt{2\pi}} e^{-S(\varphi)} \tag{65}$$

which is well defined for a finite – possibly very large – number of integration variables. This is achieved e.g. by taking a finite system in a box of volume $\Omega = L^d$. The size L will be sent to infinity at the end of the calculations. This operation (*thermodynamic limit*) will always be implicitly assumed. The correlation function between two fields $\varphi(\mathbf{x})$ and $\varphi(\mathbf{0})$ takes the same form as the propagator $G(\mathbf{x}, \lambda)$ for Brownian motion,

$$G(\mathbf{x}, \lambda) = \langle \varphi(\mathbf{x})\varphi(\mathbf{0}) \rangle = \int_{-\pi}^{\pi} \frac{d^d \mathbf{k}}{(2\pi)^d} \frac{e^{i\mathbf{k}\cdot\mathbf{x}}}{\lambda^{-1} - \frac{1}{d}\sum_\mu \cos k_\mu}$$

$$= Z^{-1} \int \prod_{\mathbf{x}} \frac{d\varphi_{\mathbf{x}}}{\sqrt{2\pi}} e^{-S(\varphi)} \varphi(\mathbf{x})\varphi(\mathbf{0}) \tag{66}$$

The action (64b) is diagonalized by a Fourier transformation

$$S(\varphi) = \frac{1}{2} \int \frac{d^d \mathbf{k}}{(2\pi)^d} \varphi(-\mathbf{k}) \left[\lambda^{-1} - \frac{1}{d}\sum_\mu \cos k_\mu \right] \varphi(\mathbf{k}) \tag{67}$$

with $\varphi(-\mathbf{k}) \equiv \overline{\varphi(\mathbf{k})}$, so that the denominator of the propagator is the kernel of the quadratic form in the action. Although φ can take both positive and negative values, its nonvanishing correlations are always positive, as G is. This positivity property is associated with the *ferromagnetic* character of the corresponding statistical model, which describes schematically the statistics of ferromagnetic materials. To be more precise, the action is positive definite for $0 < \lambda < 1$, while it is nonnegative for $\lambda = 1$ due to the zero mode $\varphi = Cst$.

The preceding section has shown the relation between Brownian motion and free field Green functions. We also introduced an additional variable t (the length of the path or proper time) which has no interpretation at equilibrium, but which may describe the approach to equilibrium.

To obtain the continuous limit, we rescale the lattice field φ_L to get the continuous field φ_c (a is the lattice spacing).

$$\varphi_L(\mathbf{x}/a) = \sqrt{2d}\, a^{d/2-1} \varphi_c(\mathbf{x}) \tag{68}$$

with again

$$\lambda^{-1} = 1 + \frac{m^2 a^2}{2d} \tag{69}$$

The dimensionless field φ_L is now replaced by the field φ_c, with a *canonical* dimension (in terms of the wavenumber $1/a$)

$$[\varphi_c] = \frac{1}{2}(d - 2) \tag{70}$$

We can rewrite the action as

$$S(\varphi_L) = \frac{1}{2} \sum_{\substack{x/a,x'/a \\ integers}} \varphi_L(\mathbf{x}/a) \left[\lambda^{-1} \delta_{\mathbf{x},\mathbf{x}'} - J_{\mathbf{x}/a,\mathbf{x}'/a} \right] \varphi_L(\mathbf{x}'/a)$$

$$= \frac{1}{2} \int \frac{d^d\mathbf{x}}{a^d} 2da^{d-2} \left[\left(1 + \frac{m^2 a^2}{2d} \right) \varphi_c^2(\mathbf{x}) \right.$$

$$\left. -\varphi_c(\mathbf{x}) \frac{1}{d} \sum_{\mu=1}^{d} \left(\varphi_c(\mathbf{x}) + \frac{a^2}{2} \partial_\mu^2 \varphi_c(\mathbf{x}) + \cdots \right) \right] \qquad (71)$$

$$= \frac{1}{2} \int d^d\mathbf{x}\, \varphi_c(\mathbf{x}) [m^2 - \Delta] \varphi_c(\mathbf{x}) + \mathcal{O}(a^2)$$

$$= \frac{1}{2} \int d^d\mathbf{x} \left[m^2 \varphi_c^2(\mathbf{x}) + (\nabla \varphi_c(\mathbf{x}))^2 \right] + \mathcal{O}(a^2)$$

where, in the last integral, one has performed an integration by parts and neglected the corresponding contributions at infinity.

Up to terms of order a^2, the action has a limit in terms of φ_c, which is its continuous Euclidean form. We omit now the index of φ and use a compact symbol $\mathcal{D}\varphi$ for the integration measure. The scaling transformation leads of course to a corresponding change for the correlation functions as in equation (31),

$$\langle \varphi(\mathbf{x})\varphi(\mathbf{0}) \rangle = g(\mathbf{x}, m^2) = \int_{-\infty}^{\infty} \frac{d^d\mathbf{k}}{(2\pi)^d} \frac{e^{i\mathbf{k}\cdot\mathbf{x}}}{\mathbf{k}^2 + m^2}$$

$$= Z^{-1} \int \mathcal{D}\varphi\, e^{-S(\varphi)} \varphi(\mathbf{x})\varphi(\mathbf{0}) \qquad (72)$$

More generally, averages of an odd number of fields vanish, while those with an even number are given by Wick's theorem

$$\langle \varphi(\mathbf{x}_1) \cdots \varphi(\mathbf{x}_{2n}) \rangle = \frac{1}{2^n n!} \sum_{perm.} \left\langle \varphi(\mathbf{x}_{p_1})\varphi(\mathbf{x}_{p_2}) \right\rangle \cdots \left\langle \varphi(\mathbf{x}_{p_{2n-1}})\varphi(\mathbf{x}_{p_{2n}}) \right\rangle$$

$$= \sum_{pairings} \prod \langle \varphi(\mathbf{x}_s)\varphi(\mathbf{x}_t) \rangle \qquad (73)$$

The number of pairings on the r.h.s. is $(2n)!/(2^n n!) = (2n - 1)!!$. A compact form which summarizes equations (73) is provided by the generating functional

$$\frac{Z(J)}{Z(0)} = Z^{-1} \int \mathcal{D}\varphi \exp\left\{ -S(\varphi) + \int d^d\mathbf{x} J(\mathbf{x})\varphi(\mathbf{x}) \right\}$$

$$= \exp\left\{ \frac{1}{2} \int d^d\mathbf{x}_1 d^d\mathbf{x}_2 J(\mathbf{x}_1) \langle \varphi(\mathbf{x}_1)\varphi(\mathbf{x}_2) \rangle J(\mathbf{x}_2) \right\} \qquad (74)$$

where the two kernels $\langle\varphi(\mathbf{x}_1)\varphi(\mathbf{x}_2)\rangle$ and $\delta^2 S/\delta\varphi(\mathbf{x}_1)\delta\varphi(\mathbf{x}_2)$ are inverse from each other. Correlation functions can be generated by taking functional derivatives with respect to J. This leads to an immediate proof of Wick's theorem (73).

Each term in the right-hand side of (73) appears as a sum over independent Brownian paths. Hence the free field acts, from this point of view, as a generating function for studying symmetrized sets of Brownian particles. Green functions of interacting fields are similarly expected to have an interpretation in terms of particles moving under restrictions (Symanzik, 1969), as we shall now show.

1.2.2   Interacting fields and random walks

Let us consider now a class of models with interacting fields, invariant with respect to the orthogonal group $O(n)$. The n-component vector field $\mathbf{\Phi}$ is defined on each lattice site. The lattice spacing acts as an ultraviolet cutoff and we modify the action by adding a local interaction term $(\mathbf{\Phi}^2)^2$, the simplest possible one. The action is

$$S(\mathbf{\Phi}) = -\beta \sum_{(\mathbf{x},\mathbf{x}')} \mathbf{\Phi}(\mathbf{x}) \cdot \mathbf{\Phi}(\mathbf{x}') + \sum_{\mathbf{x}} V(\mathbf{\Phi}^2(\mathbf{x})) \qquad (75)$$

with

$$V(\mathbf{\Phi}^2) = \frac{1}{2}\mu^2\mathbf{\Phi}^2 + \frac{1}{4}g(\mathbf{\Phi}^2)^2 \qquad (76)$$

Only two of the three constants β, μ^2, g are really independent as can be seen using a scaling argument. To deal with a bounded action (from below), we restrict ourselves to a positive coupling constant, $g \geq 0$; however, now we need not exclude negative values of μ^2. The point $g = 0$, which is the starting point for a perturbation theory around the free field, is singular, because the integrals cannot be defined for arbitrarily small g, including $\mathrm{Re}\, g < 0$. The notation $(\mathbf{x}, \mathbf{x}')$ refers to all pairs of neighbouring sites, and hence $\sum_{(\mathbf{x},\mathbf{x}')}$ is equivalent to $d\sum_{\mathbf{x},\mathbf{x}'} J_{\mathbf{x},\mathbf{x}'}$. For $n = 1$ the model describes a scalar field in self-interaction. In the limit $g+\mu^2 = 0$, $g \to \infty$, we obtain the Ising model, with $\mathbf{\Phi}$ constrained to take only the values ± 1. When $n > 1$, this limit constrains $|\mathbf{\Phi}|$ to be a constant and yields the *classical Heisenberg model*, the continuous counterpart of which is the *nonlinear σ-model*, referring to the model introduced in particle physics by Gell-Mann and Levy in their description of spontaneous breaking of chiral invariance. These models will be described in greater detail in following chapters.

It is often natural in formal developments to combine the local part of the action $\sum_{\mathbf{x}} V(\mathbf{\Phi}^2)$ with the integration measure. The partition function Z – with the infinite volume limit implicitly understood – reads

$$Z = \int \prod_{\mathbf{x}} \left\{ e^{-V(\mathbf{\Phi}^2)} \prod_{\alpha=1}^{n} \frac{d\Phi^{\alpha}(\mathbf{x})}{\sqrt{2\pi}} \right\} e^{\beta \sum_{(\mathbf{x},\mathbf{x}')} \mathbf{\Phi}(\mathbf{x}) \cdot \mathbf{\Phi}(\mathbf{x}')} \qquad (77)$$

Except for the propagation or kinetic term, the integrand is factorized. In this statistical version, this term therefore acts as an interaction between fields on different sites. Similarly, correlation functions read

$$\langle \Phi^{\alpha_1}(\mathbf{x}_1) \cdots \Phi^{\alpha_{2k}}(\mathbf{x}_{2k}) \rangle = Z^{-1} \int \prod_{\mathbf{x}} \left\{ e^{-V(\mathbf{\Phi}^2)} \prod_{\alpha=1}^{n} \frac{d\Phi^{\alpha}(\mathbf{x})}{\sqrt{2\pi}} \right\}$$
$$\times e^{\beta \sum_{(\mathbf{x},\mathbf{x}')} \mathbf{\Phi}(\mathbf{x}) \cdot \mathbf{\Phi}(\mathbf{x}')} \Phi^{\alpha_1}(\mathbf{x}_1) \cdots \Phi^{\alpha_{2k}}(\mathbf{x}_{2k})$$
$$(78)$$

The relation with a sum over paths appears when one expands the kinetic term in powers of β, exactly as one proceeds in the high temperature expansion of statistical mechanics, with β proportional to the inverse temperature. However, as the weight $\exp(-V(\mathbf{\Phi}^2))$ leads to unmanageable integrals, it is convenient to introduce its Fourier transform in order to perform explicitly the subsequent integrals over $\mathbf{\Phi}$, owing to their Gaussian form. Therefore one writes the Laplace transform

$$e^{-V(\mathbf{\Phi}^2)} = \int d\mu(a) e^{-a\mathbf{\Phi}^2} \qquad (79)$$

which defines the measure $d\mu(a)$. The integral implicitly contains a prescription for the integration path over the variable a, to be distinguished from the lattice spacing, running along the imaginary axis. For brevity, let us use the operator σ

$$\sigma f(\mathbf{x}) \equiv 2d(\Delta_r + 1) f(\mathbf{x}) \equiv \sum_{\mu=1}^{d} [f(\mathbf{x} + \mathbf{e}_{\mu}) + f(\mathbf{x} - \mathbf{e}_{\mu})] \qquad (80)$$

Hence

$$
\begin{aligned}
Z &= \int \prod_{\mathbf{x}} \left\{ \prod_{\alpha=1}^{n} \frac{d\Phi^\alpha(\mathbf{x})}{\sqrt{2\pi}} d\mu(a_{\mathbf{x}}) \right\} e^{-\frac{1}{2}\sum_{\mathbf{x}} \Phi(\mathbf{x})[2a_{\mathbf{x}}-\beta\sigma]\Phi(\mathbf{x})} \\
&= \int \prod_{\mathbf{x}} \frac{d\mu(a_{\mathbf{x}})}{(2a_{\mathbf{x}})^{n/2}} \det^{-n/2}\left[1 - \frac{\beta\sigma}{2a_{\mathbf{x}}}\right]
\end{aligned}
\tag{81}
$$

In the last expresion, we have performed a Gaussian integral providing an inverse square root of a determinant raised to the power n, the number of field components. Similar formulae hold for the Green functions. We only display the two-point function, the generalization being straightforward,

$$
\begin{aligned}
\langle \Phi^{\alpha_1}(\mathbf{x}_1)\Phi^{\alpha_2}(\mathbf{x}_2) \rangle &= Z^{-1} \int \prod_{\mathbf{x}} \left\{ \prod_{\alpha=1}^{n} \frac{d\Phi^\alpha(\mathbf{x})}{\sqrt{2\pi}} d\mu(a_{\mathbf{x}}) \right\} \\
&\quad \times e^{-\frac{1}{2}\sum_{\mathbf{x}} \Phi(\mathbf{x})[2a_{\mathbf{x}}-\beta\sigma]\Phi(\mathbf{x})} \Phi^{\alpha_1}(\mathbf{x}_1)\Phi^{\alpha_2}(\mathbf{x}_2) \\
&= Z^{-1} \int \prod_{\mathbf{x}} \left\{ \prod_{\alpha=1}^{n} \frac{d\Phi^\alpha(\mathbf{x})}{\sqrt{2\pi}} d\mu(a_{\mathbf{x}}) \right\} \\
&\quad \times e^{-\frac{1}{2}\sum_{\mathbf{x}} \Phi(\mathbf{x})[2a_{\mathbf{x}}-\beta\sigma]\Phi(\mathbf{x})} \delta^{\alpha_1\alpha_2} \langle \mathbf{x}_1 | \frac{1}{2a_{\mathbf{x}}-\beta\sigma} | \mathbf{x}_2 \rangle
\end{aligned}
\tag{82}
$$

In the integrand of the last expression, we recognize the Green function for free fields, generalized to the case where the term $a_{\mathbf{x}}$ depends on $\mathbf{x}$. As a result, the expressions implying the Fourier transform cannot be used anymore. However, the expansion in terms of paths going from $\mathbf{x}_1$ to $\mathbf{x}_2$ is always valid, generalizing equation (30)

$$
\langle \mathbf{x}_1 | \frac{1}{2a_{\mathbf{x}}-\beta\sigma} | \mathbf{x}_2 \rangle = \sum_{\substack{\omega \\ \partial\omega=\{\mathbf{x}_1,\mathbf{x}_2\}}} \beta^{|\omega|} \prod_{\mathbf{x}} \left(\frac{1}{2a_{\mathbf{x}}} \right)^{n_{\mathbf{x}}(\omega)}
\tag{83}
$$

The notation $n_{\mathbf{x}}(\omega) \geq 0$ stands for the number of times the path goes through $\mathbf{x}$ ($\mathbf{x}$ may coincide with one of the end points), that is the "time" spent at point $\mathbf{x}$. Of course, $\sum_{\mathbf{x}} n_{\mathbf{x}}(\omega) = |\omega| + 1$ so that most of the terms in the products on the right-hand side are equal to 1 as $|\omega| < \infty$ (only a finite number of terms contribute).

The auxiliary quantities $a_{\mathbf{x}}$ (varying mass terms) must now be eliminated. This is done by using an integral representation

$$(2a)^{-k} = \int d\nu_k(t)e^{-2at} \qquad (84)$$

with

$$d\nu_k(t) = \begin{cases} \theta(t)\dfrac{t^{k-1}}{(k-1)!}dt & k > 0 \\ \delta(t)dt & k = 0 \end{cases} \qquad (85)$$

Combining the expansion in (83) with this representation and substituting in (82), one finally obtains

$$\langle \Phi^{\alpha_1}(\mathbf{x}_1)\Phi^{\alpha_2}(\mathbf{x}_2)\rangle = \delta^{\alpha_1\alpha_2} \sum_{\substack{\omega \\ \partial\omega=\{\mathbf{x}_1,\mathbf{x}_2\}}} \beta^{|\omega|} Z^{-1}$$

$$\times \int \prod_{\mathbf{x}} \left\{ \prod_{\alpha=1}^{n} \frac{d\Phi^{\alpha}(\mathbf{x})}{\sqrt{2\pi}} d\nu_{n_{\mathbf{x}}(\omega)}(t_{\mathbf{x}}) \right\} \qquad (86)$$

$$\times e^{\beta \sum_{(\mathbf{x},\mathbf{x}')} \Phi(\mathbf{x})\cdot\Phi(\mathbf{x}') - \sum_{\mathbf{x}} V(\Phi^2(\mathbf{x})+2t_{\mathbf{x}})}$$

The factor $\delta^{\alpha_1\alpha_2}$ expresses the $O(n)$ invariance. The coefficient of $\beta^{|\omega|}$ in the sum over paths is a weight $z(\omega)$ which takes into account the number of times, $n_{\mathbf{x}}$, the path ω visits each point $\mathbf{x}$. The bigger $n_{\mathbf{x}}$, the smaller the contribution. In the free field case where $V(\Phi^2) = \frac{1}{2}\mu^2\Phi^2$, the integral over Φ can be done, it compensates the factor Z^{-1} and gives the same result as in the preceding section, up to a change of notation

$$\langle \Phi^{\alpha_1}(\mathbf{x}_1)\Phi^{\alpha_2}(\mathbf{x}_2)\rangle = \delta^{\alpha_1\alpha_2} \langle \mathbf{x}_1| \frac{1}{\mu^2 - \beta\sigma} |\mathbf{x}_2\rangle$$

$$= \delta^{\alpha_1\alpha_2}\frac{1}{\mu^2} \sum_{\substack{\omega \\ \partial\omega=\{\mathbf{x}_1,\mathbf{x}_2\}}} \left(\frac{\beta}{\mu^2}\right)^{|\omega|} \qquad (87)$$

In the general case however, the positive weight

$$z(\omega)$$

$$= Z^{-1}\int \mathcal{D}\Phi\mathcal{D}\nu \exp\left\{\beta\sum_{(\mathbf{x},\mathbf{x}')}\Phi(\mathbf{x})\cdot\Phi(\mathbf{x}') - \sum_{\mathbf{x}}V(\Phi^2(\mathbf{x})+2t_{\mathbf{x}})\right\}$$

$$\qquad (88)$$

cannot be recast in the form $\exp(\text{Cst }|\omega|)$.

For the $2k$-point function, the result appears as a sum over $(2k-1)!!$ terms corresponding to all pairings. This provides a nontrivial generalization of Wick's theorem. If $\mathcal{P}$ denotes the pairings,

$$\langle \Phi^{\alpha_1}(\mathbf{x}_1) \cdots \Phi^{\alpha_{2k}}(\mathbf{x}_{2k}) \rangle = \sum_{\mathcal{P}} \delta^{\alpha_{P_1}\alpha_{P_2}} \ldots \delta^{\alpha_{P_{2k-1}}\alpha_{P_{2k}}}$$

$$\times \sum_{\substack{\omega_1,\ldots,\omega_k \\ \partial\omega_j = \left\{ {}^{x}P_{2j-1}, {}^{x}P_{2j} \right\}}} \beta^{|\omega_1| + \cdots + |\omega_k|} z(\omega_1, \ldots, \omega_k)$$

$$(89)$$

with

$$z(\omega_1, \ldots, \omega_k) = Z^{-1} \int \prod_{\mathbf{x}} \prod_{\alpha} \frac{\mathrm{d}\Phi^\alpha(\mathbf{x})}{\sqrt{2\pi}} \mathrm{d}\nu_{\mathcal{N}_\mathbf{x}}$$

$$\times \exp \left\{ \beta \sum_{(\mathbf{x},\mathbf{x}')} \Phi(\mathbf{x}) \cdot \Phi(\mathbf{x}') - \sum_{\mathbf{x}} V(\Phi^2(\mathbf{x}) + 2t_\mathbf{x}) \right\}$$

$$(90)$$

The weight $z(\omega_1, \ldots, \omega_k)$ uses the total number $\mathcal{N}_\mathbf{x} = \sum_{j=1}^k n_\mathbf{x}(\omega_j)$ of visits to site $\mathbf{x}$ by all paths.

Fröhlich (1982) has obtained the following remarkable inequalities

$$(i) \quad \omega_1 \cap \omega_2 = \emptyset \quad \Rightarrow \quad z(\omega_1, \omega_2) \geq z(\omega_1) z(\omega_2)$$

$$(ii) \quad \sum_{\omega_2} z(\omega_1, \omega_2) \leq z(\omega_1) \sum_{\omega_2} z(\omega_2) \qquad (91)$$

The proof is based on the following idea. The average values of products of fields decrease if the parameters in the action are modified in such a way as to decrease the probability measure for large Φ. Let $\mathcal{D}\nu_\omega(t) = \prod_x \mathrm{d}\nu_{n_x(\omega)}(t_x)$. As

$$a^{-n_1 - n_2} = \int \mathrm{d}\nu_{n_1 + n_2}(t) e^{-at} = \int \int \mathrm{d}\nu_{n_1}(t_1) \mathrm{d}\nu(t_2) e^{-a(t_1 + t_2)},$$

the integral over $t_{\mathbf{x}}$ can be replaced by a double integral over $t_{1\mathbf{x}}$ and $t_{2\mathbf{x}}$ in expression (90), written for $k = 2$,

$$z(\omega_1, \omega_2) = Z^{-1} \int \mathcal{D}\Phi \mathcal{D}\nu_{\omega_1}(t_1)\mathcal{D}\nu_{\omega_2}(t_2)$$

$$\times e^{\beta \sum_{(\mathbf{x},\mathbf{x}')} \Phi(\mathbf{x})\cdot\Phi(\mathbf{x}') - \sum_{\mathbf{x}} V(\Phi^2 + 2t_{1\mathbf{x}} + 2t_{2\mathbf{x}})}$$

$$= \int \mathcal{D}\nu_{\omega_1}(t_1)\mathcal{D}\nu_{\omega_2}(t_2)z(t_1 + t_2)$$

with

$$z(t_1 + t_2) = e^{-\sum_{\mathbf{x}}\mu^2(t_{1\mathbf{x}} + t_{2\mathbf{x}}) + g(t_{1\mathbf{x}} + t_{2\mathbf{x}})^2} \left\langle e^{-\sum_{\mathbf{x}}g\Phi^2(\mathbf{x})(t_{1\mathbf{x}} + t_{2\mathbf{x}})} \right\rangle$$

the variables t_1 and t_2 being associated with the paths ω_1 and ω_2 respectively. Similar relations hold for $z(\omega_1)$ and $z(\omega_2)$, so that the inequality (i) is satisfied if

$$\omega_1 \cap \omega_2 = \emptyset \quad \Rightarrow \quad z(t_1 + t_2) \ge z(t_1)z(t_2)$$

Defining $\tilde{z}(t) = \langle \exp(-g\sum_{\mathbf{x}}\Phi^2(\mathbf{x})t_{\mathbf{x}}\rangle$, one finds

$$\ln \tilde{z}(t_1 + t_2) = \ln \tilde{z}(t_2) + \int_0^1 d\lambda \tilde{z}^{-1}(\lambda t_1 + t_2)\frac{\partial}{\partial\lambda}\tilde{z}(\lambda t_1 + t_2)$$

If one uses a notation $\langle . \rangle_t$ to indicate an additional term $g\sum_{\mathbf{x}} t_{\mathbf{x}}\Phi^2(\mathbf{x})$ in the action (with of course $\langle . \rangle_0 = \langle . \rangle$), the integrand reads $-g\sum_{\mathbf{x}} t_{1\mathbf{x}} < \Phi^2(\mathbf{x}) >_{\lambda t_1 + t_2}$. Now, we can use the remark mentioned previously; as t is positive, $\langle \Phi^2 \rangle_t$ is a decreasing function of t. More precisely, one has the following inequality due to Griffiths

$$\langle \Phi^2(\mathbf{x})\Phi^2(\mathbf{y}) \rangle_t^{\mathrm{conn}} = \langle \Phi^2(\mathbf{x})\Phi^2(\mathbf{y}) \rangle_t - \langle \Phi^2(\mathbf{x}) \rangle_t \langle \Phi^2(\mathbf{y}) \rangle_t \ge 0$$

$$\frac{\partial}{\partial\lambda'} \langle \Phi^2(\mathbf{x}) \rangle_{\lambda t_1 + \lambda' t_2} = -g\sum_{\mathbf{y}} t_{2\mathbf{y}} \langle \Phi^2(\mathbf{x})\Phi^2(\mathbf{y}) \rangle_{\lambda t_1 + \lambda' t_2}^{\mathrm{conn}} \le 0$$

$$(92)$$

The integral is therefore minimized by replacing $\langle \Phi^2 \rangle_{\lambda t_1 + t_2}$ by $\langle \Phi^2 \rangle_{\lambda t_1}$. Hence

$$\ln \tilde{z}(t_1 + t_2) \ge \ln \tilde{z}(t_2) - \int_0^1 d\lambda g\sum_{\mathbf{x}} t_{1\mathbf{x}} \langle \Phi^2(\mathbf{x}) \rangle_{\lambda t_1} = \ln \tilde{z}(t_1) + \ln \tilde{z}(t_2)$$

and for the quantities $z(t)$,

$$\exp\left(g\sum_{\mathbf{x}} t_{1\mathbf{x}}t_{2\mathbf{x}}\right) z(t_1 + t_2) \ge z(t_1)z(t_2)$$

Whenever ω_1 and ω_2 do not intersect, $t_{1\mathbf{x}}t_{2\mathbf{x}} = 0$ for all $\mathbf{x}$, and inequality (91i) is proved. For inequality (ii), taking $\partial\omega_2 =$

$\{\mathbf{x}_3,\mathbf{x}_4\}$ fixed, one has

$$\sum_{\omega_2} z(\omega_1,\omega_2) = \sum_{\omega_2}\int \mathcal{D}\nu_{\omega_1}(t_1)\mathcal{D}\nu_{\omega_2}(t_2)z(t_1+t_2)$$

$$= \int \mathcal{D}\nu_{\omega_1}(t_1)z(t_1)\langle\Phi^\alpha(\mathbf{x}_3)\Phi^\alpha(\mathbf{x}_4)\rangle_{t_1}$$

with the same notation as above. One now uses the inequality

$$\langle\Phi^\alpha(\mathbf{x}_3)\Phi^\alpha(\mathbf{x}_4)\rangle_t \le \quad \langle\Phi^\alpha(\mathbf{x}_3)\Phi^\alpha(\mathbf{x}_4)\rangle = \sum_{\omega_2}z(\omega_2) \qquad (93)$$

which completes the proof of the inequalities (91). These formulae are useful for deriving numerous inequalities for $2k$-point correlation functions.

1.2.3   Self-avoiding walks and the limit $n \to 0$

We now look at the n-dependence of correlation functions, focusing on the two-point function. When the internal indices coincide, the latter can be written, using equations (82,83), as

$$\langle\Phi^1(\mathbf{x}_1)\Phi^1(\mathbf{x}_2)\rangle = Z^{-1}\sum_{\partial\omega=\{\mathbf{x}_1,\mathbf{x}_2\}}^{\omega}\beta^{|\omega|}\int\prod_{\mathbf{x}}\left(d\mu(a_\mathbf{x})(2a_\mathbf{x})^{-\frac{1}{2}n-n_\mathbf{x}(\omega)}\right)$$

$$\times \det{}^{-n/2}\left(1-\frac{\beta\sigma}{2a_\mathbf{x}}\right)$$

$$Z = \int\prod_{\mathbf{x}}\left(d\mu(a_\mathbf{x})(2a_\mathbf{x})^{-n/2}\right)\det{}^{-n/2}\left(1-\frac{\beta\sigma}{2a_\mathbf{x}}\right)$$

$$(94)$$

The dependence in n is explicitly written everywhere, except in the integration measure $d\mu(a_\mathbf{x})$. For simple choices of $V(\mathbf{\Phi}^2)$, an analytic continuation in the variable n will be possible. We look at the limit $n \to +0$. A simplification arises, since

$$\lim_{n\to 0}\det{}^{-n/2}\left(1-\frac{\beta\sigma}{2a_\mathbf{x}}\right) = 1 \qquad (95)$$

so that the weight $z(\omega)$ becomes a sum over factorized contributions

$$z(\omega) = \sum_{\mathbf{x}}\rho_{n_\mathbf{x}(\omega)}$$

$$\rho_k = \lim_{n\to 0}\frac{\int d\mu(a)\,(2a)^{-(n/2)-k}}{\int d\mu(a)\,(2a)^{-n/2}} \qquad (96)$$

and

$$\lim_{n \to 0} \langle \Phi^1(\mathbf{x}_1)\Phi^1(\mathbf{x}_2) \rangle = \sum_{\substack{\omega \\ \partial\omega = \{\mathbf{x}_1, \mathbf{x}_2\}}} \beta^{|\omega|} \sum_{\mathbf{x}} \rho_{n_\mathbf{x}(\omega)} \qquad (97)$$

The normalization is such that $\rho_0 = 1$. With the action (75)-(76), and using $d\mu(a)$ as defined in (79), we can evaluate ρ_k. If $p > 0$,

$$\int d\mu(a)(2a)^{-p} = \int d\mu(a) \int_0^\infty dt \frac{t^{p-1}}{\Gamma(p)} e^{-2at}$$

$$= \int_0^\infty dt \frac{t^{p-1}}{\Gamma(p)} e^{-(\mu^2 t + g t^2)}$$

while

$$\int d\mu(a) = \int d\mu(a) \int_{-\infty}^\infty dt \, \delta(t) e^{-2at} = 1$$

Inserting these relations in (96), we get

$$\rho_0 = 1$$

$$\rho_k = \int_0^\infty dt \frac{t^{k-1}}{(k-1)!} e^{-(\mu^2 t + g t^2)} \qquad (k \geq 1) \qquad (98)$$

We have already observed that, as the number of visits to a site increases, the corresponding weight decreases (i.e. ρ_k decreases as k increases). Let us consider more precisely the limit $n \to 0$ for the nonlinear σ-model, which constrains $\mathbf{\Phi}^2$ to a constant value conveniently fixed to n

$$\mathbf{\Phi}^2 = n \qquad (99)$$

This is achieved by choosing

$$\mu^2 = -ng \qquad (100)$$

and taking the limit $g \to \infty$ (before $n \to 0$). Indeed, for a regular test function

$$\lim_{g \to \infty} \frac{\int \prod_\alpha \frac{d\Phi^\alpha}{\sqrt{2\pi}} e^{g[n\mathbf{\Phi}^2/2 - (\mathbf{\Phi}^2)^2/4]} f(\mathbf{\Phi})}{\int \prod_\alpha \frac{d\Phi^\alpha}{\sqrt{2\pi}} e^{g[n\mathbf{\Phi}^2/2 - (\mathbf{\Phi}^2)^2/4]}} = \frac{\int \prod_\alpha d\Phi^\alpha \delta(\mathbf{\Phi}^2 - n) f(\mathbf{\Phi})}{\int \prod_\alpha d\Phi^\alpha \delta(\mathbf{\Phi}^2 - n)}$$

$$(101)$$

Thus we have

$$\rho_k = \lim_{n \to 0} \frac{\int d\nu_{k+n/2}(t) \delta(t - n/2)}{\int d\nu_{n/2}(t) \delta(t - n/2)} = \lim_{n \to 0} (n/2)^k \frac{\Gamma(n/2)}{\Gamma(n/2 + k)}$$

and finally

$$\rho_k = \begin{cases} 1 & \text{if } k = 0, 1 \\ 0 & \text{otherwise} \end{cases} \tag{102}$$

The result is quite remarkable (de Gennes, 1972), as it shows that the $n \to 0$ limit of the nonlinear σ-model leads for the correlation function to a sum over *non-self-intersecting* paths (excluded volume limit), a model which describes dilute solutions of polymers. It is worthwhile to distinguish this sum over self-avoiding paths from a random walk where the walker would choose with equal probabilities the neighbouring sites not previously visited (and which might lead him into traps). The latter model, of interest in itself, belongs to a quite different *universality* class. This term will be defined more precisely as we proceed. In the present context, it means that the two processes have a different large-distance behaviour.

Extend the preceding result to the $2k$-point function

$$\lim_{n \to 0} \langle \Phi^1(\mathbf{x}_1) \cdots \Phi^1(\mathbf{x}_{2k}) \rangle = \sum_{\substack{pairings \\ \mathcal{P}}} \sum_{\substack{\omega_1, \dots, \omega_k \\ \partial\omega_j = \{x_{\mathcal{P}_{2j-1}}, x_{\mathcal{P}_{2j}}\}}} \beta^{|\omega_1| + \cdots + |\omega_k|}$$

$$\times \prod_{\mathbf{x}} \rho_{n_{\mathbf{x}}(\omega_1) + \cdots + n_{\mathbf{x}}(\omega_k)} \tag{103}$$

with ρ_n given by equations (98) or (102). Again, the paths do not intersect.

Looking for a continuous limit of these models, the intersection properties of Brownian curves suggest that, in dimension greater than four, the interactions of φ^4 type will have no effect on a critical limit which will be equivalent to a free field theory, i.e. a Gaussian model. Conversely, for $d = 4$ (as a limiting case) or for $d < 4$, the continuous limit will be nontrivial.

1.2.4   Comparison with the high temperature expansion

The above expansion in terms of paths has suggestive properties and is useful for theoretical purposes, but it does not lead to easy calculations, except in some particular cases. However, it looks very similar to another expansion (*high temperature*, or *strong coupling* series) which provides a practical tool for obtaining explicit series and which will be considered in greater detail in chapter 7. It is instructive to perform a comparison in the case of the Ising model ($n = 1$ "σ-model") where the field φ is restricted to the values ± 1. The integration measure is thus

$$\frac{d\varphi}{\sqrt{2\pi}} e^{-V(\varphi^2)} = 2\delta(\varphi^2 - 1)d\varphi = d\varphi \int_{-\infty}^{\infty} \frac{da}{\pi} e^{ia - ia\varphi^2} \qquad (104)$$

Consider the partition function

$$Z = \sum_{\varphi_{\mathbf{x}} = \pm 1} \exp\left(\beta \sum_{(\mathbf{x},\mathbf{x}')} \varphi_{\mathbf{x}} \varphi_{\mathbf{x}'}\right) = \sum_{\varphi_{\mathbf{x}} = \pm 1} \prod_{(\mathbf{x},\mathbf{x}')} \exp\left(\beta \varphi_{\mathbf{x}} \varphi_{\mathbf{x}'}\right) \quad (105)$$

Each pair of neighbours (associated to a link of the lattice) yields a term $\exp \beta \varphi \varphi'$, which may be expanded as a power series in β

$$e^{\beta \varphi \varphi'} = \sum_{j=0}^{\infty} \frac{\beta^j}{j!} \varphi^j \varphi'^j \qquad (106)$$

Substituting in (105) and using

$$\sum_{\varphi = \pm 1} \varphi^n = 2\delta_{n,0}^{(mod\ 2)} = \begin{cases} 2 & \text{if } n \text{ is even} \\ 0 & \text{if } n \text{ is odd} \end{cases} \qquad (107)$$

one obtains a sum of terms corresponding to configurations of non-negative integers $\{q_l\}$ assigned to each link l and satisfying the condition

$$\partial \{q_l\} = \emptyset \qquad (108)$$

where $\partial \{q_l\}$ is the set of sites $\mathbf{x}$ such that the sum of the q_l's on incident links is odd. Hence, if Ω is the total volume (the total number of lattice sites),

$$Z = 2^{\Omega} \sum_{\partial \{q_l\} = \emptyset} \frac{\beta^{\sum q_l}}{\prod_l q_l!} \qquad (109)$$

Similarly, one obtains the 2-point function

$$Z\left\langle \varphi_{\mathbf{x}} \varphi_{\mathbf{y}} \right\rangle = 2^{\Omega} \sum_{\partial \{q_l\} = \{\mathbf{x},\mathbf{y}\}} \frac{\beta^{\sum q_l}}{\prod_l q_l!} \qquad (110)$$

and analogous expressions for higher order functions. Reordering in increasing powers of β, one gets the desired strong coupling expansion, which looks similar to the sum over paths, as discussed previously. For a precise comparison, we rewrite the latter as (with $d\mu(a) = e^{ia}da/\pi$)

$$Z = \int \prod_{\mathbf{x}} d\varphi_{\mathbf{x}} d\mu(a_{\mathbf{x}}) \exp \left(\beta \sum_{(\mathbf{x},\mathbf{x}')} \varphi_{\mathbf{x}}\varphi_{\mathbf{x}'} - i \sum_{\mathbf{x}} a_{\mathbf{x}}\varphi_{\mathbf{x}}^2 \right)$$

$$= \int \prod_{\mathbf{x}} \left(d\mu(a_{\mathbf{x}}) \sqrt{\frac{\pi}{ia_{\mathbf{x}}}} \right) \exp \left(-\tfrac{1}{2} \operatorname{Tr} \ln \left(1 - \frac{\beta\sigma}{2ia_{\mathbf{x}}} \right) \right) \quad (111)$$

$$-\tfrac{1}{2} \operatorname{Tr} \ln \left(1 - \frac{\beta\sigma}{2ia_{\mathbf{x}}} \right) = \tfrac{1}{2} \sum_{\mathbf{x}} \sum_{\omega_{\mathbf{x}}} \frac{\beta^{|\omega_{\mathbf{x}}|}}{|\omega_{\mathbf{x}}|} \prod_{\mathbf{y}} \left(\frac{1}{2ia_{\mathbf{y}}} \right)^{n_{\mathbf{y}}(\omega_{\mathbf{x}})} \quad (112)$$

The last expression involves a sum over $\mathbf{x}$ and over the paths $\omega_{\mathbf{x}}$ starting from and returning to $\mathbf{x}$, and visiting $n_{\mathbf{y}}(\omega_{\mathbf{x}})$ times the site $\mathbf{y}$. Therefore,

$$Z = \int \prod_{\mathbf{x}} \left(da_{\mathbf{x}} \sqrt{\frac{2}{\pi}} e^{ia_{\mathbf{x}}} \right) \sum_{k,r_q} \frac{1}{2^{\sum_{q=1}^{k} r_q} \prod_{q=1}^{k} r_q!}$$

$$\times \sum_{\substack{\mathbf{x}_1,\dots,\mathbf{x}_k \\ \omega_{\mathbf{x}_1},\dots,\omega_{\mathbf{x}_k}}} \frac{\beta^{\sum_{q=1}^{k}|\omega_{\mathbf{x}_q}|}}{\prod_{q=1}^{k}|\omega_{\mathbf{x}_q}|} \prod_{\mathbf{y}} \left(\frac{1}{2ia_{\mathbf{y}}} \right)^{\frac{1}{2}+\sum_{q=1}^{k} n_{\mathbf{y}}(\omega_{\mathbf{x}_q})} \quad (113)$$

where the path $\omega_{\mathbf{x}}$ describes the same geometric image r times. The integration over a can be performed to give

$$\sqrt{\frac{2}{\pi}} \int da e^{ia} \left(\frac{1}{2ia} \right)^{k+\frac{1}{2}} = \frac{2}{(2k-1)!!} \quad (114)$$

(using the convention $(-1)!! = 1$). Indeed,

$$2 = \int d\varphi \, 2\delta(\varphi^2 - 1)\varphi^{2k} = \int \frac{da}{\pi} e^{ia} \int d\varphi e^{-ia\varphi^2} \varphi^{2k}$$

$$= (2k-1)!! \sqrt{\frac{2}{\pi}} \int da \frac{e^{ia}}{(2ia)^{k+\frac{1}{2}}}$$

Finally

$$Z = 2^{\Omega} \sum_{k,r_q} \frac{1}{2^{\sum_{q=1}^{k} r_q} \prod_{q=1}^{k} r_q!} \sum_{\substack{\mathbf{x}_1,\dots,\mathbf{x}_k \\ \omega_{\mathbf{x}_1},\dots,\omega_{\mathbf{x}_k}}} \frac{\beta^{\sum_{q=1}^{k}|\omega_{\mathbf{x}_q}|}}{\prod_{q=1}^{k}|\omega_{\mathbf{x}_q}|}$$

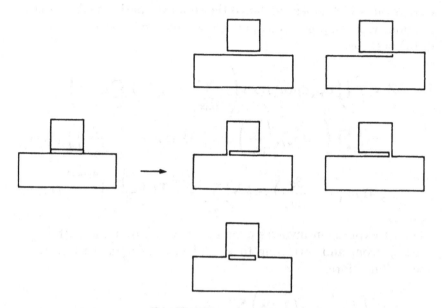

Fig. 5 Example of the relation between high temperature diagrams and closed paths, according to the discussion given in the text.

$$\times \prod_{\mathbf{y}} \frac{1}{\left(2\sum n_{\mathbf{y}}(\omega_{\mathbf{x}_q}) - 1\right)!!} \qquad (115)$$

This expression can be simplified, observing that the denominator takes partially into account the number of choices of an origin $\mathbf{x}_q$ on a closed path, as well as the two possible orientations of these paths (except for some exceptional paths, e.g. those of length 2). A set of closed paths (without a choice of an origin nor an orientation) defines a configuration in the sense of equation (109), which corresponds to the contribution of several terms in the expansion (115).

Let us analyze an example, taking the $\{q_l\}$ depicted on figure 5. The high temperature expansion assigns a contribution $\beta^{12}/2$ to this configuration. In the path decomposition, two sites are visited twice, yielding a factor $[\prod_{\mathbf{y}}(2\sum n_{\mathbf{y}}(\omega_{\mathbf{x}_q}) - 1)!!]^{-1} = 1/3^2$. The power of β is of course 12. The last term in the path decomposition yields an additional factor $\frac{1}{2}$, due to the path of length two which has only one possible orientation (with a fixed origin). Adding all contributions, we recover the expected value $\frac{1}{2} = (4 + \frac{1}{2}) \times (\frac{1}{3^2})$.

The reader may convince himself in the general case that the two methods give identical results.

For the Ising model – as well as for other similar models – the high temperature expansion as presented above is not the most economical one. From the fact that $\varphi\varphi' = \pm 1$, we have the relation

$$e^{\beta\varphi\varphi'} = \cosh\beta + \varphi\varphi'\sinh\beta = \cosh\beta(1 + t\varphi\varphi') \qquad (116)$$
$$\text{with } t = \tanh\beta$$

Inserting this identity in the expression (105) for the partition function, and summing over $\varphi_{\mathbf{x}}$, one is led to restricted configurations for which $q_l = 0$ or 1. If $L = \Omega d$ stands for the total number of links, we get

$$Z = 2^\Omega (\cosh\beta)^L \sum_{\substack{\partial\{q_l\}=0\\q_l=0,1}} t^{\Sigma q_l} \qquad (117)$$

This defines again a set of closed curves with, however, an ambiguous separation between connected parts. For correlation functions, we have a similar expansion

$$Z \left\langle \varphi_{\mathbf{x}_1} \cdots \varphi_{\mathbf{x}_{2k}} \right\rangle = 2^\Omega (\cosh\beta)^L \sum_{\substack{\partial\{q_l\}=\{\mathbf{x}_1,\ldots,\mathbf{x}_{2k}\}\\q_l=0,1}} t^{\Sigma q_l} \qquad (118)$$

We shall study these formulae in great depth later.

1.2.5 The one-dimensional case

One-dimensional models lead to simple functional integrals. They can be interpreted as solving quantum mechanical problems in terms of stochastic fields, after analytic continuation to imaginary time. As an example, we consider an $O(n)$ symmetric model, using periodic boundary conditions $\Phi_{L+x} = \Phi_x$, with a partition function

$$Z = \int \prod_{x=1}^{L} \left(\frac{d^n\Phi_x}{(2\pi)^{n/2}} e^{-V(\Phi_x^2)} \right) \exp\beta \sum_x \Phi_x \cdot \Phi_{x+1} \qquad (119)$$

Let us introduce the notion of transfer matrix which will be generalized to higher dimensions later. It is an operator $\mathcal{T}$ acting on the states $\psi(\Phi)$ (or wave functions) where Φ denotes a point in n-dimensional space. Its integral kernel satisfies

$$\mathcal{T}\psi(\Phi') = \int \frac{d^n\Phi}{(2\pi)^{n/2}} e^{-V(\Phi^2)} e^{\beta\Phi\cdot\Phi'} \psi(\Phi) \qquad (120)$$

so that the partition function may be written as the trace of the Lth iterate of T, assuming periodic boundary conditions

$$Z_L = \text{Tr}\, T^L = t_0^L + t_1^L + \cdots \qquad (121)$$

The eigenvalues $\{t_i\}$ of T have been ordered according to decreasing magnitude. In general, the highest eigenvalue t_0 is not degenerate. This corresponds in quantum mechanics to the uniqueness of the ground state. In the present context, this property can be deduced from a theorem due to Frobenius, which states that a symmetric real matrix with positive coefficients has a nondegenerate highest eigenvalue, and that all coordinates of the corresponding eigenvector have the same sign. The operator T defined by equation (120) is not symmetric, but we can substitute another equivalent symmetric operator, by replacing $V(\mathbf{\Phi}^2)$ by $\frac{1}{2}(V(\mathbf{\Phi}^2) + V(\mathbf{\Phi}'^2))$ in such a way that equation (121) is still fulfilled. Strictly speaking, Frobenius' theorem only applies to finite matrices, while T operates here in an infinite dimensional space.

In the thermodynamic limit $L \to \infty$, Z_L is dominated by the highest eigenvalue, so that the free energy is

$$\mathcal{F} = \lim_{L \to \infty} \frac{1}{L} \ln Z_L = \ln t_0 \qquad (122)$$

In the case of the nonlinear σ-model, the integration measure is

$$\frac{\mathrm{d}^n \mathbf{\Phi}}{(2\pi)^{n/2}} \mathrm{e}^{-V(\mathbf{\Phi}^2)} = \mathrm{d}^n \mathbf{\Phi}\, 2\delta(\mathbf{\Phi}^2 - n) \qquad (123)$$

The wave functions are defined on a sphere of radius $\sqrt{n}$. Rescaling $\mathbf{\Phi}$ so that it takes its values on the unit sphere, with a measure $\mathrm{d}^{n-1}\hat{\mathbf{\Phi}}$ normalized to

$$\int \mathrm{d}^{n-1}\hat{\mathbf{\Phi}} = S_n = \frac{2\pi^{n/2}}{\Gamma(n/2)}$$

equation (120) takes the form

$$T\psi(\hat{\mathbf{\Phi}}') = n^{(n/2)-1} \int \mathrm{d}^{n-1}\hat{\mathbf{\Phi}}\, \mathrm{e}^{\beta n \hat{\mathbf{\Phi}} \cdot \hat{\mathbf{\Phi}}'} \psi(\hat{\mathbf{\Phi}}) \qquad (124)$$

As T commutes with rotations in $\mathbf{\Phi}$ space, its eigenvectors are obtained by an expansion in spherical harmonics. The highest eigenvalue corresponding to $\psi = \text{Cst}$ is thus

$$t_0 = n^{n/2-1} S_{n-1} \int_0^\pi \mathrm{d}\theta \sin^{n-2}\theta\, \mathrm{e}^{\beta n \cos\theta} = (2\pi)^{n/2} \frac{I_{n/2-1}(n\beta)}{\beta^{n/2-1}}$$

$$(= 2\cosh\beta \qquad \text{for } n = 1) \tag{125}$$

Here $I_\nu(z)$ is the modified Bessel function

$$I_\nu(z) = (z/2)^\nu \sum_{k=0}^{\infty} \frac{(z/2)^{2k}}{k!\Gamma(\nu + k + 1)} \qquad (\nu \neq -1, -2, \ldots) \tag{126}$$

$$I_{\nu-1} = \frac{2\nu}{z} I_\nu + I_{\nu+1}, \qquad I_\nu' = \frac{\nu}{z} I_\nu + I_{\nu+1}$$

The eigenvalue t_0 is therefore an entire even function of β, positive whenever n is an integer. Its expression allows an analytic continuation in n. In particular, $t_0(\beta, n)$ tends to 1 as $n \to 0$, as expected from section 2.3. For large n, we find

$$t_0(\beta, n) \underset{n \to \infty}{\longrightarrow} \left(\frac{2\pi}{\beta}\right)^{\frac{1}{2}(n-1)} \frac{e^{\beta n}}{\sqrt{n}} \tag{127}$$

Let us turn now to correlation functions. Due to rotational invariance,

$$G(1, 2) = \langle \Phi^{\alpha_1}(x_1)\Phi^{\alpha_2}(x_2)\rangle = \delta^{\alpha_1\alpha_2}\frac{1}{n}\langle \Phi(x_1) \cdot \Phi(x_2)\rangle \tag{128}$$

The eigenvectors (*spherical harmonics*) of the symmetrized T are orthogonal, since this operator is Hermitian. Denoting them by $|l, m\rangle$ (m being a degeneracy index), one obtains

$$G(1, 2) = \frac{\delta^{\alpha_1\alpha_2}}{n}\sum_{l,m}\left(\frac{t_l}{t_0}\right)^{|x_1 - x_2|}\langle 0|\, \Phi\, |l, m\rangle\langle l, m|\, \Phi\, |0\rangle \tag{129}$$

where the ground state $|0\rangle$ corresponds to a constant $\psi_0(\Phi) = S_n^{-\frac{1}{2}}$. This expression is typical of transfer matrix methods. For a large distance $|x_1 - x_2|$, the correlation function is dominated by the leading term in $(t_1/t_0)^{|x_1 - x_0|}$, where t_1 is the second largest eigenvalue, provided the associated matrix elements do not vanish. One observes an exponential behaviour of the correlations, as long as the first two eigenvalues are not degenerate, which is the case here. Applying one component of Φ to the ground state leads to a first order harmonic (let us recall that harmonics of order p are of the form $s_{\alpha_1..\alpha_p}\Phi^{\alpha_1}..\Phi^{\alpha_p}$, where $s_{\alpha_1..\alpha_p}$ is a symmetric tensor with vanishing partial traces). Hence, in the present case, the sum (129) reduces to the contributions from the terms with $l = 1$, $m = 1, 2, \ldots, n$, and one gets an exponentially decreasing correlation function

$$G(1,2) = \delta^{\alpha_1\alpha_2} \left(\frac{t_1}{t_0}\right)^{|x_1-x_2|} = \delta^{\alpha_1\alpha_2} \exp\left(-\frac{|x_1-x_2|}{\xi}\right) \quad (130)$$

whatever x_1 and x_2, and ξ has the meaning of a *correlation length*, governing the size of the region over which the fields are correlated. In particular, $t_1/t_0 = n^{-1}\langle \mathbf{\Phi}_x \cdot \mathbf{\Phi}_{x+1}\rangle = (nL)^{-1}\sum_{x=1}^L \langle \mathbf{\Phi}_x \cdot \mathbf{\Phi}_{x+1}\rangle$ for two neighbouring sites. This last quantity is nothing but

$$\frac{1}{nL}\frac{\partial}{\partial\beta}\ln Z_L = \frac{1}{n}\frac{\partial}{\partial\beta}\ln t_0$$

Thus,

$$e^{-1/\xi} = \frac{t_1}{t_0} = \frac{1}{n}\frac{\partial}{\partial\beta}\ln t_0 \quad (131)$$

which yields

$$e^{-1/\xi} = \frac{I_{n/2}(n\beta)}{I_{n/2-1}(n\beta)} = \beta - \frac{n}{n+2}\beta^3 + \frac{2n^2}{(n+2)(n+4)}\beta^5 + \mathcal{O}(\beta^7) \quad (132)$$

Note the two particular cases of the Ising model ($n=1$)

$$\xi^{-1} = -\ln\tanh\beta \quad (133)$$

and of the rigid rotator ($n=2$)

$$\xi^{-1} = -\ln\frac{I_1(2\beta)}{I_0(2\beta)} \quad (134)$$

In the Ising case, formula (133) can be established directly in a straightforward way. For any integer $n \geq 1$, the correlation length ξ is finite for all β and diverges at zero temperature ($\beta \to \infty$). For example, for $n=2$, $\exp(-1/\xi) \sim 1-1/4\beta+\mathcal{O}(1/\beta^2)$ and therefore $\xi \sim 4\beta$ as $\beta \to \infty$. In some sense, we can say that $\beta = \infty$ is the only *critical point*, namely the value of the temperature near which one can substitute a continuous model to the discrete one for the description of the large-distance behaviour. A similar property holds for all one-dimensional models with short-range interactions (Landau).

Two limiting cases are interesting. First, as n vanishes, t_0 tends to unity and one verifies easily that

$$e^{-1/\xi} = \frac{I_{n/2}(n\beta)}{I_{n/2-1}(n\beta)} = \beta\frac{I_{n/2}(n\beta)}{I_{n/2}(n\beta) + \beta I_{n/2+1}(n\beta)} \xrightarrow[n\to 0]{} \beta \quad (135)$$

Hence

$$\lim_{n \to 0} \langle \Phi^1(x_1)\Phi^1(x_2) \rangle = \beta^{|x_1 - x_2|} \qquad (136)$$

as expected since there is only one path without self-intersection going from x_1 to x_2. Positivity properties are lost whenever one performs analytic continuations; this is reflected here by a correlation function which can exceed unity.

Another limiting case is the one corresponding to an infinite number of components, which will be treated in greater detail later. One can use the asymptotic behaviour of Bessel functions

$$I_\nu(\nu z) \underset{\nu \to \infty}{\simeq} \frac{1}{\sqrt{2\pi\nu}} \exp \left\{ \nu \left[\sqrt{1 + z^2} + \ln \frac{z}{1 + \sqrt{1 + z^2}} \right] \right\} \qquad (137)$$

or remark that $t_1/t_0 = n^{-1} \langle \Phi_x \cdot \Phi_{x+1} \rangle$ is the value of $\cos\theta$ at the saddle point of the integrand $\exp n(\beta\cos\theta + \ln | \sin\theta |)$ in equation (125), yielding

$$-\beta\sin\theta_{\rm sp} + \cotan\theta_{\rm sp} = 0$$

As a result,

$$\frac{t_1}{t_0} = \cos\theta_{\rm sp} = \beta\left(1 - \frac{t_1^2}{t_0^2}\right) \qquad \text{as } n \to \infty$$

Hence

$$\frac{t_1}{t_0} = e^{-1/\xi} = \frac{\sqrt{1 + 4\beta^2} - 1}{2\beta} = \beta - \beta^3 + 2\beta^5 + \mathcal{O}(\beta^7) \qquad (138)$$

This expansion coincides with the large n limit of equation (132), as it should. The algebraic form of the equation giving t_1/t_0 in the infinite n limit can be derived in another instructive way. From the expression

$$\frac{1}{n} \langle \Phi \cdot \Phi' \rangle = \frac{\int d^n \Phi\, 2\delta(\Phi^2 - n) e^{\beta\Phi \cdot \Phi'}\, \Phi \cdot \Phi'/n}{\int d^n \Phi\, 2\delta(\Phi^2 - n) e^{\beta\Phi \cdot \Phi'}} \qquad (139)$$

the $\langle\rangle$ bracket can be interpreted as an average on the vector Φ keeping Φ' fixed of length $\sqrt{n}$. One can also average over the orientations of Φ'. Let us perform an infinitesimal rotation of the integration variables $\Phi \to (1 + \delta A)\Phi$, where δA is an infinitesimal antisymmetric $n \times n$ real matrix, in the numerator of equation (139). It is clear that the result is unchanged and, owing to the invariance of the measure, one finds

$$\delta\langle \Phi'^\alpha \Phi^\beta \rangle = \delta A_{\beta\gamma} \langle \Phi'^\alpha \Phi^\gamma \rangle + \beta\,\delta A_{\gamma\delta} \langle \Phi'^\alpha \Phi^\beta \Phi'^\gamma \Phi^\delta \rangle = 0$$

Identifying the coefficient of the expansion of δA in the basis $A_{(\mu\nu)\alpha\beta} = \delta_{\alpha\mu}\delta_{\beta\nu} - \delta_{\alpha\nu}\delta_{\beta\mu}$, one finds

$$0 = \delta_{\beta\mu}\langle\Phi'^{\alpha}\Phi^{\nu}\rangle - \delta_{\beta\nu}\langle\Phi'^{\alpha}\Phi^{\mu}\rangle + \beta\langle\Phi'^{\alpha}\Phi^{\beta}(\Phi'^{\mu}\Phi^{\nu} - \Phi'^{\nu}\Phi^{\mu})\rangle$$

Summing on the indices $\mu = \alpha$, $\nu = \beta$, one obtains the following relation, valid for any n

$$\left(1 - \frac{1}{n}\right)\frac{1}{n}\langle\Phi\cdot\Phi'\rangle = \beta\left(1 - \frac{1}{n^2}\langle(\Phi\cdot\Phi')^2\rangle\right) \qquad (140)$$

For $n = 1$, the relation is trivial, since $(\Phi\cdot\Phi')^2 = 1$. As n increases, one can use the steepest descent method, and the fluctuations of the invariant quantity $\Phi\cdot\Phi'/n$ become negligible. In this limit, mean values of invariant observables factorize. In particular,

$$\lim_{n\to\infty}\frac{1}{n^2}\langle(\Phi\cdot\Phi')^2\rangle = \left(\frac{1}{n}\langle\Phi\cdot\Phi'\rangle\right)^2 \qquad (141)$$

and hence

$$\lim_{n\to\infty}\left\{\frac{1}{n}\langle\Phi\cdot\Phi'\rangle + \beta\left[\left(\frac{1}{n}\langle\Phi\cdot\Phi'\rangle\right)^2 - 1\right]\right\} = 0 \qquad (142)$$

which is nothing but the algebraic equation leading to the result (138).

Absence of ordered phase in one dimension.

The very simple expressions obtained for the Ising model can be used to illustrate Landau's proof of the absence of phase transitions in one-dimensional models with short-range interactions. The partition function, normalized to 1 as $\beta \to \infty$, can be written for a system of size L in the form

$$\tilde{Z}_L(\beta) = e^{-L\beta}Z_L(\beta) = \left(1 + e^{-2\beta}\right)^L = \sum_{k=0}^{L}\binom{L}{k}e^{-2k\beta}$$

The last expression is interpreted as a partition of the system by k walls separating pure phases (i.e. with all spins up or down). Each wall leads to a variation -2β of the action, and there are $\binom{L}{k}$ ways to choose the positions of these walls. The partition function is a sum over all possible values of k. In the general case of the coexistence of two phases, the term -2β would be replaced by the interface free energy, and the entropy factor $\binom{L}{k}$ would be the same. Now, let us find the average value of k. If this average is

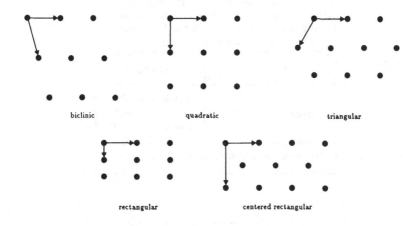

Fig. 6   Two-dimensional lattices.

small, the existence of pure phases over large intervals is favored. However, using Stirling's formula (which is a good approximation to the factorial, even for small values of the argument), we have

$$\tilde{F}(\beta) = \frac{1}{L} \ln \tilde{Z}_L(\beta) = -(x \ln x + (1-x)\ln(1-x) + 2\beta x) \qquad x = \frac{k}{L}$$

Hence, $\partial \tilde{F}/\partial x = 0$ yields $k = L/(1 + \exp(2\beta))$. For all finite β, k is proportional to L and disorder will be present in the thermodynamic limit. As $\beta \to \infty$ (zero temperature limit), k/L vanishes and this explains why the correlation length (L/k) increases indefinitely. One verifies that the substitution $x = k/L = (1 + \exp(2\beta))^{-1}$ gives $\tilde{F} = \ln(1 + e^{-2\beta})$ and that $\xi \sim \frac{1}{2}e^{2\beta}$ (as $\beta \to \infty$) agrees with $L/k \sim e^{2\beta}$.

This reasoning, valid for the coexistence of a finite number of phases, should be modified for a continuous symmetry, since there is a continuum of possible phases. The behaviour of the correlation length is only linear and no longer exponential, the coefficient of β varying from 4 to 2 as n grows from 2 to infinity.

Appendix 1.A   Lattices

In this appendix, we remind the reader of some characteristics of two- and three-dimensional regular lattices. Such a lattice is

Table II. Fundamental two-dimensional
lattices. The lengths of some conventional
translation vectors are denoted by a and b,
and the angle between them by φ.

lattice	definition
biclinic	$a \neq b,\ \varphi \neq 90°$
square	$a = b,\ \varphi = 90°$
triangular	$a = b,\ \varphi = 120°$
rectangular	$a \neq b,\ \varphi = 90°$
centered rectangular	$a \neq b,\ \varphi = 90°$

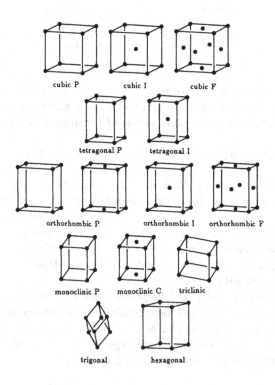

Fig. 7   three-dimensional lattices.

Table III. Fundamental three-dimensional lattices. The lengths of some conventional translation vectors are denoted a, b, c, and their respective angles α, β, γ.

lattice	definition
triclinic	$a \neq b \neq c\ \alpha \neq \beta \neq \gamma$
monoclinic $\begin{cases} P \\ C \end{cases}$	$a \neq b \neq c\ \alpha = \gamma = 90° \neq \beta$
orthorhombic $\begin{cases} P \\ C \\ I \\ F \end{cases}$	$a \neq b \neq c\ \alpha = \beta = \gamma = 90°$
tetragonal $\begin{cases} P \\ I \end{cases}$	$a = b \neq c\ \alpha = \beta = \gamma = 90°$
cubic $\begin{cases} P \text{ (simple cubic)} \\ I \text{ (centered cubic)} \\ F \text{ (face centered cubic)} \end{cases}$	$a = b = c\ \alpha = \beta = \gamma = 90°$
trigonal	$a = b = c\ \alpha = \beta = \gamma \neq 90°$
hexagonal	$a = b \neq c\ \alpha = \beta = 90°\ \gamma = 120°$

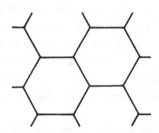

Fig. 8   Honeycomb lattice.

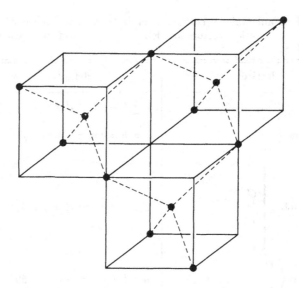

Fig. 9   Diamond lattice.

characterized by a group of translations

$$\mathbf{r} \rightarrow \mathbf{r} + \sum n_i \mathbf{e}_i$$

n_i being integers and $\mathbf{e}_i$ being a set of generators $(i = 1, \ldots, d)$. Besides translational invariance, there may also exist point symmetries. For instance, lattices may be invariant under cyclic rotation groups of order $n = 2$, 3, 4 or 6. Other values of n are excluded due to incompatibility with translational symmetry. The existence of a rotational invariance provides relations between the fundamental generators, leading to a classification of lattice types. Figures 6 and 7 display the five two-dimensional and fourteen three-dimensional types, while their characteristics are given in tables II and III.

More complex lattices are obtained by translating a group of sites. Among the various possibilities, let us give the examples of the honeycomb two-dimensional lattice (figure 8) and diamond lattice (figure 9).

We focus on lattices which present a maximal rotational symmetry. This is the case of the triangular and square lattices in two dimensions, and of the simple cubic, body centered cubic and face centered cubic in three dimensions. More generally, in higher dimensions, one can consider three general types of such lattices. They are the hypercubic lattice, with sites at integral position in an orthonormal frame, the body centered hypercubic lattice composed of two hypercubic lattices displaced from each other by the vector $(\frac{1}{2}, \frac{1}{2}, \ldots, \frac{1}{2})$, and the generalized fcc lattice, which is the cross section of a $(d+1)$-dimensional hypercubic lattice by the hyperplane $x_1 + \cdots + x_{d+1} = 0$.

Notes

Our presentation of Brownian motion is very schematic. A more complete treatment can be found in the article by E.W. Montroll and B.J. West in *Studies in Statistical Mechanics VII, Fluctuation Phenomena*, E.W. Montroll and J.L. Lebowitz eds, North Holland, Amsterdam (1979). The complete volume is very instructive reading and provides a wealth of references. Among the numerous mathematical references, we quote the two volume text by W. Feller *An introduction to probability theory and its applications*, second edition, John Wiley and Sons, New York (1971) and *Principles of Random Walk* by F. Spitzer, Springer Verlag, Heidelberg (1976).

Euclidean quantum field theory and its statistical interpretation is presented in a classical course of K. Symanzik in *Rendiconti della Scuola Internazionale di Fisica "Enrico Fermi"* XLV Corso, Teorica quantistica locale, R. Jost ed., Academic Press, New York (1969). The observation that the $O(n)$-symmetric model describes self-avoiding chains in the $n \to 0$ limit is due to P.G. de Gennes, *Phys. Lett.* **38A**, 339 (1972). For an extensive treatment of polymer physics, see J. des Cloizeaux and G. Jannink, *Les polymères en solution: leur modélisation et leur structure*, Les Editions de Physique (1987). Section 2 on interacting field theory as a weighted sum over paths is patterned after J. Fröhlich, *Nucl. Phys.* **B200** [FS4], 281 (1982) who discusses the triviality of $\lambda \phi^4$ theories in dimension larger or equal to four. Similar work is described by M. Aizenman in *Comm. Math. Phys.* **86**, 1 (1982). For the ferromagnetic inequalities, we refer to R.B. Griffiths, *J. Math. Phys.* **8**, 478 (1967).

2

GRASSMANNIAN INTEGRALS AND THE TWO-DIMENSIONAL ISING MODEL

Up to now we have considered models arising from classical statistical physics which only involve commuting variables. It is a fascinating aspect of Onsager's solution of the two-dimensional Ising model, that he was able to turn it into a problem involving fermions, indeed into a fermionic free field theory. The natural tools to deal with such systems combine path integral methods with anticommuting variables, described in the first section of this chapter. Such variables originate in the work of 19th century geometers (Grassmann) but their systematic use in a statistical or quantum context is however rather recent (Berezin). We present an overview of Onsager's solution in the second section, leading to the continuum (lattice independent) limit studied in the third section.

2.1 Grassmannian integrals

2.1.1 Anticommuting variables

Consider a finite number of N symbols η_i to be combined linearly using c-number coefficients, which satisfy the following (associative) multiplication rule

$$\eta_i \eta_j + \eta_j \eta_i = 0 \qquad (1)$$

i.e. they anticommute, which implies in particular that $\eta_i^2 = 0$. More generally $(\Sigma f^i \eta_i)^2 = 0$, for c-number coefficients f^i. The Grassmann algebra generated by these symbols contains all expressions of the form

$$f(\eta) = f^o + \sum_i f^i \eta_i + \sum_{i<j} f^{ij} \eta_i \eta_j + \sum_{i<j<k} f^{ijk} \eta_i \eta_j \eta_k + \cdots \ (2a)$$

48

$$= \sum_{0 \le k \le N} \frac{1}{k!} \sum_{\{i\}} f^{i_1 \ldots i_k} \eta_{i_1} \eta_{i_2} \cdots \eta_{i_k} \tag{2b}$$

where we extend the coefficients as antisymmetric tensors with k indices, each ranging from 1 to N. Since there are $\binom{N}{k} = N!/k!(N-k)!$ such linearly independent tensors, summing over k from 0 to N produces a 2^N-dimensional algebra. The word algebra is used here, since equation (1) allows us to define an associative product

$$
\begin{aligned}
f(\eta)g(\eta) = f^0 g^0 &+ \sum_i \left(f^0 g^i + f^i g^0 \right) \eta_i \\
&+ \tfrac{1}{2} \sum_{i,j} \left(f^{ij} g^0 + f^i g^j - f^j g^i + f^0 g^{ij} \right) \eta_i \eta_j + \cdots
\end{aligned}
\tag{3}
$$

In general note that fg is not equal to $\pm gf$. Nevertheless the subalgebra containing terms with an even number (possibly zero) of η variables commutes with any element f.

We can draw an analogy with the commuting case, where the η_i's would be replaced by commuting indeterminates X_i, i ranging from 1 to N, and the functions by finite power series (at first) of arbitrary high degree. The coefficients in the latter can be considered as symmetric tensors (of arbitrary high rank). Whereas in the antisymmetric case the number of coefficients of rank k is obtained by expanding $(1+A)^N$ and looking for the term in A^k, i.e. $\binom{N}{k}$, in the symmetric case we have to expand $(1-A)^{-N}$ with a coefficient of A^k given by $\binom{N+k-1}{k} = (N+k-1)!/k!(N-1)!$.

One can present the above expansions in yet another guise which clearly exhibits the relation with Fermi statistics, namely

$$ f = \sum_{n_i = 0,1} f_{n_1, n_2, \ldots, n_N} \eta_1^{n_1} \eta_2^{n_2} \cdots \eta_N^{n_N} \tag{2c}$$

The integers $n_i = 0, 1$ can be thought of as occupation numbers of "states" described by η_i. In the commuting case we have a similar expression with X_i replacing η_i and integers n_i unbounded.

A possible choice for a particle-hole-correspondence (in the description given by (2b)) is $f \to {}^*f$ where

$$
{}^*f_{\mu_1 \ldots \mu_k} = \frac{1}{(N-k)!} \sum_{\mu_{k+1}, \ldots, \mu_N} \varepsilon_{\mu_1 \ldots \mu_k \mu_{k+1} \ldots \mu_N} f_{\mu_{k+1} \ldots \mu_N} \tag{4a}
$$

and ε is the totally antisymmetric (Levi Civita) symbol such that $\varepsilon_{12 \ldots N} = 1$. Show that

(i) in the description (2c) ${}^*f_{n_1 \ldots n_N} = \pm f_{n_1+1, \ldots, n_N+1}$ where the occupation numbers are understood modulo two. Discuss the corresponding sign.

(ii) $\qquad\qquad {}^{**}f_{\mu_1 \ldots \mu_k} = (-1)^{k(N-k)} f_{\mu_1 \ldots \mu_k} \tag{4b}$

(iii) the coefficient of $\eta_1 \ldots \eta_N$ in ${}^*f(\eta) f(\eta)$ is

$$
\sum_k \left(\frac{1}{k!} \sum_{\mu_1 \ldots \mu_k} f^2_{\mu_1 \ldots \mu_k} \right) \tag{4c}
$$

In the commuting case one introduces a Hilbert space structure, such that two distinct monomials are orthogonal. An orthonormal basis is given by the elements

$$
X_{\{n\}} \equiv \frac{X_1^{n_1}}{(n_1!)^{1/2}} \frac{X_2^{n_2}}{(n_2!)^{1/2}} \cdots \frac{X_N^{n_N}}{(n_N!)^{1/2}} \quad \to | \{n\} > \tag{5}
$$

Correspondingly the finite series are complemented by infinite ones, defining the Fock–Bargmann space of entire functions. The orthonormality property of the state vectors (5) is expressed using integrals over products of complex planes in the form

$$
\delta_{\{n\}, \{m\}} = \langle \{n\} \mid \{m\} \rangle = \int \prod_{j=1}^N \left(\frac{\mathrm{d}^2 X_j}{2\pi} e^{-X_j \bar{X}_j} \right) \bar{X}_{\{n\}} X_{\{m\}} \tag{6}
$$

The bar is complex conjugation, and $\mathrm{d}^2 X$ stands for $\mathrm{d}\,\mathrm{Re}X\,\mathrm{d}\,\mathrm{Im}X$. A general vector is represented by an entire series

$$
B = \sum_{\{n\}} b_{\{n\}} X_{\{n\}} \tag{7a}
$$

with a square norm

$$
\langle B \mid B \rangle = \sum_{\{n\}} | b_{\{n\}} |^2 = \int \prod_{j=1}^N \left(\frac{\mathrm{d}^2 X_j}{2\pi} e^{-X_j \bar{X}_j} \right) \overline{B(X)}\, B(X) \tag{7b}
$$

An analogous formula holds for scalar products. This construction is supplemented by the definition of creation and annihilation operators satisfying the Bose commutation rules

$$[a_i, a_j^\dagger] = \delta_{ij} \tag{8a}$$

represented by multiplications and derivations when acting on functions

$$a_i^\dagger \rightarrow X_i \qquad\qquad a_i \rightarrow \frac{\partial}{\partial X_i} \tag{8b}$$

That these are adjoint operators follows from the exponential factor in the norm.

As Berezin showed, this structure has its exact parallel in the fermionic case. Having already defined sums and products in the Grassmann algebra we now define a left derivative $\partial/\partial\eta_i$. The latter gives zero on a monomial which does not contain the variable η_i. If the monomial contains η_i, the latter is moved to the left, sometimes at the price of a minus sign, then is suppressed. The operation is then extended by linearity to any element of the algebra. One could similarly define a right derivative. It is easily verified that

$$\frac{\partial}{\partial\eta_i}\frac{\partial}{\partial\eta_j} + \frac{\partial}{\partial\eta_j}\frac{\partial}{\partial\eta_i} = 0 \tag{9a}$$

$$\frac{\partial}{\partial\eta_i}\eta_j + \eta_j\frac{\partial}{\partial\eta_i} = \delta_{ij} \tag{9b}$$

The first relation is equivalent to the fact that any linear combination (with c-number coefficients) has a vanishing square. It also shows that derivatives $\partial/\partial\eta_i$ generate in turn a Grassmann algebra, isomorphic with the one expressed in terms of η_i. These quantities may then be viewed as (right-)derivatives on the $\partial/\partial\eta_i$ algebra. Equation (9b) allows one to define fermionic creation and annihilation operators

$$a_i^\dagger \rightarrow \eta_i \qquad\qquad a_i \rightarrow \frac{\partial}{\partial\eta_i}$$

$$\{a_i, a_j\} = \{a_i^\dagger, a_j^\dagger\} = 0 \qquad \{a_i, a_j^\dagger\} = \delta_{ij} \tag{10}$$

The bracket $\{A, B\}$ is the symmetric anticommutator $AB + BA$. The algebra of functions $f(\eta)$ is turned into a Hilbert space through the definition of a scalar product

$$\langle g, f \rangle = \sum_{k=1}^{N} \frac{1}{k!} \sum_{i_1,\ldots,i_k} \bar{g}^{i_1 \ldots i_k} f^{i_1 \ldots i_k} \tag{11}$$

where the bar stands again for complex conjugation. If $\bar{\eta}_i$ is a second family of N anticommuting variables (also anticommuting with the η_i's), one can introduce an antilinear one-to-one map

$$f \to \bar{f} \qquad \sum_{k} \sum_{\{i\}} f^{i_1 \ldots i_k} \eta_{i_1} \cdots \eta_{i_k} \to \sum_{k} \sum_{\{i\}} \bar{f}^{i_1 \ldots i_k} \bar{\eta}_{i_k} \cdots \bar{\eta}_{i_1} \tag{12}$$

Note the reverse ordering. This map extends to the 2^{2N} dimensional algebra including products of η's and $\bar{\eta}$'s.

Returning to the creation and annihilation operators, consider the $2N$ combinations

$$\Gamma_j = a_j + a_j^\dagger \qquad \Gamma_{N+j} = i(a_j - a_j^\dagger) \qquad 1 \le j \le N \tag{13}$$

These can be also viewed as $2^N \times 2^N$ matrices, and they satisfy the relations

$$\{\Gamma_s, \Gamma_{s'}\} = 2\delta_{s,s'} \qquad 1 \le s, s' \le 2N \tag{14}$$

characteristic of a Clifford algebra. It can be checked that the N pairs $(a_i, a_i^\dagger)$ or the $2N$ combinations Γ_s generate by sums and products all the 2^{2N} linearly independent operators on the original Grassmann algebra. The case $N = 1$ corresponds to quaternions (or Pauli matrices), while if $N = 2$ we have the (four-dimensional) algebra of Dirac matrices.

(i) For $N = 1, 2$ write explicitly the operators $a, a^\dagger, \Gamma$ in the basis of polynomials in η.

(ii) Study the effect of the duality operation, given by equation (4).

(iii) Describe the linear canonical group on creation and annihilation operators in the Bose or Fermi case (symplectic or orthogonal group).

2.1.2   Integrals

The parallel with the Fock–Bargmann construction can be extended by defining anticommuting integrals as linear operations on the functions $f(\eta)$, with the seemingly paradoxical property that they can be identified with the (left) derivatives! Correspondingly

$$\int d\eta f(\eta) = \frac{\partial f}{\partial \eta}, \qquad \int d\eta_1 d\eta_2 f(\eta_1, \eta_2) = \frac{\partial}{\partial \eta_1} \frac{\partial}{\partial \eta_2} f(\eta_1, \eta_2), \qquad \cdots$$
(15)

This yields for the scalar product $\langle g, f \rangle$ defined by equation (11) the following integral representation

$$\langle g, f \rangle = \int d\eta_N d\bar{\eta}_N \cdots d\eta_1 d\bar{\eta}_1 \exp\left(-\sum_1^N \eta_i \bar{\eta}_i\right) \bar{g}(\bar{\eta}) f(\eta) \quad (16)$$

This can be verified by expanding the exponential and using the definition (15)

$$\langle g, f \rangle = \sum \frac{\bar{g}^{j_1 \cdots j_q}}{q!} \frac{f^{i_1 \cdots i_k}}{k!}$$

$$\frac{\partial}{\partial \eta_N} \cdots \frac{\partial}{\partial \bar{\eta}_1} (1 + \bar{\eta}_1 \eta_1) \cdots (1 + \bar{\eta}_N \eta_N) \, \bar{\eta}_{j_q} \cdots \bar{\eta}_{j_1} \eta_{i_1} \cdots \eta_{i_k}$$

Nonvanishing terms arise only when $k = q$ and when the indices j are a permutation of the i's. Taking into account the antisymmetry of the coefficients, this means that the above expression reduces to

$$\langle g, f \rangle = \sum_k \frac{1}{k!} \sum_{\{i\}} \bar{g}^{i_1 \cdots i_k} f^{i_1 \cdots i_k}$$

$$\frac{\partial}{\partial \eta_N} \cdots \frac{\partial}{\partial \bar{\eta}_1} (1 + \bar{\eta}_1 \eta_1) \cdots (1 + \bar{\eta}_N \eta_N) \, \bar{\eta}_{i_n} \cdots \bar{\eta}_{i_1} \eta_{i_1} \cdots \eta_{i_k}$$

with the indices i all distinct. The last factor can be rewritten $\bar{\eta}_{i_1} \eta_{i_1} \cdots \bar{\eta}_{i_k} \eta_{i_k}$. This shows that all other missing pairs $\bar{\eta}\eta$ arise from the other factor $\prod_1^N (1 + \bar{\eta}_s \eta_s)$ to produce once and only once $\bar{\eta}_1 \eta_1 \cdots \bar{\eta}_N \eta_N$, the derivative of which is one. The final result is indeed the combination given in equation (11).

Given this integral expression for the scalar product, we can now proceed to the corresponding representation of operators through integral kernels. The fermionic case (for N finite) is even simpler than its bosonic counterpart, everything being now algebraic as opposed to analytic, which requires us, in principle, to worry about domains of definitions for operators... Only signs have to be watched carefully.

To any linear operator A we associate a kernel $\mathcal{A}$. To comply with a convenient convention we will henceforth denote the members of the Grassmann algebra on which A operates, $f(\bar{\xi})$ or $f(\bar{\eta})$... In other words we permute the roles of η and $\bar{\eta}$. Hence

$$(Af)(\bar{\xi}) = \int d\bar{\eta}_N d\eta_N \cdots d\bar{\eta}_1 d\eta_1 \exp\left(-\sum_1^N \bar{\eta}_i \eta_i\right) \mathcal{A}(\bar{\xi}, \eta) f(\bar{\eta}) \tag{17}$$

For brevity we shall use the compact notation $d\bar{\eta}d\eta$ for $d\bar{\eta}_N d\eta_N \cdots d\bar{\eta}_1 d\eta_1$ and $\bar{\eta}\eta$ for $\bar{\eta}_1\eta_1 + \cdots + \bar{\eta}_N\eta_N$.

A typical example is the integral representation of the identity operator, in analogy with the commutative case

$$f(\bar{\xi}) = \int d\bar{\eta}d\eta\, e^{-\bar{\eta}\eta} e^{\bar{\xi}\eta} f(\bar{\eta}) \tag{18}$$

Check relation (18) and ascertain the numerical factor (π^{-N}) for the corresponding identity in the Bose case (hint: use the same reasoning as the one leading to equation (16) with $e^{\bar{\xi}\eta}$ playing the role of $\bar{g}(\eta)$).

Equation (18) enables one to represent any operator by its corresponding kernel $\mathcal{A}(\bar{\xi}, \eta) = A e^{\bar{\xi}\eta}$, where the right-hand side is interpreted as the action of the operator A on the kernel of the identity considered as a function of $\bar{\xi}$. Thus we have for instance (remember the $\eta \leftrightarrow \bar{\eta}$ interchange)

$$a_i \to e^{\bar{\xi}\eta}\eta_i \qquad a_i^\dagger \to \bar{\xi}_i e^{\bar{\xi}\eta} \tag{19}$$

and more generally, omitting indices,

$$aA \to \frac{\partial}{\partial \bar{\xi}} \mathcal{A}(\bar{\xi}, \eta) \qquad Aa \to \mathcal{A}(\bar{\xi}, \eta)\eta$$

$$a^\dagger A \to \bar{\xi}\mathcal{A}(\bar{\xi}, \eta) \qquad Aa^\dagger \to \mathcal{A}(\bar{\xi}, \eta)\frac{\overleftarrow{\partial}}{\partial \eta} \tag{20}$$

In the last relation $\overleftarrow{\partial}/\partial\eta$ means that $\partial/\partial\eta$ operates as a right derivative. Let : : be the symbol for Wick ordering an expression in a and $a^\dagger$'s. It requires writing the annihilation operators a to the right of the creation ones $a^\dagger$, with due respect for the sign of the corresponding permutation. For instance $: a^\dagger a : = a^\dagger a$, $: aa^\dagger : = -a^\dagger a$, etc... Observe that we do not use the commutation rules for otherwise we would be led to the absurd result that $I =: I :=: aa^\dagger + a^\dagger a := -a^\dagger a + a^\dagger a = 0$! Then, for any monomial

$$: a_{\alpha_1}^\dagger \ldots a_{\alpha_s}^\dagger a_{\beta_1} \ldots a_{\beta_t} :\to \bar\xi_{\alpha_1} \ldots \bar\xi_{\alpha_s} e^{\bar\xi\eta} \eta_{\beta_1} \ldots \eta_{\beta_t} \qquad (21)$$

Check formulae (20) and (21).

Therefore to obtain the kernel corresponding to an operator in normal order form, we simply substitute $\bar\xi$ for $a^\dagger, \eta$ for a and supply a (commuting) factor $e^{\bar\xi\eta}$.

Using the scalar product we endow the Grassmann algebra with a Hilbert space structure. Corresponding to Hermitian conjugation for operators, we have the following operation on kernels

$$A \to A^\dagger \qquad \mathcal{A}(\bar\eta, \eta) \to \overline{\mathcal{A}(\bar\eta, \eta)} \qquad (22)$$

where the bar over $\mathcal{A}$ implies complex conjugation of coefficients and substitutes for a monomial $\bar\eta_{\alpha_1} \ldots \bar\eta_{\alpha_s} \eta_{\beta_1} \ldots \eta_{\beta_t}$ its conjugate $\bar\eta_{\beta_t} \ldots \bar\eta_{\beta_1} \eta_{\alpha_s} \ldots \eta_{\alpha_1}$ implying the exchange $\bar\eta \leftrightarrow \eta$ together with a reverse ordering. The unit operator being Hermitian, the corresponding kernel is self-conjugate according to (22). The kernels corresponding to a pair $a_i, a_i^\dagger$ are also conjugate.

The kernel for a product is obtained by convolution as

$$A \to \mathcal{A} \qquad B \to \mathcal{B} \qquad AB \to \mathcal{C}$$
$$\mathcal{C}(\bar\xi, \xi) = \int d\bar\eta d\eta \, \mathcal{A}(\bar\xi, \eta) e^{-\bar\eta\eta} \mathcal{B}(\bar\eta, \xi) \qquad (23)$$

Moreover the trace can also be written as

$$\text{Tr}\, A = \int d\bar\eta d\eta \, e^{-\bar\eta\eta} \mathcal{A}(-\bar\eta, \eta) \qquad (24)$$

Notice the characteristic minus sign in the argument of $\mathcal{A}$. Changing $\eta, \bar\eta$ into $-\eta, -\bar\eta$, which does not affect the integral, one could instead use $\mathcal{A}(\bar\eta, -\eta)$. From equations (18) and (24) it follows that

$$\text{Tr}\, I = \int d\bar\eta d\eta \, e^{-2\bar\eta\eta} = 2^N \qquad (25)$$

consistent with the dimension 2^N of the underlying vector space, the Grassmann algebra.

Let us look at some specific properties of anticommuting integrals. Changes of coordinate have to preserve the anticommuting structure of the Grassmann algebra. This allows non-singular linear transformations (with c-number coefficients) of the form

$$\bar{\eta}_i = \sum_{j=1}^{N} J_i^j \bar{\xi}_j \qquad \det J = \det J_i^j = \frac{D(\bar{\eta})}{D(\bar{\xi})} \qquad (26a)$$

One verifies that by setting

$$f(\bar{\eta}) = F(\bar{\xi}) \qquad (26b)$$

one obtains the following relation

$$\int d\bar{\xi}\, F(\bar{\xi}) = \det J \int d\bar{\eta}\, f(\bar{\eta}) \qquad (26c)$$

in contradistinction to the commuting case where the coefficient on the right-hand side would be $|\det J|^{-1}$ involving both an absolute value sign and an inversion. On reflection it is realized that (26c) results in the present case from the identification of derivation and integration. As another consequence we have

$$\int d\bar{\eta}\, \frac{\partial}{\partial \bar{\eta}} f(\bar{\eta}) = 0 \qquad (27)$$

where the commutative analog is a boundary contribution. An important application of these formulae is the one concerning Gaussian integrals, the only nontrivial ones that can be done in practice. First we have to look for the anticommutative analog of quadratic forms. These are obviously antisymmetric forms. Omitting bars on the variables, such a form reads $\sum_{ij} \eta_i Q^{ij} \eta_j$ with $Q^{ij} + Q^{ji} = 0$. The expansion of $\exp \frac{1}{2} \sum_{ij} \eta_i Q^{ij} \eta_j$ only involves even monomials. Consequently an integral over all variables vanishes unless the total number of variables is even: $N = 2n$. In this case

$$\int d\eta_{2n} \ldots d\eta_1 \exp \frac{1}{2} \sum_{ij} \eta_i Q^{ij} \eta_j$$

$$= \int d\eta_{2n} \ldots d\eta_1 \prod_{i<j} \left(1 + Q^{ij} \eta_i \eta_j \right)$$

$$= \frac{1}{2^n n!} \sum_{\sigma} (-1)^\sigma Q^{\sigma_1 \sigma_2} \ldots Q^{\sigma_{2n-1} \sigma_{2n}} \equiv \mathrm{Pf}(Q) \qquad (28)$$

Here σ runs over permutations of the $2n$ indices, $(-1)^\sigma$ is the parity of the permutation. The prefactor $2^{-n}(n!)^{-1}$ is the number of those permutations which leave any one of the terms invariant. Omitting this factor leads to a sum of distinct pairings (Wick's theorem), each one with the corresponding $\pm$ sign. Finally the whole expression is called a Pfaffian, hence the symbol $\mathrm{Pf}(Q)$. According to (26), a linear change of variables yields

$$\mathrm{Pf}(^{T}\!JQJ) = \det(J)\,\mathrm{Pf}(Q) \qquad (29)$$

As a particular case of (28) and (29) suppose that we split the $2n$ variables $\eta_i, 1 \le i \le N = 2n$ into two sets $\xi_a, \bar{\xi}_a$ of n variables in such a way that the matrix Q only connects ξ and $\bar{\xi}$ variables. In other words $\{\eta_1 \ldots \eta_{2N}\} \equiv \{\bar{\xi}_1 \ldots \bar{\xi}_n \xi_1 \ldots \xi_n\}$ and $\frac{1}{2}\eta Q\eta = -\bar{\xi}K\xi$, meaning that the matrix Q has the form

$$Q = \begin{pmatrix} 0 & -K \\ K^T & 0 \end{pmatrix} \qquad (30)$$

where K is any $n \times n$ matrix. Then

$$(-)^{n(n+1)/2}\mathrm{Pf}(Q) = \int \mathrm{d}\bar{\xi}_n \mathrm{d}\xi_n \cdots \mathrm{d}\bar{\xi}_1 \mathrm{d}\xi_1\, \mathrm{e}^{-\bar{\xi}K\xi} = \det K \qquad (31)$$

To see this it is sufficient to make the change of variables $\xi' = K\xi$, $\bar{\xi}' = \bar{\xi}$ and to note that $\int \mathrm{d}\bar{\eta}\mathrm{d}\eta\,\mathrm{e}^{-\bar{\eta}\eta} = 1$. Equation (31) is a widely used formula and has the following consequence. Squaring both sides of equation (28) we get

$$\mathrm{Pf}(Q)^2 = \int \mathrm{d}\eta_{2n} \cdots \mathrm{d}\eta_1 \mathrm{d}\xi_{2n} \cdots \mathrm{d}\xi_1 \exp \tfrac{1}{2}(\eta Q\eta + \xi Q\xi)$$

Set $\xi_j = (\bar{\rho}_j - \rho_j)/\mathrm{i}\sqrt{2}$, $\eta_j = (\bar{\rho}_j + \rho_j)/\sqrt{2}$, for $1 \le j \le 2n$. The Jacobian is unity. Applying (31) we get for any antisymmetric even-dimensional matrix Q

$$\mathrm{Pf}(Q)^2 = \int \mathrm{d}\bar{\rho}_{2n} \cdots \mathrm{d}\rho_1 \exp -\bar{\rho}Q\rho = \det Q \qquad (32)$$

Of course, if Q is odd-dimensional, $\det Q = 0$.

Mean values with a Gaussian weight satisfy Wick's theorem, again as in the commutative case. This follows from yet another invariance property of the measure $\mathrm{d}\eta$ under a translation $\eta \to \eta + \xi$ with ξ another set of anticommuting variables

$$\int \mathrm{d}\eta\, f(\eta) = \int \mathrm{d}\eta\, f(\eta + \xi) \qquad (33)$$

Assuming $\mathrm{Pf}(Q) \neq 0$ (hence $\det Q \neq 0$ and the number of variables is even) define the mean value of any expression $O(\eta)$ as

$$\langle O \rangle = \frac{\int d\eta \, O(\eta) \exp \frac{1}{2}\eta Q \eta}{\int d\eta \exp \frac{1}{2}\eta Q \eta} \tag{34}$$

Then, using translational invariance, we can derive the analog of the generating functional (1.74)

$$\langle \exp \xi \eta \rangle = \exp -\tfrac{1}{2}\xi Q^{-1}\xi \tag{35}$$

which is the compact form of Wick's theorem. If we denote by $(Q^{-1})_{ij}$ the ij component of Q^{-1}, it follows from differentiating (35) with respect to the ξ_i's that

$$\langle \eta_1 \eta_2 \rangle = (Q^{-1})_{12}$$
$$\langle \eta_1 \eta_2 \eta_3 \eta_4 \rangle = (Q^{-1})_{12}(Q^{-1})_{34} - (Q^{-1})_{13}(Q^{-1})_{24} + (Q^{-1})_{14}(Q^{-1})_{23}$$
$$\langle \eta_1 \ldots \eta_{2p} \rangle = \mathrm{Pf}(Q^{-1})_{[1\ldots 2p]} \tag{36}$$

where the last notation implies a restriction of Q^{-1} to the subspace $[1 \ldots 2p]$, for $2p \leq 2n$. This leads to a sum over products of all distinct contractions affected by the sign of the corresponding permutation of indices. A particular case corresponds to a matrix Q of the form indicated in equation (30).

2.2   The two-dimensional Ising model

Let us apply Grassmannian integrals to the solution of the two-dimensional Ising model (Onsager 1944). The model is intended to simulate the Curie transition from a ferro- to a paramagnetic phase. We can also interpret it as a model of the liquid gas transition (the lattice-gas model) where sites are either occupied or empty, or to binary mixtures or alloys. We shall use the magnetic interpretation. The starting point is the partition function in the absence of a magnetic field on a large but finite (regular) lattice with V sites. Pairs of neighbouring sites are denoted (ij). At each site is attached a "spin variable" σ_i taking the values ± 1. Boltzmann's constant as well as other energetic units are lumped together in the factor β interpreted as inverse "temperature" in dimensionless units. The partition function is written

$$Z_V(\beta) = \frac{1}{2^V} \sum_{\sigma_i = \pm 1} \exp \beta \sum_{(ij)} \sigma_i \sigma_j \qquad (37)$$

It may seem at first extraordinary that such a simple looking expression is so difficult to compute, as experience has taught us! By going to the infinite volume (or thermodynamic) limit we define the free energy (without a $-\beta$ factor), as

$$F(\beta) = \lim_{V \to \infty} \frac{1}{V} \ln Z_V(\beta) \qquad (38)$$

In the following the infinite volume limit will be implicit most of the time. In the remainder of this chapter, we restrict ourselves to the two-dimensional case, so volume should properly be replaced by area. Also we use a square lattice, each site having four neighbours. Such a lattice admits (geometrically) a notion of duality with the roles of sites and plaquettes (i.e. elementary squares) being interchanged. This leads to an important duality transformation for the Ising model. A generalization is discussed in chapter 6 on lattice gauge fields.

2.2.1   Duality

We have seen in Chapter 1 that $Z_V(\beta)$ can be expanded in powers of β, or even better in powers of

$$t = \tanh \beta \qquad (39)$$

after factoring out $(\cosh \beta)^{2V}$. The exact exponent is the number of links belonging to V which may differ from $2V$ by a term negligible compared to V in the thermodynamic limit (it is precisely $2V$ for periodic boundary conditions). The expansion assumes the form

$$Z_V(\beta) = (\cosh \beta)^{2V} \sum_{\partial \{n_l\} = \emptyset} t^{\sum n_l} \qquad (40)$$

Each integer $n_l = 0, 1$ refers to a link between neighbours. The boundary $\partial \{n_l\}$ of such a configuration is the set of sites where the sum of n's attached to incident links is odd. If $\partial \{n_l\} = \emptyset$, each site belongs to an even number (0, 2, 4) of links in the configuration (those for which $n_l = 1$).

Let us now use the dual (plaquette) lattice with corresponding binary variables $\tau_\kappa = \pm 1$. Each link of the new lattice (pair of neighbouring plaquettes) is in one-to-one correspondence with the link of the original lattice on which it is incident. Implicitly we already think in terms of the infinite volume limit and neglect boundary effects. We set up a one-to-two correspondence between configurations $\{n_l\}$ with empty boundaries $\partial \{n_l\} = \emptyset$ and assignments ± 1 to plaquette variables τ_κ, by requiring that adjacent plaquettes carry the same value for τ's if they are incident on a link l such that $n_l = 0$, or opposite values of τ's if $n_l = 1$. Obviously reversing all the τ's leads to a second configuration with the same property, while $\partial \{n_l\} = \emptyset$ insures the consistency of the assignment as one will check by turning around a point of the original lattice. The overall duplication is irrelevant in the calculation of the free energy (38). According to the above procedure

$$\sum n_l = \sum_{(ab)} \tfrac{1}{2}(1 - \tau_a \tau_b) \qquad (41)$$

On the left, the sum runs over the links of the original lattice, on the right, on the dual set of pairs of adjacent plaquettes. Finally, it is realized that the dual lattice is isomorphic to the original one, again in the infinite volume limit. Let us define variables $\tilde{\beta}$ and $\tilde{t} = \tanh \tilde{\beta}$ according to

$$t = \tanh \beta = \mathrm{e}^{-2\tilde{\beta}} = \frac{1 - \tilde{t}}{1 + \tilde{t}} \qquad \tilde{t} = \tanh \tilde{\beta} \qquad (42)$$

Of course $\tilde{\beta}$ runs from zero to infinity and $\tilde{t}$ from zero to one. In the duality transformation, small and large temperatures are interchanged and $\tilde{t} = t$. Using this notation, we rewrite equation (40) as

$$Z_V(\beta) = (\cosh \beta)^{2V} (2\mathrm{e}^{-2\tilde{\beta}})^V \frac{1}{2^{V+1}} \sum_{\tau_a} \exp\left(\tilde{\beta} \sum_{(ab)} \tau_a \tau_b \right) \qquad (43)$$

Up to

$$\left(2 \cosh^2 \beta \mathrm{e}^{-2\tilde{\beta}} \right)^V = (\sinh 2\beta)^V = \left(\sinh 2\tilde{\beta} \right)^{-V}$$

and inessential nonthermodynamic factors, we have reconstructed the original partition function on the dual lattice evaluated at the dual value $\tilde{\beta}$. Hence we obtain the duality formula (Kramers–Wannier)

$$F(\beta) = \ln \sinh 2\beta + F(\tilde{\beta}) \tag{44}$$

The relation

$$\sinh 2\beta \, \sinh 2\tilde{\beta} = 1 \tag{45}$$

shows both the consistency of equation (44) and its involutory nature. Duality relates observables and their dual counterparts in the high and low temperature regimes.

The fixed point of the transformation (42), i.e. the one for which $\beta = \tilde{\beta} = \beta_c$, satisfies

$$t_c = \sqrt{2} - 1 \qquad \sinh 2\beta_c = 1 \qquad \cosh 2\beta_c = \sqrt{2}$$
$$\beta_c = \tfrac{1}{2}\ln(\sqrt{2}+1) = 0.440686794\ldots \tag{46}$$

The forthcoming solution will confirm that β_c is a critical value that separates a disordered high temperature phase $(\beta < \beta_c)$ from an ordered low temperature one $(\beta > \beta_c)$. The latter is characterized by a spontaneous magnetization M which can be recovered, even in the absence of an external field, from the large distance behaviour of correlations. For instance

$$\lim_{|i-j|\to\infty} \left\langle \sigma_i \sigma_j \right\rangle = M^2 \tag{47}$$

with $M^2 > 0$ for $\beta > \beta_c$.

(i) For a doubly periodic finite square lattice with V points, L links and P plaquettes, supply in the above derivation of duality the exact nonthermodynamic factors. Hint: use Euler's relation $V - L + P = 0$. Discuss the case of free boundary conditions.

(ii) Observe that (a) a triangular and an hexagonal lattice with twice as many points, are (geometrically) dual to each other (b) a decimation procedure may be set up on the latter which maps an Ising model back on a triangular lattice. Therefore find the critical temperature in both cases

$$\begin{array}{ll} \text{triangular} & \mathrm{e}^{2\beta_c} = \sqrt{3} \\ \text{hexagonal} & \mathrm{e}^{2\beta_c} = 2 + \sqrt{3} \end{array} \tag{48}$$

Since the number of neighbours of a site is 3, 4 and 6 in the hexagonal, square and triangular lattices respectively, one expects for equal densities that the corresponding critical temperatures are increasing in the same order. Check this fact.

2.2.2   Transfer matrix

To compute the partition function we use, as in the one-dimensional case, a transfer matrix T. Properly speaking it is in fact an operator. The latter acts on spin configurations along a row of length L (the x-direction) repeated M times (in the y- or "time" direction), thus enclosing $V = LM$ sites. To simplify future expressions the partition function is written as

$$Z_{LM}(\beta) = \left(\tfrac{1}{2}\sinh 2\beta\right)^{LM/2} \operatorname{Tr} T^M \tag{49}$$

and therefore assumes periodic boundary conditions in the y-direction as dictated by the trace occuring on the right-hand side. State vectors are described by values $\sigma_x = \pm 1$ with x ranging from 0 to $L-1$. We consider these values as eigenvalues of two by two Pauli matrices $\sigma_x^{(1)}$ with the usual definitions

$$\sigma^{(1)} = \begin{pmatrix} 0 & 1 \\ 1 & 0 \end{pmatrix} \qquad \sigma^{(2)} = \begin{pmatrix} 0 & -i \\ i & 0 \end{pmatrix} \qquad \sigma^{(3)} = \begin{pmatrix} 1 & 0 \\ 0 & -1 \end{pmatrix}$$

$$\sigma^{(1)}\sigma^{(2)} = i\sigma^{(3)}, \qquad \sigma^{(2)}\sigma^{(3)} = i\sigma^{(1)}, \qquad \sigma^{(3)}\sigma^{(1)} = i\sigma^{(2)} \tag{50}$$

The choice of $\sigma^{(1)}$ rather than $\sigma^{(3)}$, which would seem more natural at this stage, is dictated by future convenience. The transfer matrix T is then factorized into a contribution θ collecting the interactions of spins along a row, and $\tilde{\theta}$ involving interactions between spins located on two successive rows

$$T = \theta\tilde{\theta} \tag{51}$$

This immediately gives

$$\theta(\beta) = \exp\beta \sum_x \sigma_x^{(1)}\sigma_{x+1}^{(1)} \tag{52}$$

where the sum over x runs from 0 to $L-2$ for open boundary conditions or from 0 to $L-1$ for periodic boundary conditions in which case $\sigma_L^{(k)} = \sigma_0^{(k)}$. As for $\tilde{\theta}$ it is factorized into commuting contributions from each site along the row. Let V be a matrix which diagonalizes $\sigma^{(1)}$, i.e.

$$V^{-1}\sigma^{(1)}V = -\sigma^{(3)} \qquad V^{-1}\sigma^{(2)}V = \sigma^{(2)} \qquad V^{-1}\sigma^{(3)}V = \sigma^{(1)} \tag{53}$$

Remembering the factor extracted in (49) as well as the one included in the normalisation of the partition function, the contribution of site x to $\tilde\theta_x$ is,

$$V^{-1}\tilde\theta_x V = \frac{1}{2(\sinh\beta\cosh\beta)^{1/2}}\begin{pmatrix} e^\beta & e^{-\beta} \\ e^{-\beta} & e^\beta \end{pmatrix}$$

$$= \frac{e^\beta}{2(\sinh\beta\cosh\beta)^{1/2}}\left[1+\tanh\tilde\beta\sigma_x^{(1)}\right]$$

Remarking that $e^\beta(4\sinh\beta\cosh\beta)^{-1/2}$ is $\cosh\tilde\beta$, and using (53) we deduce that

$$\tilde\theta(\tilde\beta) = e^{\tilde\beta\sum_x \sigma_x^{(3)}} \tag{54}$$

completing the construction of the transfer matrix T. One observes a parallelism between the roles of β and $\tilde\beta$. There remains however a difference between the operators $\sigma_x^{(1)}\sigma_{x+1}^{(1)}$ and $\sigma_x^{(3)}$, both Hermitian and with a square equal to one (since $\sigma_x^{(k)}$ and $\sigma_{x+1}^{(k')}$ commute) involved respectively in θ and $\tilde\theta$. This difference will be eliminated by the fermionic formalism to which we turn.

Show that $T^\dagger = \tilde\theta\theta$ is different from T but nevertheless the partition function is both real (and positive in a finite volume) for real β.

2.2.3   Fermionic representation

On each site the operators $\frac{1}{2}\left(\sigma_x^{(1)}\pm i\sigma_x^{(2)}\right)$ form a pair of fermionic creation and annihilation operators, but they commute from site to site. There exists a transformation (Jordan–Wigner) which turns them into bona fide fermionic creators and annihilators, after ordering the sites. In the present case, such an ordering is directly afforded by the ordering of the x-axis. Using notations due to Perk, we introduce in addition to the integer labelling sites of the original one-dimensional lattice, a second set of points on the one-dimensional dual lattice with half integral coordinates. We then assign to each pair α consisting of two neighbours, one from each lattice, an operator Γ_α belonging to a Clifford algebra (hence a fermionic one) in such a way that

$$\left\{\Gamma_\alpha,\Gamma_\beta\right\} = 2\delta_{\alpha\beta} \tag{55}$$

To be precise, starting from a fixed origin, call it $x = 0$, we write

$$\Gamma_{-\frac{1}{2},0} = \sigma_0^{(1)} \qquad \Gamma_{x-\frac{1}{2},x} = \left(\prod_0^{x-1} \sigma_{x'}^{(3)}\right) \sigma_x^{(1)} \qquad x \geq 1$$

$$\Gamma_{0,\frac{1}{2}} = \sigma_0^{(2)} \qquad \Gamma_{x,x+\frac{1}{2}} = \left(\prod_0^{x-1} \sigma_{x'}^{(3)}\right) \sigma_x^{(2)} \qquad x \geq 1 \tag{56}$$

As a result Γ_α is Hermitian. The starting point $x = 0$ could be made to recede to $-\infty$ at the price of introducing an ill-defined infinite product. The string factor $\prod_0^{x-1} \sigma_{x'}^{(3)} = \exp \frac{1}{2} i\pi \sum_0^{x-1} \left(\sigma_{x'}^{(3)} - 1\right)$ reverses the sign of all states with $\sigma^{(3)} = -1$ up to $x - 1$. It is readily checked that this factor produces the desired anticommutation rules (55). Should we require periodic boundary conditions $\sigma_0^{(k)} \equiv \sigma_L^{(k)}$, the previous definitions would still make sense up to $x = L - 1$ at the price of introducing a slight complication in the transfer matrix (see below). We have now the symmetric expressions

$$\sigma_x^{(3)} = i^{-1}\Gamma_{x-\frac{1}{2},x}\Gamma_{x,x+\frac{1}{2}}$$

$$\sigma_x^{(1)}\sigma_{x+1}^{(1)} = i^{-1}\Gamma_{x,x+\frac{1}{2}}\Gamma_{x+\frac{1}{2},x+1} \tag{57}$$

The second relation would fail for periodic boundary conditions in the subspace $\prod_0^{L-1} \sigma_x^{(3)} = 1$ (it holds in its orthogonal subspace where $\prod_0^{L-1} \sigma_x^{(3)} = -1$). For the particular bond $\sigma_{L-1}^{(1)}\sigma_L^{(1)} \equiv \sigma_{L-1}^{(1)}\sigma_0^{(1)}$ we therefore have to write with our conventions

$$\sigma_{L-1}^{(1)}\sigma_0^{(1)} = i^{-1}\gamma\Gamma_{L-1,L-\frac{1}{2}}\Gamma_{-\frac{1}{2},0}$$

$$\gamma = -i^{-L}\prod_0^{L-1}\Gamma_{x-\frac{1}{2},x}\Gamma_{x,x+\frac{1}{2}} \tag{58}$$

We leave it to the reader to amend the forthcoming expressions appropriately if it is desired to impose periodic boundary conditions. We shall not do so, having ultimately in mind only the discussion of the thermodynamic limit, except for the finite expressions in equations (81) and (82) below.

The two factors of the transfer matrix are then expressed as

$$\theta(\beta) = \exp -\mathrm{i}\beta \sum_x \Gamma_{x,x+\frac{1}{2}}\Gamma_{x+\frac{1}{2},x+1}$$

$$\tilde{\theta}(\tilde{\beta}) = \exp -\mathrm{i}\tilde{\beta} \sum_x \Gamma_{x-\frac{1}{2},x}\Gamma_{x,x+\frac{1}{2}} \tag{59}$$

The transition from $\theta(\beta)$ to $\tilde{\theta}(\tilde{\beta})$ implies interchanging both β with $\tilde{\beta}$ and the two one-dimensional dual lattices, exhibiting the full symmetry of the problem.

We are now in a position to obtain in a straightforward way a path integral formulation of the partition function. Indeed both θ and $\tilde{\theta}$ appear as exponentials of quadratic forms in fermionic operators. The latter generate a group, the so called spin-group, namely a double covering of the orthogonal group which is familiar in the context of the spin representations of the ordinary three-dimensional rotation group, or of its relativistic four-dimensional extension.

We can carry out the discussion either in operator or path integral language. We use the second option, but we still need to introduce creation and annihilation operators by pairing the Γ's. This has the unfortunate drawback of destroying the explicit duality. Set for instance

$$\Gamma_{x-\frac{1}{2},x} = a_x + a_x^\dagger \qquad \Gamma_{x,x+\frac{1}{2}} = \mathrm{i}(a_x - a_x^\dagger) \tag{60}$$

Consequently

$$\tilde{\theta}(\tilde{\beta}) = \exp \tilde{\beta} \sum_x \left(a_x^\dagger a_x - a_x a_x^\dagger \right) = \prod_x \left(\mathrm{e}^{-\tilde{\beta}} + 2\sinh \tilde{\beta}\, a_x^\dagger a_x \right) \tag{61}$$

Accordingly, using the same symbol, the corresponding integral kernel is

$$\tilde{\theta}(\tilde{\beta}; \bar{\xi}, \xi) = \prod_x \left(\mathrm{e}^{-\tilde{\beta}} + 2\sinh \tilde{\beta}\bar{\xi}_x \xi_x \right) \mathrm{e}^{\bar{\xi}_x \xi_x} = \prod_x \left(\mathrm{e}^{-\tilde{\beta}} + \mathrm{e}^{\tilde{\beta}}\bar{\xi}_x \xi_x \right)$$

$$= \exp \left\{ -L\tilde{\beta} + \sum_x t^{-1}\bar{\xi}_x \xi_x \right\} \tag{62}$$

recalling that $t^{-1} = \mathrm{e}^{2\tilde{\beta}}$.

To compute the kernel of $\theta(\beta)$ we use the unitary operator U which shifts the coordinate x by a half integral unit

$$U^{-1}\Gamma_{x,x+\frac{1}{2}}U = \Gamma_{x+\frac{1}{2},x+1} \qquad\qquad U^{-1}\Gamma_{x-\frac{1}{2},x}U = \Gamma_{x,x+\frac{1}{2}}$$

$$U^{-1}a_x^\dagger U = b_x^\dagger = \tfrac{1}{2}\mathrm{i}\left(a_{x+1}^\dagger - a_x^\dagger + a_{x+1} + a_x\right)$$

$$U^{-1}a_x U = b_x = \tfrac{1}{2}\mathrm{i}\left(-a_{x+1}^\dagger - a_x^\dagger - a_{x+1} + a_x\right) \tag{63}$$

This assumes implicitly a limit $L \to \infty$. Defining Fourier transforms as

$$\xi_x = \int_{-\pi}^{+\pi} \frac{\mathrm{d}p}{2\pi} e^{\mathrm{i}px}\xi_p \qquad\qquad \bar{\xi}_x = \int_{-\pi}^{+\pi} \frac{\mathrm{d}p}{2\pi} e^{-\mathrm{i}px}\bar{\xi}_p \tag{64}$$

one verifies that the kernels corresponding to U and its adjoint U^{-1} read

$$U(\bar{\xi},\xi) = 2^{-L/2} \exp \sum_{x \geq x'} 2\mathrm{i}\bar{\xi}_x\xi_{x'} - \bar{\xi}_x\bar{\xi}_{x'} + \xi_x\xi_{x'}$$

$$= 2^{-L/2} \exp \int_{-\pi}^{+\pi} \frac{\mathrm{d}p}{2\pi}\left[\frac{2\mathrm{i}}{1-e^{-\mathrm{i}p}}\bar{\xi}_p\xi_p\right.$$

$$\left. +\frac{1+e^{-\mathrm{i}p}}{2(1-e^{-\mathrm{i}p})}\left(\bar{\xi}_{-p}\bar{\xi}_p + \xi_{-p}\xi_p\right)\right] \tag{65a}$$

$$U^{-1}(\bar{\xi},\xi) = 2^{-L/2} \exp \int_{-\pi}^{+\pi} \frac{\mathrm{d}p}{2\pi}\left[\frac{-2\mathrm{i}}{1-e^{\mathrm{i}p}}\bar{\xi}_p\xi_p\right.$$

$$\left. +\frac{1+e^{\mathrm{i}p}}{2(1-e^{\mathrm{i}p})}\left(\bar{\xi}_p\bar{\xi}_{-p} + \xi_p\xi_{-p}\right)\right] \tag{65b}$$

The kernel for $\theta(\beta) = U^{-1}\tilde{\theta}(\beta)U$ follows from applying twice the convolution formula (23) to Gaussian kernels with the result

$$\theta(\beta;\bar{\xi},\xi) = (\cosh\beta)^L$$

$$\times \exp \int_{-\pi}^{+\pi} \frac{\mathrm{d}p}{2\pi} \frac{(1-t^2)\bar{\xi}_p\xi_p + \mathrm{i}t\sin p\left(\bar{\xi}_{-p}\bar{\xi}_p + \xi_{-p}\xi_p\right)}{1+t^2+2t\cos p} \tag{66}$$

In order to obtain the partition function we now need to convolute these expressions M times and to take the trace according to equation (24). Neglecting again a boundary term, and calling y the temporal variable, we obtain

$$Z = \left(\tfrac{1}{2}\sinh 2\beta\right)^{LM} \int \prod_{x,y} \mathrm{d}\bar{\eta}_{xy}\mathrm{d}\eta_{xy}\mathrm{d}\bar{\rho}_{xy}\mathrm{d}\rho_{xy} \exp S \tag{67a}$$

$$S = \sum_y \int_{-\pi}^{\pi} \frac{dp}{2\pi} \left[-\bar{\eta}_{p,y}\eta_{p,y+1} - \bar{\rho}_{p,y}\rho_{p,y} + t^{-1}\bar{\rho}_{p,y}\eta_{p,y} + \right.$$

$$\left. \frac{(1-t^2)\bar{\eta}_{p,y}\rho_{p,y} + it\sin p \left(\bar{\eta}_{-p,y}\bar{\eta}_{p,y} + \rho_{-p,y}\rho_{p,y} \right)}{1 + t^2 + 2t\cos p} \right] \tag{67b}$$

According to the definitions (64), the variable p is conjugate to x. This seems a very nonlocal form for the action S due to the denominator. It simplifies, however, considerably if we set

$$\begin{array}{ll} \bar{\eta}_{py} = \bar{\xi}_{py}^V & \bar{\rho}_{py} = -\bar{\xi}_{py}^H - \xi_{py}^V + \xi_{-py}^H \\ \eta_{py} = -t\xi_{-py}^V & \rho_{py} = -\bar{\xi}_{py}^V + (1 + te^{ip})\,\xi_{py}^H \end{array} \tag{68a}$$

It is even clearer to return to configuration space, where the equations (68a) now read

$$\begin{array}{ll} \bar{\eta}_{xy} = \bar{\xi}_{xy}^V & \bar{\rho}_{xy} = -\bar{\xi}_{xy}^H - \xi_{xy}^V + \xi_{xy}^H \\ \eta_{xy} = -t\xi_{xy}^V & \rho_{xy} = -\bar{\xi}_{xy}^V + \xi_{xy}^H + t\xi_{x+1,y}^H \end{array} \tag{68b}$$

The Jacobian is t per site, hence we have to divide by t^{LM}. Therefore

$$Z = (\cosh \beta)^{2LM} \int \mathcal{D}(\xi, \bar{\xi}) \exp S$$

$$S = \sum_{xy} \left\{ t \left(\bar{\xi}_{x,y}^V \xi_{x,y+1}^V + \bar{\xi}_{x,y}^H \xi_{x+1,y}^H \right) \right. \tag{69}$$

$$\left. + \bar{\xi}_{xy}^H \xi_{xy}^H + \bar{\xi}_{xy}^V \xi_{xy}^V + \bar{\xi}_{xy}^V \xi_{xy}^H + \bar{\xi}_{xy}^V \bar{\xi}_{xy}^H + \bar{\xi}_{xy}^H \xi_{xy}^V + \xi_{xy}^V \xi_{xy}^H \right\}$$

an expression first derived by S. Samuel. The action has a suggestive form. Terms proportional to t correspond to propagation from one site to a neighbouring one, while the six remaining terms establish connections between incident links at a site. We have lost explicit duality due to the particular choice made in equation (60). Finally we have two pairs of fermionic variables per site $(\xi^V, \bar{\xi}^V)$ $(\xi^H, \bar{\xi}^H)$ since we kept the variables involved in both kernels for θ and $\bar{\theta}$ separately. We can now turn to the computation of the free energy per site.

2.2.4 Free energy

From equation (69), the partition function is obtained as a Pfaffian, i.e. the square root of the determinant of an antisymmetric matrix. We can take advantage of the translational invariance by using two-dimensional Fourier transforms, setting $\mathbf{x} \equiv (x, y)$, $\mathbf{p} \equiv \left(p_x, p_y \right)$ and letting ξ stand for the vector $\left(\bar{\xi}^V \xi^V \bar{\xi}^H \xi^H \right)$

$$\xi_{\mathbf{x}} = \int_{-\pi}^{+\pi} \frac{d^2 p}{(2\pi)^2} e^{i\mathbf{p} \cdot \mathbf{x}} \xi_{\mathbf{p}}$$

$$\xi_{\mathbf{p}} \equiv \begin{pmatrix} \bar{\xi}^V_{-\mathbf{p}} \\ \xi^V_{\mathbf{p}} \\ \bar{\xi}^H_{-\mathbf{p}} \\ \xi^H_{\mathbf{p}} \end{pmatrix} \tag{70}$$

The action is now block diagonal

$$S = \tfrac{1}{2} \int_{-\pi}^{+\pi} \frac{d^2 p}{(2\pi)^2} \xi^T_{-\mathbf{p}} K_{\mathbf{p}} \xi_{\mathbf{p}} \tag{71a}$$

$$K_{\mathbf{p}} = \begin{pmatrix} 0 & 1 + te^{ip_y} & 1 & 1 \\ -(1 + te^{-ip_y}) & 0 & -1 & 1 \\ -1 & 1 & 0 & 1 + te^{ip_x} \\ -1 & -1 & -(1 + te^{-ip_x}) & 0 \end{pmatrix} \tag{71b}$$

We have

$$\ln Z = 2LM \ln(\cosh \beta) + \tfrac{1}{2} \operatorname{Tr} \ln K \tag{72a}$$

$$\operatorname{Tr} \ln K = LM \int_{-\pi}^{+\pi} \frac{d^2 p}{(2\pi)^2} \ln \det K_{\mathbf{p}} \tag{72b}$$

$$\det K_{\mathbf{p}} = (1 + t^2) - 2t(1 - t^2)(\cos p_x + \cos p_y)$$
$$= (\cosh \beta)^{-4} \left[\cosh^2 2\beta - \sinh 2\beta (\cos p_x + \cos p_y) \right] \tag{72c}$$

Thus follows Onsager's celebrated result for the free energy per site

$$F(\beta) = \tfrac{1}{2} \int_{-\pi}^{+\pi} \frac{d^2 p}{(2\pi)^2} \ln \left[\cosh^2 2\beta - \sinh 2\beta (\cos p_x + \cos p_y) \right] \tag{73}$$

which indeed satisfies the duality relation (44). There exist many alternative derivations of this expression, some of them much shorter, and perhaps more illuminating than the previous one. The latter however suits our purposes in providing a motivation for the continuous theory in the critical region.

The argument of the logarithm in equation (73) appears in the denominator of the "propagator" $K_{\mathbf{p}}^{-1}$ for the fermionic variables

ξ. In the Brillouin zone $|p_x|, |p_y| \leq \pi$ the combination $(\cos p_x + \cos p_y)$ reaches its maximum value 2 at the origin. Therefore the effective square mass is $\cosh^2 2\beta - 2\sinh 2\beta$ and is always positive (for real positive β) except for the unique critical value β_c, where $\cosh 2\beta_c = \sqrt{2}$, $\sinh 2\beta_c = 1$ as expected. For β negative corresponding to the antiferromagnetic regime not discussed here, one would have to go to the boundary of the Brillouin zone where $\cos p_x + \cos p_y = -2$ with β_c replaced by $-\beta_c$. Here we limit ourselves to the ferromagnetic system. The free energy F is always regular except possibly at β_c, where its second derivative has a logarithmic singularity. We adopt for brevity the following notation

$$
\begin{array}{ll}
C = \cosh 2\beta & \tilde{C} = \cosh 2\tilde{\beta} = C/S \\
S = \sinh 2\beta & \tilde{S} = \sinh 2\tilde{\beta} = 1/S
\end{array}
\tag{74}
$$

Half the internal energy, i.e. the mean value of a neighbouring pair of spins $\langle \sigma\sigma' \rangle$, is continuous at $\beta = \beta_c$ and is given by

$$
\frac{1}{2}\frac{d}{d\beta}F(\beta) = \langle \sigma\sigma' \rangle = \frac{\tilde{C}}{2}\left[1 + \int_{-\pi}^{+\pi}\frac{d^2p}{(2\pi)^2}\frac{S - \tilde{S}}{C\tilde{C} - \cos p_x - \cos p_y}\right]
\tag{75}
$$

Its behaviour close to β_c is

$$
\langle \sigma\sigma' \rangle = \frac{4\sqrt{2}}{\pi}(\beta - \beta_c)\ln\frac{1}{|\beta - \beta_c|} + \text{regular part}
\tag{76}
$$

Show by performing one of the integrals in (75) that

$$
\frac{1}{2}\frac{dF}{d\beta}(\beta) = \langle \sigma\sigma' \rangle = \frac{\tilde{C}}{2}\left[1 + \left(S - \tilde{S}\right)\phi(\beta)\right]
\tag{77a}
$$

$$
\phi(\beta) = \int_0^{2\pi}\frac{d\theta}{2\pi}\left[\left(\cosh 2(\beta + \tilde{\beta}) - \cos\theta\right)\right.
$$

$$
\left.\times \left(\cosh 2(\beta - \tilde{\beta}) - \cos\theta\right)\right]^{-1/2}
$$

$$
= \frac{k}{2}\cdot\frac{2}{\pi}\int_0^{\pi/2}d\varphi\,(1 - k^2\sin^2\varphi)^{-1/2}
$$

$$
= \frac{k}{2}F\left(\tfrac{1}{2}, \tfrac{1}{2}; 1; k^2\right)
\tag{77b}
$$

with $F(a, b; c; x)$ the hypergeometric function, and $k = 2/C\tilde{C} = 2S/C^2$. One recognizes a complete elliptic integral of the first kind.

The specific heat, namely the second derivative of F, has a logarithmic divergence at $\beta = \beta_c$

$$\frac{1}{2}\frac{d^2 F(\beta)}{d\beta^2} = \sum_{(x,x')} \{\langle\sigma\sigma'\sigma_x\sigma_{x'}\rangle - \langle\sigma\sigma'\rangle\langle\sigma_x\sigma_{x'}\rangle\}$$

$$\simeq \frac{4\sqrt{2}}{\pi}\ln\frac{1}{(\beta - \beta_c)} + \text{reg. part} \tag{78}$$

showing that at criticality energy–energy correlations have an infinite range. Here we call energy the observable $\sigma\sigma'$ product of spins at neighbouring sites. This is our first encounter with second order phase transitions where correlation lengths become infinite and in an appropriate scale the discrete theory can be replaced by a continuous one, ignoring the fine details of the lattice. As it will turn out, close to criticality, the two-dimensional Ising model is nothing but a free Fermi field theory. What is particular to the statistical mechanics aspect is the emphasis on certain observables which derive from the original formulation and interpretation of the model.

(1) Diagonalize $K_{\mathbf{p}} = -K_{\mathbf{p}}^\dagger = -K_{-\mathbf{p}}^T$. The eigenvalues are of the form $\pm i\lambda$, $\pm i\lambda'$, with $\pm\lambda$ and $\pm\lambda'$ solutions of

$$\lambda^4 - 2\lambda^2 \left[2 + (1+t^2) + t(\cos p_x + \cos p_y)\right]$$
$$+ (1+t^2)^2 - 2t(1-t^2)(\cos p_x + \cos p_y) = 0 \tag{79a}$$

Let λ be the eigenvalue which vanishes at criticality for $\mathbf{p} = 0$, while λ' remains finite,

$$\pm i\lambda = \pm i \left[2 + (1+t^2) + t(\cos p_x + \cos p_y)\right.$$
$$\left. - \sqrt{4 + t(\cos p_x + \cos p_y) + 4t^2 - 8}\right]^{1/2}$$

$$\pm i\lambda' = \pm i \left[2 + (1+t^2) + t(\cos p_x + \cos p_y)\right.$$
$$\left. + \sqrt{4 + t(\cos p_x + \cos p_y) + 4t^2 - 8}\right]^{1/2} \tag{79b}$$

If the eigenvectors are such that

$$K_{\mathbf{p}}n_{\mathbf{p}}^{(\pm)} = \pm i\lambda n_{\mathbf{p}}^{(\pm)} \qquad K_{\mathbf{p}}n_{\mathbf{p}}'^{(\pm)} = \pm i\lambda' n_{\mathbf{p}}'^{(\pm)}$$

one may choose $n_{-\mathbf{p}}^{(-)} = n_{\mathbf{p}}^{(+)*}$ and similarly for n', and adjust the relative phase of n and n' in such a way that the matrix

$$U_{\mathbf{p}} = \left(n_{\mathbf{p}}^{(+)}, n_{\mathbf{p}}^{(-)}, n_{\mathbf{p}}'^{(+)} n_{\mathbf{p}}'^{(-)}\right)$$

be unitary, unimodular, such that

$$
U_{\mathbf{p}}^{\dagger} = \begin{pmatrix} 0 & 1 & 0 & 0 \\ 1 & 0 & 0 & 0 \\ 0 & 0 & 0 & 1 \\ 0 & 0 & 1 & 0 \end{pmatrix} U_{-\mathbf{p}}^{T}
$$

and

$$
K_{\mathbf{p}} = U_{\mathbf{p}} \begin{pmatrix} 0 & i\lambda_{\mathbf{p}} & 0 & 0 \\ -i\lambda_{\mathbf{p}} & 0 & 0 & 0 \\ 0 & 0 & 0 & i\lambda_{\mathbf{p}}' \\ 0 & 0 & -i\lambda_{\mathbf{p}}' & 0 \end{pmatrix} U_{-\mathbf{p}}^{T}
$$

If we now change variables in (71) by defining

$$
\xi_{\mathbf{p}}' = U_{-\mathbf{p}}^{T} \xi_{\mathbf{p}}
$$

we obtain the canonical form of the action

$$
\xi_{-\mathbf{p}} K_{\mathbf{p}} \xi_{\mathbf{p}} = i\xi_{-\mathbf{p}}' \begin{pmatrix} 0 & \lambda & 0 & 0 \\ -\lambda & 0 & 0 & 0 \\ 0 & 0 & 0 & \lambda' \\ 0 & 0 & -\lambda' & 0 \end{pmatrix} \xi_{\mathbf{p}}' \tag{80}
$$

The last two components of ξ' are decoupled from the first two and correspond to massive modes for any value of β positive. They can safely be ignored if one is only interested in the critical region.

(2) Find the partition function on a finite $L \times M$ lattice, with periodic boundary conditions on the spins as

$$
Z = (\cosh \beta)^{2LM} \frac{1}{2} \left(\int_{-,-} + \int_{-,+} + \int_{+,-} + \int_{+,+} \right) \mathcal{D}(\xi, \bar{\xi}) \, \exp S \tag{81}
$$

Here S is the action (69), with x and y restricted to the range $(0, L-1)$ and $(0, M-1)$ respectively, while the signs refer to the boundary conditions $\xi_{L,y}^{H} = \pm \xi_{0,y}^{H}$ and $\xi_{x,M}^{V} = \pm \xi_{x,0}^{V}$. Hence obtain the expressions

$$
Z = (\cosh \beta)^{2LM} \frac{1}{2} \sum_{\varepsilon_1, \varepsilon_2 = 0, \frac{1}{2}} \eta_{\varepsilon_1, \varepsilon_2}
$$

$$
\times \prod_{q=0}^{M-1} \prod_{p=0}^{L-1} \left\{ (1 + t^2)^2 \right.
$$

$$
\left. - 2t(1 - t^2) \left[\cos \frac{2\pi(p + \varepsilon_1)}{L} + \cos \frac{2\pi(q + \varepsilon_2)}{M} \right] \right\}^{1/2} \tag{82}
$$

where $\eta_{\frac{1}{2},0} = \eta_{0,\frac{1}{2}} = \eta_{\frac{1}{2},\frac{1}{2}} = 1$, $\eta_{0,0} = \varepsilon(T - T_c)$. Of course, the $(0,0)$ term vanishes at criticality. Generalize this expression when the couplings in the horizontal and vertical directions are distinct.

2.2.5   *Spontaneous magnetization*

For $\beta > \beta_c$, when we consider far away spins σ_0 and $\sigma_{\mathbf{x}}$, the correlation $G(\mathbf{x}) = \langle \sigma_0 \sigma_{\mathbf{x}} \rangle$ does not vanish in the limit where $\mathbf{x}$ goes to infinity, but rather tends to a positive constant, the square of the spontaneous magnetization, which we denote by $\mathcal{M}$. Thus it is as if the mean value of a spin were equal to $\mathcal{M}$, although in the absence of any symmetry breaking perturbation, such as one induced by a small magnetic field or a specific boundary condition, one should expect the mean value of a spin to vanish. This means in reality that such a symmetric situation beyond β_c is in fact a superposition of two pure states, each one corresponding to a well-defined spontaneous magnetization $\pm\mathcal{M}$. As a convenience in the calculation, we reach $\mathcal{M}$ indirectly by computing the asymptotic behaviour of $G(\mathbf{x})$, the same in any of these two pure states, hence the same in any mixture,

$$\mathcal{M}^2 = \lim_{\mathbf{x} \to \infty} G(\mathbf{x}) \qquad (83)$$

We say that symmetry is spontaneously broken whenever $\mathcal{M} \neq 0$. It is not *a priori* obvious that $\mathcal{M}$ should be independent of the direction in which we let $\mathbf{x}$ go to infinity. We presuppose this property when, as a further matter of convenience, we let $\mathbf{x}$ go to infinity along one of the two principal axes, for instance along the first one. Then $\mathbf{x} \equiv (x,0)$ and we note that $\sigma_0 \sigma_x = (\sigma_0 \sigma_1)(\sigma_1 \sigma_2) \cdots (\sigma_{x-1} \sigma_x)$ using the fact that at any site $\sigma_i^2 = 1$. Now we recall that in the high temperature expansion (1.116) each link of the lattice contributed a factor $(1 + t\sigma\sigma')$ to the Boltzmann weight. When computing the numerator in the average value $\langle \sigma_0 \sigma_x \rangle$ we therefore modify each of the links $(01)(12)...(x-1,x)$ through $(1 + t\sigma_i \sigma_{i+1}) \to \sigma_i \sigma_{i+1}(1 + t\sigma_i \sigma_{i+1}) = t(1 + t^{-1}\sigma_i \sigma_{i+1})$. We therefore conclude that, apart from a factor $t^{|x|}$, the required mean value takes the form of a ratio between a "frustrated" partition function Z' and the original one Z

$$G(x) = t^{|x|} \frac{Z'}{Z} \qquad (84)$$

where in Z' all the links joining 0 to x one has substituted t^{-1} for t. Thus following the steps which lead to the integral (69), we have a similar representation for Z' with a new kernel $K' = K + Q$ and

$$G(x) = t^{|x|} \left[\det \left(1 + K^{-1}Q \right) \right]^{1/2} \qquad (85a)$$

with the (antisymmetric) matrix Q given by

$$\tfrac{1}{2}\xi Q \xi = \frac{1 - t^2}{t} \sum_{x'=0}^{x-1} \bar{\xi}^H_{x',0} \xi^H_{x'+1,0} \qquad (85b)$$

The matrix $K^{-1}Q$ has lines and columns indexed by the two coordinates x, y, as well as a fourfold symbol $\bar{\xi}^V, \xi^V, \bar{\xi}^H, \xi^H$. Obviously all matrix elements but those in the columns such that $y = 0$, $x = 0, 1, ..., x-1$, and the internal index is either $\bar{\xi}^H$ or ξ^H, vanish. As a result, in the computation of the above determinant we can delete all lines but those with the same characterization. This leaves a $(2x) \times (2x)$ determinant which involves the inverse of K. Restricted to the above subspace, the latter, which we denote by $(K^{-1})^H_{x_1,x_2}$, has the form

$$(K^{-1})^H_{x_1,x_2} = \int_{-\pi}^{+\pi} \frac{\mathrm{d}^2 p}{(2\pi)^2} \frac{e^{ip_x(x_1 - x_2)} \begin{bmatrix} a_{\mathbf{p}} & b_{\mathbf{p}} \\ c_{\mathbf{p}} & d_{\mathbf{p}} \end{bmatrix}}{(1 + t^2)^2 - 2t(1 - t^2)(\cos p_x + \cos p_y)} \qquad (86a)$$

$$a_{\mathbf{p}} = -d_{\mathbf{p}} = -2\mathrm{i}t \sin p_y$$
$$b_{\mathbf{p}} = -c^*_{\mathbf{p}} = 2(1 + t \cos p_y) - (1 + te^{ip_x})(1 + t^2 + 2t \cos p_y) \qquad (86b)$$

Let $\chi(u)$ denote the characteristic function of the interval 1 to x i.e. $\chi(u) = 1$ if $n = 1, 2, ..., x$ and zero otherwise. Then

$$[K^{-1H}Q]_{x_1 x_2} = \frac{1 - t^2}{t} \int_{-\pi}^{+\pi} \frac{\mathrm{d}^2 p}{(2\pi)^2}$$

$$\frac{e^{ip_x(x_1 - x_2)}}{(1 + t^2)^2 - 2t(1 - t^2)\left(\cos p_x + \cos p_y \right)}$$

$$\times \begin{bmatrix} -b_{\mathbf{p}} e^{-ip_x} \chi(x_2 + 1) & a_{\mathbf{p}} e^{ip_x} \chi(x_2) \\ -d_{\mathbf{p}} e^{-ip_x} \chi(x_2 + 1) & c_{\mathbf{p}} e^{ip_x} \chi(x_2) \end{bmatrix} \qquad (87)$$

Since $a_{\mathbf{p}}$ and $d_{\mathbf{p}}$ are odd in p_y, their contribution vanishes. Using $c_{\mathbf{p}} = -b^*_{\mathbf{p}}$ one discovers that $\det(1 + K^{-1}Q)$ is the square of an

$x \times x$ determinant of the form

$$G(x)^2 = t^{2|x|} \det\left(1 + K^{-1}Q\right) = (\det A)^2 \qquad (88a)$$

$$A = \begin{vmatrix} a_0 & a_{-1} & a_{-2} & \cdots & a_{-(x-1)} \\ a_1 & a_0 & a_{-1} & \cdots & a_{-(x-2)} \\ \vdots & \vdots & \vdots & \ddots & \vdots \\ a_{x-1} & a_{x-2} & a_{x-3} & \cdots & a_0 \end{vmatrix} \qquad (88b)$$

$$a_k = \int_{-\pi}^{+\pi} \frac{\mathrm{d}^2 p}{(2\pi)^2} \mathrm{e}^{-ikp_x}$$

$$\frac{2t\left(1+t^2\right) - t^2\left(1-t^2\right)\mathrm{e}^{-\mathrm{i}p_x} - \left(1-t^2\right)\mathrm{e}^{\mathrm{i}p_x}}{\left(1+t^2\right)^2 - 2t\left(1-t^2\right)\left(\cos p_x + \cos p_y\right)} \qquad (88c)$$

We pause to note that one has therefore in "closed form" the value of the spin–spin correlation function at any separation along the principal axes. In particular if $x = 1$, the matrix A reduces to a_0 and (88) to the previous result (75). However, in the general case $x > 1$ equation (88) is still not very manageable. We proceed to evaluate it in the limit $x \to \infty$. First the integral over p_y, can be performed in (88c) and

$$a_k = \int_{-\pi}^{+\pi} \frac{\mathrm{d}p}{2\pi} \mathrm{e}^{-ikp} \left[\frac{\left(t - \tilde{t}\mathrm{e}^{\mathrm{i}p}\right)\left(t\tilde{t} - \mathrm{e}^{\mathrm{i}p}\right)}{\left(t\mathrm{e}^{\mathrm{i}p} - \tilde{t}\right)\left(t\tilde{t}\mathrm{e}^{\mathrm{i}p} - 1\right)} \right]^{1/2} \qquad (88d)$$

The quantity between brackets in the integral is of unit modulus. To extract the square root, its phase is chosen by continuity assuming it is zero for $p = 0$ when $t > t_c$ (i.e. in the low temperature phase). We observe again the very similar roles played by t and $\tilde{t}$. A matrix such as A given in (88b) is referred to as a Toeplitz matrix, and the corresponding determinant as a Toeplitz determinant. To obtain its limiting value when $x \to \infty$, we use a result due to Szegö, which goes as follows. Let a function $f(z)$ analytic in the vicinity of the unit circle be expanded in a Laurent (or Fourier) series as

$$f(z) = \sum_{-\infty}^{+\infty} a_q z^q \qquad |z| = 1 \qquad (89a)$$

We require furthermore that as one rotates around the unit circle the phase of f returns to its original value, making its logarithm well-defined. Thus the image of the unit circle through f does not encircle the origin in the (complex) f plane. Expand the logarithm $h(z) = \ln f(z)$ in Laurent series

$$h(z) = \ln f(z) = \sum_{-\infty}^{+\infty} h_q z^q \qquad |z| = 1 \qquad (89b)$$

Then Szegö's lemma states that

$$\lim_{x \to \infty} e^{-x h_0} \det \begin{vmatrix} a_0 & a_{-1} & \cdots & a_{(x-1)} \\ \vdots & \vdots & \ddots & \vdots \\ a_{x-1} & a_{x-2} & \cdots & a_0 \end{vmatrix} = \exp \sum_1^\infty q h_q h_{-q} \quad (89c)$$

We present below an interpretation of this formula which makes it sound plausible. For the time being, we apply it to the function appearing in (88d), namely with $z = e^{ip}$

$$f(z) = \left(\frac{t - \tilde{t}z}{tz - \tilde{t}} \frac{t\tilde{t} - z}{t\tilde{t}z - 1} \right)^{1/2} \qquad (90)$$

Now $f(z)$ has two branch points inside the unit circle, one located at $z_1 = t\tilde{t}$, the second one z_2 is either at $t/\tilde{t}$ or its inverse, according to whichever one has a modulus smaller than one. Two other branch points lie outside the unit circle. Let us look at the low temperature regime, corresponding to $t > t_c > \tilde{t}$. It is readily seen that under such circumstances the total variation of the argument of f as one describes the unit circle vanishes, and as a result $h = \ln f$ is well-defined. Moreover

$$h_0 = 0, \qquad h_q = -h_{-q} = \frac{1}{2q} \left((t\tilde{t})^{|q|} - \left(\frac{\tilde{t}}{t} \right)^{|q|} \right) \qquad q \neq 0 \quad (91)$$

Applying (89c) we conclude that in the low temperature phase

$$\mathcal{M}^2 = \lim_{x \to \infty} \langle \sigma_0 \sigma_x \rangle = \left[1 - \frac{\tilde{t}^2 (1 - t^2)^2}{t^2 (1 - \tilde{t}^2)^2} \right]^{1/4} = \left[1 - (\sinh 2\beta)^{-4} \right]^{1/4}$$

$$(92a)$$

Therefore the spontaneous magnetization $\mathcal{M}$ is (Onsager, Yang)

$$\mathcal{M} = \left[1 - (\sinh 2\beta)^{-4} \right]^{1/8} \qquad (92b)$$

At zero temperature, or $\beta \to \infty$, $\mathcal{M}$ takes the value 1, then decreases down to zero as β recedes to β_c where $\sinh 2\beta_c = 1$.

More significant is the nontrivial power law behaviour as β comes close to β_c

$$\mathcal{M} \sim (\beta - \beta_c)^{1/8} \qquad (92c)$$

What is somehow surprising is the simple appearance of the expression (92b), considering the complexity of the intermediate stages in the calculation. A simpler derivation is still awaiting. Furthermore, the power law behaviour in $(\beta - \beta_c)^{1/8}$ or $(1 - T/T_c)^{1/8}$ (with $T = \beta^{-1}$) is not at all expected. We shall see later that a mean field approximation, ignoring fluctuations, would have led to a $(1 - T/T_c)^{1/2}$ behaviour. In the present case the departure from zero at the critical point is much steeper. Finally it is customary to characterize the behaviour of the spontaneous magnetization close to T_c as $(1 - T/T_c)^{\beta}$ and hence

$$\beta = \frac{1}{8} \qquad (92d)$$

The reader will notice a conflict in notation between the inverse temperature β and the critical exponent for the magnetization. We can only offer the apology that both are in common use and should not lead to confusion. That the index β comes out as a rational fraction is part of a much more involved story which will be described in chapter 9. Below T_c, two pure phases with magnetizations $\pm \mathcal{M}$ may coexist, and one can compute the associated surface tension (it would be more appropriate to speak of perimeter tension in a two-dimensional world). This is discussed below.

Szegö's lemma and averages over unitary matrices   Instead of reproducing the original proof of Szegö and Kac, let us present an heuristic proof, the interest of which is to rely on averages over the unitary group. Let A denote the Toeplitz matrix built from the Fourier coefficients of the function f of size $x \times x$. Expressing these coefficients as integrals we find

$$\det A = \det \left[a_{k_1 - k_2} \right] = \int \prod_{k=0}^{x-1} \frac{\mathrm{d}\varphi_k}{2\pi} f\left(e^{i\varphi_k} \right)$$

$$\times \det \begin{vmatrix} 1 & e^{i\varphi_0} & \dots & e^{i(x-1)\varphi_0} \\ e^{-i\varphi_1} & 1 & \dots & e^{i(x-2)\varphi_1} \\ \vdots & \vdots & \ddots & \vdots \\ e^{-i(x-1)\varphi_{x-1}} & e^{-i(x-2)\varphi_{x-1}} & \dots & 1 \end{vmatrix}$$

The determinant under the integral sign is recognized as a Vandermonde determinant up to a factor

$$\exp\left(-i\sum_{k=0}^{x-1} k\varphi_k\right) \prod_{0\le k_1 < k_2 \le x-1} \left(e^{i\varphi_{k_1}} - e^{i\varphi_{k_2}}\right)$$

Since the measure is invariant under any permutation of the k-indices, the latter expression can be replaced by its average over permutations. Under such permutation, the above product picks a sign, the signature of the permutation, and the combination of prefactors builds up the complex conjugate Vandermonde determinant. Hence the equality

$$\det A = \frac{1}{x!}\int \prod_{k=0}^{x-1} \frac{d\varphi_k f\left(e^{i\varphi_k}\right)}{2\pi} \prod_{0\le k_1 < k_2 \le x-1} \left|e^{i\varphi_{k_1}} - e^{i\varphi_{k_2}}\right|^2$$

If f is replaced by unity, the determinant is obviously one, hence the measure is normalized. Now let U stand for any unitary $x \times x$ matrix with eigenvalues $e^{i\varphi_k}$ and set

$$F(U) = \sum a_k U^k$$

Then $\det F(U) = \prod_{k=0}^{x-1} f\left(e^{i\varphi_k}\right)$ and

$$\frac{1}{x!}\prod_{k=0}^{x-1} \frac{d\varphi_k}{2\pi} \prod_{0\le k_1 < k_2 \le x-1} \left|e^{i\varphi_{k_1}} - e^{i\varphi_{k_2}}\right|^2$$

is the unique normalized measure on the conjugacy classes of the unitary group. Denoting by brackets averages over the unitary group, our expression reads

$$\det A = \langle \det F(U)\rangle = \langle \exp \mathrm{Tr}\, H(U)\rangle$$

where in analogy with $f = \exp h$ we have $F = \exp H$ and the expansion of H in powers of U involves the same coefficients as those of h, as follows readily once we bring U to a diagonal form. We then evaluate the last average using the standard cumulant expansion

$$\det A = \exp\left\{ \langle \mathrm{Tr}\, H\rangle + \tfrac{1}{2}\left[\left\langle (\mathrm{Tr}\, H)^2\right\rangle - \langle \mathrm{Tr}\, H\rangle^2\right]\right.$$
$$\left. + \tfrac{1}{6}\left[\left\langle (\mathrm{Tr}\, H)^3\right\rangle - 3\left\langle (\mathrm{Tr}\, H)^2\right\rangle \langle \mathrm{Tr}\, H\rangle + 2\langle \mathrm{Tr}\, H\rangle^3\right] + \cdots\right\}$$

The mean values of the powers of $\operatorname{Tr} H$ are readily computed, leading to

$$\det A = \exp x h_0 + \sum_{p=1}^{\infty} \operatorname{Min}(p, x) h_p h_{-p} +$$

$$+ \tfrac{1}{2} \sum_{\substack{q,r=1 \\ q+r \geq x}}^{\infty} \operatorname{Min}(q, r, q+r-x, x) \left(h_q h_r h_{-q-r} + h_{-q} h_{-r} h_{q+r} \right) + \cdots$$

If h_k tends to zero fast enough as $x \to \infty$, the generic term tends to zero and we are left with the desired result

$$\lim_{x \to \infty} e^{-x h_0} \det A = \exp \sum_{p=1}^{\infty} p h_p h_{-p}$$

2.2.6   *Correlation function in the high temperature phase*

We follow here the derivation given by McCoy and Wu. We return to the expressions (88) and assume $t < t_c < \tilde{t}$, in which case the phase of f varies by 2π as we describe the unit circle. By factoring out z, we write $f(z) = zg(z)$, with $\ln g(z)$ well-defined for $|z| = 1$ and

$$a_k = \int_{-\pi}^{\pi} \frac{d\varphi}{2\pi} e^{-i(k-1)\varphi} g(e^{i\varphi}) = g_{k-1}$$

$$g^2(z) = \frac{\tilde{t} - tz^{-1}}{\tilde{t} - tz} \times \frac{1 - t\tilde{t}z^{-1}}{1 - t\tilde{t}z} \tag{93}$$

If B stands for the $(x+1) \times (x+1)$ matrix

$$B = \begin{vmatrix} g_0 & g_{-1} & \cdots & g_{-x} \\ g_1 & g_0 & \cdots & g_{-(x-1)} \\ \vdots & \vdots & \ddots & \vdots \\ g_x & g_{x-1} & \cdots & g_0 \end{vmatrix} \tag{94}$$

the determinant we wish to evaluate is the minor of the element in the last row and first column, i.e.

$$\det A = \det B \left(B^{-1} \right)_{0x} \tag{95}$$

From the above, as $x \to \infty$

$$\det B \xrightarrow[x \to \infty]{} \left\{ (1 - t^2)^2 \, (1 - t^2 \tilde{t}^2) \, (1 - t^2 \tilde{t}^{-2}) \right\}^{1/4} \tag{96}$$

To compute the inverse matrix B^{-1}, we have to invert the equation

$$\mathbf{u}B = \mathbf{v} \tag{97}$$

Given the vector $\mathbf{v} \equiv \{v_0, \ldots, v_x\}$ we want to find the vector $\mathbf{u} \equiv \{u_0, \ldots, u_x\}$. If $\mathbf{v}$ reduces to $v_k = \delta_{k,0}$, then $u_x = (B^{-1})_{0x}$. We can rephrase equation (97) by introducing the two polynomials

$$u(z) = \sum_0^x u_k z^k \qquad\qquad v(z) = \sum_0^x v_k z^k \tag{98}$$

and compare it with the expression $u(z)g(z^{-1}) - v(z)$. The remainder can be written for $|z| = 1$

$$u(z)g(z^{-1}) - v(z) = z^x \varphi(z) + \psi(z^{-1}) \tag{99a}$$

where

$$\varphi(z) = \sum_{q=0}^x \sum_{k=1}^\infty u_q g_{q-k-x} z^k$$

$$\psi(z) = \sum_{q=0}^x \sum_{k=1}^\infty u_q g_{q+k} z^k \tag{99b}$$

The two terms on the r.h.s. of (99a) are analytic respectively inside and outside the unit circle, assuming the coefficients to decrease fast enough, while u and v being polynomials are analytic in the entire complex plane. Let us now factor out $g(z^{-1})$ into a function analytic inside the unit circle times a function analytic outside, in the form

$$g(z^{-1}) = \frac{1}{P(z)Q(z^{-1})}$$

$$P(z) = \frac{1}{\sqrt{\left(1 - \frac{t}{\tilde z}z\right)(1 - t\tilde t z)}} \qquad Q(z) = \sqrt{\left(1 - \frac{t}{\tilde t}z\right)(1 - t\tilde t z)}$$

$$\tag{100}$$

Of course, $P(z)Q(z) = 1$. As a consequence, $P(z)^{-1}$ is analytic inside the unit circle. It then follows that, for $|z| = 1$,

$$\frac{u(z)}{P(z)} = Q(z^{-1})v(z) + Q(z^{-1})\left[z^x \varphi(z) + \psi(z^{-1})\right] \tag{101}$$

In the last term, let us split the series into two parts. One involves only non-negative powers of z, the other negative powers, denoted respectively by brackets with a subscript $+$ or $-$. As a result

$$\frac{u(z)}{P(z)} - [Q(z^{-1})v(z) + Q(z^{-1})z^x\varphi(z)]_+ =$$
$$Q(z^{-1})\psi(z^{-1}) + [Q(z^{-1})v(z) + Q(z^{-1})z^x\varphi(z)]_- \tag{102}$$

The left-hand side admits an analytic continuation inside the unit circle, the right-hand side a continuation outside and vanishes at infinity. Consequently, both sides represent the same analytic function which, according to Liouville's theorem, vanishes. The conclusion is therefore that

$$u(z) = P(z)\left[Q(z^{-1})v(z) + Q(z^{-1})z^x\psi(z)\right]_+$$
$$\psi(z^{-1}) = -\frac{1}{Q(z^{-1})}\left[Q(z^{-1})v(z) + Q(z^{-1})z^x\varphi(x)\right]_- \tag{103}$$

The argument can be repeated for $z^x u(z^{-1})$ analytic inside the unit disk with the result that

$$z^x u(z^{-1}) = Q(z)\left[P(z^{-1})z^x v(z^{-1}) + P(z^{-1})z^x\psi(z)\right]_+$$
$$\varphi(z^{-1}) = -\frac{1}{P(z^{-1})}\left[P(z^{-1})z^x(z^{-1}) + P(z^{-1})z^x\psi(x)\right]_- \tag{104}$$

In the case of interest, $v(z) = 1$. As $x \to \infty$, we can neglect in the second equation (103) the term $[Q(z^{-1})z^x\varphi(z)]_-$. Furthermore $[Q(z^{-1})v(z)]_- = [Q(z^{-1})]_- = Q(z^{-1}) - 1$. Thus $\psi(z) \simeq 1/Q(z) - 1$. This value is then inserted in (104) with the result that, for large x,

$$z^x u(z^{-1}) \simeq Q(z)\left[\frac{P(z^{-1})z^x}{Q(z)}\right]_+$$

Finally the quantity of interest is u_x, i.e. the value of the above function at $z = 0$, and referring to (95) and (96)

$$\langle\sigma_0\sigma_x\rangle \underset{x\to\infty}{\longrightarrow} \left[(1-t^2)^2\,(1-t^2\tilde{t}^2)\,(1-t^2\tilde{t}^{-2})\right]^{1/4} \times$$

$$\oint \frac{dz}{2i\pi}z^{x-1}\left[\left(1-\frac{t}{\tilde{t}}z\right)\left(1-\frac{t}{\tilde{t}}z^{-1}\right)(1-t\tilde{t}z)(1-t\tilde{t}z^{-1})\right]^{-1/2} \tag{105}$$

with exponentially small corrections for $x \to \infty$. The evaluation of the integral is obtained by contour deformation and leads to the final result that in the high temperature phase the correlation function decreases asymptotically with the distance as

$$\langle\sigma_0\sigma_x\rangle \underset{x\to\infty}{\simeq} \left(\frac{1-t^2\tilde{t}^2}{1-t^2\tilde{t}^{-2}}\right)^{1/4} \frac{1}{(1-\tilde{t}^2)^{1/2}} \frac{1}{(\pi|x|)^{1/2}} \left(\frac{t}{\tilde{t}}\right)^x [1+O(1/x)]$$

(106)

We recall that here 0 and x are points on the same crystallographic axis. It is not expected that far from criticality we find an isotropic law. However, if we are close to T_c, such should be the case, and (106) can be compared to the behaviour of a massive boson propagator

$$\int \frac{d^2k}{(2\pi)^2} \frac{e^{i\mathbf{k}\cdot\mathbf{x}}}{\mathbf{k}^2 + m^2} \simeq \left(\frac{2}{\pi m|\mathbf{x}|}\right)^{1/2} e^{-m|\mathbf{x}|} \left(1 + O\left(\frac{1}{m|\mathbf{x}|}\right)\right)$$

(107)

In both cases we find an exponential decay law with characteristic length $\xi = m^{-1}$, corrected by a prefactor in $|\mathbf{x}|^{-1/2}$.

At any rate, we conclude that the $(x-x)$ or $(y-y)$ correlation length is given by

$$\exp -1/\xi = t/\tilde{t} = \exp[-2(\tilde{\beta}-\beta)]$$
$$\xi = 1/2(\tilde{\beta}-\beta)$$

(108)

an expression only valid of course for $\beta < \beta_c$ or $t < \sqrt{2}-1$. As a consequence the correlation length diverges linearly as $T \to T_c$ from above

$$\xi \sim \left(\frac{T}{T_c}-1\right)^{-1}$$

(109)

The critical exponent ν is defined through $\xi \sim [(T/T_c)-1]^{-\nu}$, meaning in the present case that

$$\nu = 1$$

(110)

The expression (106) describes bosonic excitations with a mass $m_B = 2(\tilde{\beta}-\beta)$. Let us compare this scale with the fermionic propagator, the denominator of which is proportional to

$$\left[2\sinh\left(\tilde{\beta}-\beta\right)\right]^2 + 2(1-\cos p_x) + 2\left(1-\cos p_y\right)$$

the normalization being chosen in such a way that, for long wavelengths, the behaviour is $m_F^2 + p_x^2 + p_y^2$. Therefore

$$m_F^2 = \left[2\sinh\left(\tilde{\beta}-\beta\right)\right]^2$$

Thus as $\beta \to \tilde{\beta}$, we find $m_F \sim m_B$. Finally the relation $\exp(-1/\xi) = t/\tilde{t} = t(1+t)/(1-t)$ is to be compared with the analogous one valid in one dimension, $\exp(-1/\xi_1) = t = \exp -2\tilde{\beta}$, $\xi_1 = 1/2\tilde{\beta}$, which diverges as $\frac{1}{2}\exp 2\beta$ as $\beta \to \infty$ or $T \to 0$.

The relation $\exp(-1/\xi) = t(1+t)/(1-t) = t + 2t^2 + 2t^3 + \cdots$ suggests the following approximation for the correlation function. Recall the high temperature expansion

$$\langle \sigma_0 \sigma_x \rangle = \frac{\Sigma_{\partial\{n\}=(0,x)} t^{\Sigma n_l}}{\Sigma_{\partial\{n\}=\emptyset} t^{\Sigma n_l}}$$

with $n_l = 0$ or 1. Let us consider an anisotropic model with t_x assigned to horizontal links (parallel to the direction $0x$) and t_y to the vertical ones. Assume a limit (the so called "solid-on-solid" limit) where $t_x/t_y \to 0$. In the denominator no term different from unity survives in this limit. In the numerator the dominant terms have t_x^x as a factor, the corresponding contributions being depicted as a broken curve with no overhangs nor disconnected parts of the type shown on figure 1. It is easy to show that in this approximation

$$\langle \sigma_0 \sigma_x \rangle_{\text{sos}} = \left(t_x \frac{1+t_y}{1-t_y} \right)^x$$

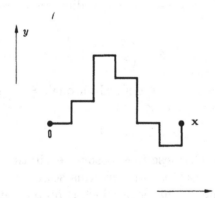

Fig. 1   A typical "solid-on-solid" curve joining 0 and x

It is intriguing to find that, if one is only interested in computing $\exp(-1/\xi)$, i.e. in the large x limit, it remains valid up to the isotropic case $t_x = t_y$.

Having found that the mass, or inverse correlation length, vanishes linearly with $T - T_c$, and assuming that, close to criticality, the correlation function becomes isotropic, we can represent its asymptotic behaviour up to a factor by the integral (107). This prefactor is in fact proportional to $m^{1/4}$. Thus

$$\langle \sigma_0 \sigma_{\mathbf{x}} \rangle \simeq \text{cst } m^{1/4} \int \frac{d^2 k}{(2\pi)^2} \frac{e^{i\mathbf{k}\cdot\mathbf{x}}}{\mathbf{k}^2 + m^2} \qquad m \equiv m_B \sim T - T_c \tag{111}$$

The susceptibility χ is the integral of $\langle \sigma_0 \sigma_{\mathbf{x}} \rangle$ over $\mathbf{x}$. Thus

$$\chi = \sum_{\mathbf{x}} \langle \sigma_0 \sigma_{\mathbf{x}} \rangle \sim m^{-7/4} \sim (T - T_c)^{-7/4} \tag{112}$$

It diverges at criticality as a power law $(T - T_c)^{-\gamma}$ and one finds in the present case

$$\gamma = \tfrac{7}{4} \tag{113}$$

In concluding this section, let us consider the correlation function in the limit $T \to T_c + 0$. We shall discuss more general expressions for multipoint correlation functions in section 3. Let ξ and x be much larger than the lattice spacing, but x much smaller than ξ, in such a way that $\exp(-|x|/\xi)$ can safely be replaced by unity. We then proceed to let $T \to T_c$. This means that x scales as $1/(1 - t/\tilde{t}\,)$ and, returning to equation (106), we find

$$\langle \sigma_0 \sigma_{\mathbf{x}} \rangle \sim \frac{1}{|\mathbf{x}|^{1/4}} \tag{114}$$

Assuming again isotropy at criticality, the scaling behaviour for such a correlation is parametrized as $1/|\mathbf{x}|^{d-2+\eta}$. Thus the critical exponent η is

$$\eta = \tfrac{1}{4} \tag{115}$$

This result can be recovered directly from the determinantal expression for the correlation function. The function $g(z)$ defined in (93) reduces at criticality to

$$g^2(z) = \frac{1 - z^{-1}}{1 - z} \times \frac{1 - t_c^2 z^{-1}}{1 - t_c^2 z}$$

The Fourier coefficients of g are asymptotically dominated by those of the first factor with a phase discontinuity of π as we travel around the unit circle. Let us altogether neglect the second factor. It can be shown that this amounts to replacing the computation of $\langle \sigma_0 \sigma_x \rangle$ by the one corresponding to a diagonal correlation $\langle \sigma_{00} \sigma_{xx} \rangle$. Within this approximation, we have $g\left(e^{ip}\right) = -ie^{ip/2}$. The corresponding Fourier coefficients are

$$a_k = -i \int_0^{2\pi} \frac{dp}{2\pi} \exp[-i(k-1)p - ip/2] = \frac{2}{\pi(1-2k)}$$

To obtain the corresponding determinant, we use a result of Cauchy which states that

$$\det \frac{1}{(a_i + b_j)} = \prod_{i<j} \frac{(a_i - a_j)(b_i - b_j)}{\prod_{i,j}(a_i + b_j)}$$

Thus

$$\langle \sigma_{00} \sigma_{xx} \rangle = \left(\frac{2}{\pi}\right)^x \prod_{0<k<q\le x-1} [4(k-q)(q-k)] \frac{1}{\prod_{k,q=0}^{x-1}(1-2(k-q))}$$

Grouping terms by pairs in the denominator, this simplifies to

$$\langle \sigma_{00} \sigma_{xx} \rangle = \left(\frac{2}{\pi}\right)^x \prod_{k=1}^{x-1}\left(1 - \frac{1}{4k^2}\right)^{k-x}$$

Recall that

$$\frac{\sin \pi a}{\pi a} = \prod_{k=1}^{\infty}\left(1 - \frac{a^2}{k^2}\right)$$

and set $a = 1/2$ to find

$$\frac{2}{\pi} = \prod_{k=1}^{\infty}\left(1 - \frac{1}{4k^2}\right).$$

Thus

$$\langle \sigma_{00} \sigma_{xx} \rangle = \prod_{k=1}^{\infty}\left(1 - \frac{1}{4k^2}\right)^x \prod_{k=1}^{x-1}\left(\frac{1}{1-4k^2}\right)^{k-x}$$

$$= \exp\left\{\sum_{k=1}^{x-1} k \ln\left(1 - \frac{1}{4k^2}\right) + x \sum_{k=x}^{\infty} \ln\left(1 - \frac{1}{4k^2}\right)\right\}$$

$$\simeq \exp - \sum_1^{x-1} \frac{1}{4k} + \text{cst} \approx \frac{\text{cst}}{x^{1/4}}$$

which is the expected behaviour.

As suggested by the previous examples, numerous quantities can be analytically evaluated for the two-dimensional Ising model, at the price of an ever increasing amount of mathematical sophistication. An important point for our future development is that one can check in detail the restoration of continuous (geometrical) symmetries as one enters the critical domain. For instance, correlation functions become isotropic on the scale of the correlation length, when the latter becomes larger and larger compared to the lattice spacing, as studied in detail by B. McCoy, T.T. Wu and their collaborators. Further elaboration is to be found in the work of Sato, Jimbo and Miwa, who used powerful mathematical tools relating the computation of correlation functions to the Riemann–Hilbert problem. The crucial result is the existence of a critical continuous domain, where the underlying field theory is a free fermionic one, with scaling laws characterized by critical exponents. We return to the continuous theory in the last section of this chapter.

2.2.7 Surface tension

At low temperature, when two macroscopic pure phases with opposite magnetization are brought into thermal equilibrium, one expects an additional contribution to the free energy proportional to the area (in three dimensions) or to the length (in two dimensions) of the interface. To compute this quantity in the present case, we can proceed as follows. Figure 2 shows a strip of width L under two sets of boundary conditions. In case I, these are uniform, say $\sigma = +1$ on the boundary. In case II, we impose mixed boundary conditions, $\sigma = +1$ above a certain line at angle ϕ, $\sigma = -1$ below. Case I will correspond in the thermodynamic limit ($L \to \infty$) to an homogeneous pure phase, while we expect an inhomogeneous mixture in case II.

We define the surface tension through the thermodynamic limit of the ratio Z_{II}/Z_I. More precisely, and restoring the conventional definition,

$$-\beta\Sigma(\varphi) = \lim_{L\to\infty} \frac{\cos\varphi}{L} \ln \frac{Z_{II}}{Z_I} \qquad (116a)$$

The tension depends on the angle φ and we note that $L/\cos\varphi$ is the length of a presumed flat interface. We can replace

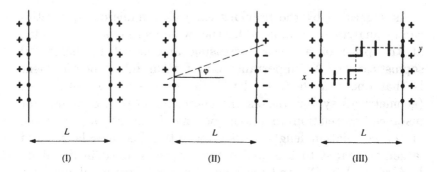

Fig. 2   Comparison of boundary conditions on a strip of width L.

the calculation of Z_{II} with mixed boundary conditions by an equivalent frustrated partition function with uniform boundary conditions. This Z_{III} is obtained by drawing an arbitrary line between two sites $\mathbf{x}$ and $\mathbf{y}$ of the dual lattice, chosen at the left (respectively right) of the link with opposite boundary conditions. One reverses the sign of the coupling, $\beta \to -\beta$, on each link of the original lattice crossing the frustration line. These are the links shown in case III on figure 2. Alternatively, this would follow starting from case II if we change all variables $\sigma \to -\sigma$ below the frustration line in the computation of Z_{II}. Except from being fixed at its end points, the line joining $\mathbf{x}$ to $\mathbf{y}$ is arbitrary. Thus we also have

$$-\beta\Sigma(\varphi) = \lim_{L\to\infty} \frac{\cos\varphi}{L} \ln \frac{Z_{III}}{Z_I} \qquad (116b)$$

This suggests a more general definition of the surface tension Σ in the two-dimensional case. Consider in the infinite volume limit two points $\mathbf{x}$ and $\mathbf{y}$ on the dual lattice and assume that $\mathbf{y} - \mathbf{x}$ is at an angle φ with the x-axis. Join these two points by a line drawn on the dual lattice and change $\beta \to -\beta$ on each link crossing this line. This is described as having set two disorder operators at $\mathbf{x}$ and $\mathbf{y}$. Denote these operators as $\tilde{\sigma}_{\mathbf{x}}$ and $\tilde{\sigma}_{\mathbf{y}}$, observing that except for its fixed boundary the location of the frustration line is again arbitrary. Then, in the thermodynamic limit, if $Z_{\mathbf{xy}}$ denotes the frustrated partition function

$$\langle \tilde{\sigma}_{\mathbf{x}} \tilde{\sigma}_{\mathbf{y}} \rangle = \frac{Z_{\mathbf{xy}}}{Z} \qquad (117)$$

and

$$-\beta\Sigma(\varphi) = \lim_{|x-y|\to\infty} \frac{\ln\left\langle \tilde{\sigma}_x \tilde{\sigma}_y \right\rangle}{|\mathbf{x} - \mathbf{y}|} \qquad (118a)$$

The above expression implies that $\beta\Sigma(\varphi)$ is identified with the inverse correlation length for disorder operators (in the ordered phase $T < T_c$). Asymptotically

$$\left\langle \tilde{\sigma}_x \tilde{\sigma}_y \right\rangle \sim \exp -\beta\Sigma(\varphi)\,|\mathbf{x} - \mathbf{y}| \qquad (118b)$$

The terminology of disorder operators is due to Kadanoff and Ceva. Off the critical region $\Sigma(\varphi)$ is expected to depend on φ (and on the lattice, here a square one) exhibiting the general anisotropy of the underlying setting. Of course, $\Sigma(\varphi)$ reflects the symmetries of this lattice, i.e. in our case

$$\Sigma(\varphi) = \Sigma(-\varphi) = \Sigma(\pi - \varphi) = \Sigma\left(\tfrac{1}{2}\pi - \varphi\right) \qquad (119)$$

The last equality might not at first be obvious in the strip geometry where it was first defined.

Let us insist on the fact that the frustration line is arbitrary except for its boundary. It can therefore be chosen in such a way as to minimize the number of frustrated links. One can indeed change a spin σ into $-\sigma$ in the computation of any partition function, thus frustrating four lines originating from the relevant site, or equivalently introducing a closed frustration curve around a plaquette of the dual lattice. Since two successive frustrations amount to no frustration at all, this means that any closed contour of frustration (on the dual lattice) can be introduced at no cost and therefore two distinct frustration lines having as common boundary the two points $\mathbf{x}, \mathbf{y}$ are equivalent. As readily verified, this extends to the introduction of an even number of frustration operators (and corresponding frustration lines).

As $T \to 0$, or $\beta \to \infty$, the energy "cost" of a frustrated link is -2β. Hence in the limit $|\mathbf{x} - \mathbf{y}| \to \infty$, $\Sigma(\varphi)$ tends to twice the ratio of the minimal number of frustrated links to the distance $|x - y|$. Per unit distance, the number of vertical and horizontal frustrated links in the minimal case are respectively $|\sin\varphi|$ and $|\cos\varphi|$. Thus

$$\lim_{T\to0} \Sigma(\varphi) = 2\left(|\cos\varphi| + |\sin\varphi|\right) \qquad (120)$$

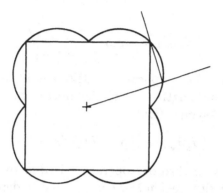

Fig. 3 Surface tension $\Sigma(\varphi)$ at zero temperature. The surface tension is the external envelope of these circles with four cusps along the horizontal and vertical axis.

In polar coordinates, this is the equation of four circles through the origin and centered at the points (± 1, ± 1) (Figure 3).

The duality property of the two-dimensional Ising model on a square lattice described in section 2.1 allows one to relate correlation functions of disorder operators $\tilde{\sigma}$ to those of the spins σ (the order operators of course, according to the same terminology) at the dual temperature. Indeed one has only to compare their low and high temperature expansions respectively to find

$$\left\langle \tilde{\sigma}_{\mathbf{x}'}\tilde{\sigma}_{\mathbf{y}'} \right\rangle_{\beta} = \left\langle \sigma_{\mathbf{x}}\sigma_{\mathbf{y}} \right\rangle_{\tilde{\beta}} \tag{121}$$

both at equal relative coordinates $\mathbf{x} - \mathbf{y} = \mathbf{x}' - \mathbf{y}'$. Comparing the asymptotic behaviour at large separation, we get

$$\beta\Sigma(\varphi) = \frac{1}{\xi_{\tilde{\beta}}(\varphi)} \tag{122}$$

where ξ stands for the correlation length for spins. In particular it follows that the surface tension vanishes at β_c with the same exponent ν with which ξ diverges. Up to now we had only computed ξ along the main crystallographic axis. But the same method extends to arbitrary directions (Cheng and Wu, 1967). From its symmetry relations, it is sufficient to compute $\Sigma(\varphi)$ for φ between 0 and $\pi/4$

$$\beta\Sigma(\varphi) = \rho(\varphi)\cos\varphi + \rho(\pi/2 - \varphi)\sin\varphi$$

$$\cos 2\varphi \cosh \rho(\varphi) = 2\cosh^2(\beta - \tilde{\beta})\cos^2\varphi$$

$$- \left(\cosh^4(\beta - \tilde{\beta})\sin^2 2\varphi + \cos^2 2\varphi\right)^{1/2} \quad (123)$$

When φ is equal to 0 or $\pi/2$ we recover $\xi_\beta^{-1} = 2(\tilde{\beta} - \beta)$, while for $\varphi = \pi/4$,

$$\cosh \frac{\beta\Sigma}{\sqrt{2}} = \cosh \frac{1}{\sqrt{2}\xi_{\tilde{\beta}}} = \cosh^2\left(\beta - \tilde{\beta}\right)$$

As $\beta \to \infty$, $\rho(\varphi) \sim \rho(\pi/2 - \varphi) \sim 2\beta$ and equation (123) reduces to the previous result (120). It is remarkable that, for any finite temperature, $\rho(\pi/2)$ vanishes and the cusps in $\Sigma(\varphi)$ disappear. Finally, in the vicinity of β_c, $\rho(\varphi) \sim 2\left(\beta - \tilde{\beta}\right)\cos\varphi$, and the surface tension (in the low temperature phase) as well as the correlation length (in the high temperature phase) become isotropic

$$\beta \to \beta_c \qquad \beta\Sigma(\varphi) = 1/\xi_{\tilde{\beta}} \sim 2\left(\beta - \tilde{\beta}\right) \qquad (124)$$

The same is true of the correlation functions in the scaling regime, showing the emergence of an isotropic (i.e. rotationally invariant) continuous theory. This is illustrated by the equilibrium shape of crystals, here two-dimensional. A side remark is that equilibrium shapes of crystals are seldom observed in nature. But under favorable conditions, they can be obtained in the laboratory for specific materials, generally at rather low temperature.

The equilibrium shape results from a competition between the bulk free energy and the surface tension. In two dimensions and at a fixed temperature, one wants therefore to minimize $\int \Sigma(\varphi)ds$ given the total area. Choosing an arbitrary origin, the element of area is $\frac{1}{2}|\mathbf{x} \wedge ds| = \frac{1}{2}\mathbf{x} \cdot \mathbf{n}ds$, $\mathbf{n}$ standing for the normal to the bounding curve (Figure 4).

Should $\Sigma(\varphi)$ be independent of the direction φ, the equilibrium shape would be a circle. The dependence on φ or equivalently on the normal $\mathbf{n}$, can then be accounted for by minimizing

$$\int \Sigma(\mathbf{n})\, ds - \lambda \int \mathbf{x} \cdot \mathbf{n}\, ds \qquad (125)$$

using a Lagrange multiplier λ for the constraint of fixed area. The quantity to be minimized is thus the integral of $\Sigma - \lambda\mathbf{x} \cdot \mathbf{n}$. Let us describe a geometrical construction (Wulff 1901) yielding the desired result. Let us draw a representation of Σ in polar

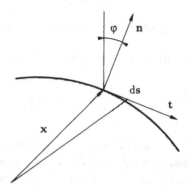

Fig. 4   Geometry for the minimization of the surface tension.

coordinates by following the point $\mathbf{n}\Sigma(\mathbf{n})$. From a point on this curve, the straight line $\Sigma - \mathbf{x} \cdot \mathbf{n}$ represents a perpendicular to the radius vector, absorbing the Lagrange multiplier in a rescaling of coordinates (or of Σ). The equilibrium shape is the envelope of these successive perpendicular lines (figure 5) up to a scale. At a point where the curve Σ is regular, let φ be the angle of $\mathbf{n}$ with the y-axis $\mathbf{n} = (-\sin\varphi, \cos\varphi)$. Let $\mathbf{t}$ denote the orthogonal unit vector tangent to the envelope, $\mathbf{t} = (\cos\varphi, \sin\varphi)$.

By minimizing (125), we find that the equation of the latter is

$$\mathbf{x} = -\mathbf{t}\frac{d\Sigma}{d\varphi} + \mathbf{n}\Sigma \tag{126}$$

When Σ is stationary, $\mathbf{x}$ is along $\mathbf{n}$. If Σ is a constant, the shape is a circle. Finally, if the slope of the Σ-curve is discontinuous (due to the presence of cusps), $d\Sigma/d\varphi$ has a discontinuity and $\Delta(\mathbf{x}\cdot\mathbf{t}) = \Delta|d\Sigma/d\varphi|$. The location of the point $\mathbf{x}$ on the line perpendicular to $\mathbf{n}$ has a jump proportional to $\Delta|d\Sigma/d\varphi|$ corresponding to a planar face on the crystal. Thus cusps in $\Sigma(\varphi)$ lead to planar faces, an example of which was given at zero temperature. The envelope was then given by the four points $(\pm 2, \pm 2)$ and four faces joining them, yielding a square equilibrium shape. The disappearance of cusps is then a signal of the disappearance of the corresponding faces. The conventional name is that the face becomes rough. At any finite temperature smaller than T_c in the case of the two-dimensional Ising model, interfaces are rough (at a

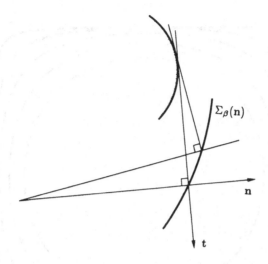

Fig. 5 The Wulff construction.

molecular scale) i.e. on a macroscopic scale the equilibrium shape is a continuous curve which does not (except in its overall shape) reveal the direction of the crystallographic axes. The equilibrium curves shown on figure 6 provide a visual demonstration for the progressive rounding off of the crystal as it tends to a circle when $T \rightarrow T_c$.

Study the curvature of the equilibrium shapes close to the would be corners as $T \rightarrow 0$.

2.3   Critical continuous theory

2.3.1   Effective action

Let us return to the Grassmannian path integral with action (69) and ask how to simplify it in the vicinity of the critical temperature. We saw that out of the four sets of modes, two remained massive at T_c. We want to decouple them from the genuine critical ones. This is accomplished if we perform a linear

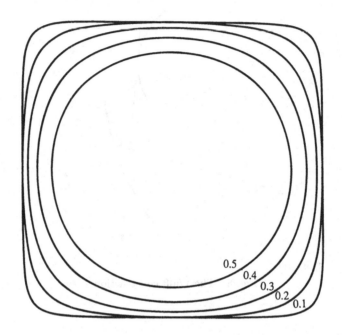

Fig. 6   Successive equilibrium shapes as a function of T/T_c.

transformation of the variables ξ of the following type

$$\bar{\xi}^H = \tfrac{1}{2} \sum_{s=\pm 1, \pm 3} e^{is\pi/4} \psi_s \qquad \bar{\xi}^V = -\tfrac{1}{2} \sum_{s=\pm 1, \pm 3} e^{2is\pi/4} \psi_s$$

$$\xi^H = -\tfrac{1}{2} \sum_{s=\pm 1, \pm 3} e^{3is\pi/4} \psi_s \qquad \xi^V = \tfrac{1}{2} \sum_{s=\pm 1, \pm 3} e^{4is\pi/4} \psi_s$$

$$(127)$$

As $t \to t_c = \sqrt{2}-1$, the ψ_s modes with $s = \pm 3$ remain massive and do not contribute to the critical singularities. We can therefore define an effective action in two steps. First neglect $\psi_{\pm 3}$, and arrange the remaining components in a two component spinor to obtain

$$S \to S' = \sum_{\mathbf{x}} t_c \tfrac{1}{2} \psi(\mathbf{x}) \sigma_2 \psi(\mathbf{x}) - \tfrac{1}{4} t \psi(\mathbf{x} + \mathbf{u}_1)(1 + \sigma_2) \psi(\mathbf{x})$$

$$- \tfrac{1}{4} t \psi(\mathbf{x} + \mathbf{u}_2)(i\sigma_3 + \sigma_2) \psi(\mathbf{x})$$

$$(128)$$

with $\mathbf{u}_1$ and $\mathbf{u}_2$ unit vectors (in lattice spacing units) along the x and y axes. Then we proceed to a formal continuous limit where sums are replaced by integrals, ψ by a continuous differentiable spinor and $\psi(\mathbf{x} + \mathbf{u})$ is expanded up to first order as $\psi(\mathbf{x}) + \mathbf{u} \cdot \nabla \psi(\mathbf{x}) + \dots$ This second step amounts to the replacement $S' \to S_{\text{eff}}$. Recall that ψ's are anticommuting variables, and use instead of $\mathbf{x} = (x, y)$ a complex notation $\mathbf{x} = (z, \bar{z})$ with

$$z = x + iy \qquad \text{and} \qquad \partial \equiv \frac{\partial}{\partial z} = \frac{1}{2} \left(\frac{\partial}{\partial x} - i\frac{\partial}{\partial y} \right)$$

the Laplacian being $4\partial\bar{\partial}$. We find

$$S' \to S_{\text{eff}} \tag{129}$$
$$= \tfrac{1}{2} t \int d^2 z \left\{ \psi_+ \bar{\partial} \psi_+ + \psi_- \partial \psi_- + \left(\frac{t - t_c}{t} \right) (i\psi_+ \psi_- - i\psi_- \psi_+) \right\}$$

The field (ψ_+, ψ_-) is referred to as a real (or Majorana) two-dimensional Fermi field. At criticality the two components decouple, while in the vicinity of this point they are coupled by a "mass" term, the mass being proportional to $t - t_c$. Finally, we rescale the field by absorbing a $\sqrt{t}$ factor in ψ. It is convenient to set

$$\sqrt{\pi t}\psi_+ = \psi \qquad \sqrt{\pi t}\psi_- = \bar{\psi} \qquad \frac{t - t_c}{t} = -\frac{m}{2} \tag{130}$$

with ψ and $\bar{\psi}$ standing now for one component fields, and m agreeing with our previous definition of m_F as $t \to t_c$. Then

$$S_{\text{eff}} = \int \frac{d^2 z}{2\pi} \left\{ (\psi\bar{\partial}\psi + \bar{\psi}\partial\bar{\psi}) + im\bar{\psi}\psi \right\} \tag{131}$$

The corresponding equation of motion, i.e. the two-dimensional analog of the Dirac equation, is the pair

$$\bar{\partial}\psi - \tfrac{1}{2}im\bar{\psi} = 0 \qquad\qquad \partial\bar{\psi} + \tfrac{1}{2}im\psi = 0 \tag{132a}$$

and as a consequence each component satisfies

$$\left(-4\partial\bar{\partial} + m^2 \right) \psi \equiv \left(-\Delta + m^2 \right) \psi = 0 \tag{132b}$$

The corresponding $\psi, \bar{\psi}$ propagator is the inverse of the matrix valued operator

$$\frac{1}{\pi} \begin{pmatrix} \bar{\partial} & -\frac{im}{2} \\ \frac{im}{2} & \partial \end{pmatrix},$$

that is

$$\begin{pmatrix} \langle \psi_1 \psi_2 \rangle & \langle \psi_1 \bar{\psi}_2 \rangle \\ \langle \bar{\psi}_1 \psi_2 \rangle & \langle \bar{\psi}_1 \bar{\psi}_2 \rangle \end{pmatrix} = \pi \begin{pmatrix} \partial_1 & \frac{1}{2}im \\ -\frac{1}{2}im & \bar{\partial}_1 \end{pmatrix} \cdot \left(\frac{1}{\partial\bar{\partial} - \frac{1}{4}m^2} \right)_{12}$$

$$= -\begin{pmatrix} \partial_1 & \frac{1}{2}im \\ -\frac{1}{2}im & \bar{\partial}_1 \end{pmatrix} \int \frac{d^2p}{\pi} \frac{e^{ip\cdot(x_1 - x_2)}}{p^2 + m^2}$$

$$(133)$$

At criticality, $m = 0$, and one has, as kernel of the Laplace operator in an arbitrary length scale,

$$-4\partial_1\bar{\partial}_1 \left(\frac{1}{2\pi} \ln \frac{1}{|x_1 - x_2|} \right) = \delta^{(2)}(x_1 - x_2) \qquad (134)$$

Therefore the propagator $\langle \psi_1 \psi_2 \rangle$ $(\langle \bar{\psi}_1 \bar{\psi}_2 \rangle)$ becomes a function of $z_{12} \equiv z_1 - z_2$ $(\bar{z}_{12})$ only

$$\langle \psi_1 \psi_2 \rangle = \frac{1}{z_{12}} \qquad\qquad \langle \bar{\psi}_1 \bar{\psi}_2 \rangle = \frac{1}{\bar{z}_{12}} \qquad (135)$$

as if the field $\psi(\bar{\psi})$ were a function of $z(\bar{z})$ alone.

Obtain the complete expression of the propagator for $m \neq 0$, and discuss its short-distance behaviour.

More general correlation functions involving several ψ and $\bar{\psi}$ fields follow from Wick's theorem. For instance, recalling the definition (28) of the Pfaffian of an antisymmetric matrix (in even dimension)

$$\langle \psi(z_1) \dots \psi(z_{2n}) \rangle = \mathrm{Pf} \frac{1}{\left(z_i - z_j \right)} \qquad (136)$$

A remarkable relation specific to this two-dimensional massless case is the following. It appears that the square of fermionic correlation functions can be identified with the correlation functions of operators derived from a free bosonic massless field. Namely, one has the identity

$$\langle \psi(z_1) \dots \psi(z_{2n}) \rangle^2 = \left(\mathrm{Pf} \frac{1}{z_i - z_j} \right)^2 = \mathrm{Hf} \left(\frac{1}{\left(z_i - z_j \right)^2} \right) \qquad (137)$$

where the Haffnian (baptized by Caianiello in analogy with the Pfaffian) of a symmetric even-dimensional matrix A_{ij} is defined as the bosonic analog of the Pfaffian

$$\operatorname*{Hf}_{1 \leq i,j \leq 2n} A_{ij} = \frac{1}{2^n n!} \sum_{\text{Permutations}} A_{p_1 p_2} \cdots A_{p_{2n-1} p_{2n}} \qquad (138)$$

and is the sum over all distinct pairings of the $2n$ indices. It expresses the content of Wick's theorem for free bosons: if A_{12} is the elementary contraction (or mean value) of two such fields, then equation (138) gives the mean value for a product of $2n$ fields.

Returning to equation (137), a proof can be obtained recursively starting from the obvious case $n = 1$. The first nontrivial instance for $n = 2$ reads

$$\left(\frac{1}{z_{12} z_{34}} - \frac{1}{z_{13} z_{24}} + \frac{1}{z_{14} z_{23}} \right)^2 = \frac{1}{z_{12}^2 z_{34}^2} + \frac{1}{z_{13}^2 z_{24}^2} + \frac{1}{z_{14}^2 z_{23}^2} \qquad (139)$$

which shows that the cross terms obtained when squaring the left-hand side cancel.

A direct proof is as follows. The Pfaffian of $1/(z_i - z_j)$ is odd under the interchange of any two variables. Consider it as a function of $z \equiv z_1$ the other variables being fixed and distinct. Therefore its Laurent expansion in z around any other z_i will have a simple pole and no constant term. By squaring, we observe that this entails the compensation of cross terms (which would generate terms of order $(z - z_i)^{-1}$). Yet another derivation can be given by studying the limit of a Cauchy determinant.

2.3.2 Correlation functions

In attempting to study the model in the vicinity of the critical point we encounter the "thermal" or "energy density" operator, or simply energy operator, coupled to the quantity m which measures the departure from T_c

$$\varepsilon(z, \bar{z}) = i \bar{\psi} \psi \qquad (140)$$

Its vacuum expectation value vanishes as well as all its odd correlation functions. As for the even ones, they follow from Wick's theorem. For instance

$$\langle \varepsilon_1 \varepsilon_2 \rangle = - \langle \bar{\psi}_1 \psi_1 \bar{\psi}_2 \psi_2 \rangle = \frac{1}{z_1 - z_2} \cdot \frac{1}{\bar{z}_1 - \bar{z}_2} \qquad (141)$$

Measured in units of inverse length, this means that the dimension of ε is 1. More generally we have

$$\left\langle \varepsilon_1 \cdots \varepsilon_{2p} \right\rangle = (-1)^p \left\langle \bar{\psi}_1 \psi_1 \cdots \bar{\psi}_p \psi_p \right\rangle = \mathrm{Pf}\frac{1}{z_i - z_j} \, \mathrm{Pf}\frac{1}{\bar{z}_i - \bar{z}_j} \tag{142}$$

Using the identity (137) we can obtain an interesting interpretation of this expression by squaring it, namely

$$\left\langle \varepsilon_1 \cdots \varepsilon_{2n} \right\rangle^2 = \mathrm{Hf}\frac{1}{\left(z_i - z_j\right)^2} \mathrm{Hf}\frac{1}{\left(\bar{z}_i - \bar{z}_j\right)^2} \tag{143}$$

This invites us to compare the model obtained by taking two distinct copies of the Ising model with a massless bosonic theory. In analogy with the effective fermionic action (131), we could consider a bosonic one, written omitting a mass term, as

$$S_{\text{bosonic}} = \int \frac{d^2 z}{8\pi} \varphi(-\Delta\varphi) = \int \frac{d^2 z}{2\pi} \bar{\partial}\varphi \partial\varphi \tag{144}$$

Using an arbitrary unit of length, the corresponding propagator

$$\left\langle \varphi_1 \varphi_2 \right\rangle = \ln\left[\frac{1}{(z_1 - z_2)(\bar{z}_1 - \bar{z}_2)}\right] \tag{145}$$

is of course unbounded at large distance and cannot therefore describe a decent observable. But φ should merely be understood as an intermediate quantity enabling one, through derivation or exponentiation, to construct bona fide observables. From (145), it follows that

$$\left\langle \partial\varphi_1 \partial\varphi_2 \right\rangle = -\frac{1}{(z_1 - z_2)^2} \qquad \left\langle \bar{\partial}\varphi_1 \bar{\partial}\varphi_2 \right\rangle = -\frac{1}{(\bar{z}_1 - \bar{z}_2)^2}$$

$$\left\langle \partial\varphi_1 \bar{\partial}\varphi_2 \right\rangle = 0 \tag{146}$$

so that

$$\left\langle \varepsilon_1 \varepsilon_2 \right\rangle^2 = \left\langle (\partial\varphi\bar{\partial}\varphi)_1 \, (\partial\varphi\bar{\partial}\varphi)_2 \right\rangle = \frac{1}{(z_1 - z_2)^2 (\bar{z}_1 - \bar{z}_2)^2} \tag{147a}$$

and more generally

$$\left\langle \varepsilon_1 \cdots \varepsilon_{2p} \right\rangle^2 = \left\langle (\partial\varphi\bar{\partial}\varphi)_1 \cdots (\partial\varphi\bar{\partial}\varphi)_{2p} \right\rangle$$

$$= \mathrm{Hf}\frac{1}{(z_i - z_j)^2} \, \mathrm{Hf}\frac{1}{(\bar{z}_i - \bar{z}_j)^2} \tag{147b}$$

The interpretation is then that by doubling the Fermi theory, the operator $\varepsilon\varepsilon'$ (prime denotes the second copy) can be thought of as $\partial\varphi\bar{\partial}\varphi$, or $\psi\psi'$ as $\partial\varphi$, both functions of z alone, and $\bar{\psi}'\bar{\psi}$ as $\bar{\partial}\varphi$. We achieve therefore a *bosonization* of the duplicated Majorana theory which may be thought of as describing a complex Fermi field. The quantities

$$\begin{aligned} J = \psi\psi' = \partial\varphi & \qquad \bar{\partial}J = 0 \\ \bar{J} = \bar{\psi}'\bar{\psi} = \bar{\partial}\varphi & \qquad \partial\bar{J} = 0 \end{aligned} \tag{148}$$

are then conserved currents (since $\bar{\partial}\partial\varphi = 0$) associated to an invariance under an Abelian rotation group in "isotopic" space (ψ, ψ') through Noether's theorem.

We record here also the analogous expressions for the spin operators, referring to the literature for a complete discussion. It is convenient to use the following normalization of the two-point function

$$\langle \sigma_1\sigma_2 \rangle = \frac{\sqrt{2}}{|z_{12}|^{1/4}} \tag{149}$$

Then the squared correlation functions may be written as

$$\langle \sigma_1 \dots \sigma_{2n} \rangle^2 = \sum_{\epsilon_i = \pm 1, \epsilon_1 + \cdots + \epsilon_{2n} = 0} \prod_{1 \le i < j \le 2n} \left| z_{ij} \right|^{\epsilon_i\epsilon_j/2} \tag{150}$$

In terms of the boson field φ we have

$$\langle : e^{i\alpha\varphi_1} :: e^{-i\alpha\varphi_2} : \rangle = \frac{1}{\left(z_1 - z_2\right)^{\alpha^2} \left(\bar{z}_1 - \bar{z}_2\right)^{\alpha^2}} \tag{151}$$

where the $::$ symbol implies the omission of self-contractions, and mean values of such exponentials are understood to vanish whenever the total "charge" i.e. the sum of coefficients of φ's, is nonzero. Choosing $\alpha = \frac{1}{2}$, we find the identification

$$\langle \sigma_1 \dots \sigma_{2n} \rangle^2 = \left\langle : e^{i\varphi_1/2} + e^{-i\varphi_1/2} : \cdots : e^{i\varphi_{2n}/2} + e^{-i\varphi_{2n}/2} : \right\rangle \tag{152}$$

or in operator form

$$\sigma\sigma' =: e^{i\varphi/2} + e^{-i\varphi/2} : \tag{153}$$

Starting from the original formulation of the Ising model and the Jordan–Wigner transformation, derive the representation (150)–(153) for the spin correlation function.

The above continuum formulation will be expanded in chapter 9 when we discuss the general two-dimensional critical theories.

Appendix 2.A   Quadratic differences and Painlevé equations

The crowning piece of the work on the two-dimensional Ising model is a set of quadratic difference equations obtained by McCoy, Perk and Wu, which determine the lattice correlation functions at any separation and any temperature. These relations reduce in the vicinity of the critical point to Painlevé equations for the two-point functions and generalizations thereof for higher correlations. A similar discussion has been given by Sato, Jimbo and Miwa, in a more mathematical setting. It would be unfair to the subject not to briefly mention these aspects, even though their developments (in the framework of integrable models) will not be pursued here. We shall limit ourselves to the case of the spin–spin correlation function, although the method extends to higher point functions. As we shall see, duality plays a crucial role in the derivation.

Let us return to the transfer matrix operating on a row of size L (tending to infinity). Consider the correlation functions of two spins located at (x_0, y_0) and (x_1, y_1). Assuming for instance $y_1 \geq y_0$

$$\left\langle \sigma_{x_1 y_1} \sigma_{x_0 y_0} \right\rangle = \lim_{M,L \to \infty} \frac{\mathrm{Tr}\left(T^{M-y_1} \sigma_{x_1}^{(1)} T^{y_1-y_0} \sigma_{x_0}^{(1)} T^{y_0} \right)}{\mathrm{Tr}\, T^M} \tag{154}$$

Recall from the definitions (56) that

$$\sigma_x^{(1)} = \sigma_{x-1/2}^{(1)} \Gamma_{x-1/2,x}$$

$$\sigma_{x-1/2}^{(1)} = \prod_0^{x-1} \frac{1}{i} \Gamma_{k-1/2,k} \Gamma_{k,k+1/2} = \prod_0^{x-1} \sigma_k^{(3)} \tag{155}$$

where $\sigma_{x-1/2}^{(1)}$ commutes with $\Gamma_{x-1/2,x}$ and can also be expressed as an exponential of a quadratic form in the Γ's since $\exp \frac{1}{2}\pi \Gamma_1 \Gamma_2 =$

$\Gamma_1 \Gamma_2$. Set

$$A = T^{y_1 - y_0} \qquad\qquad B = T^{M - (y_1 - y_0)}$$

$$\tilde{A} = \sigma^{(1)}_{x_1 - 1/2} T^{y_1 - y_0} \sigma^{(1)}_{x_0 - 1/2} \qquad AB = T^M \qquad (156)$$

Then

$$\left\langle \sigma_{x_1 y_1} \sigma_{x_0 y_0} \right\rangle = \lim_{L, M \to \infty} \frac{\mathrm{Tr}\left(\Gamma_{x_1 - 1/2, x_1} \tilde{A} \Gamma_{x_0 - 1/2, x_0} B \right)}{\mathrm{Tr}\ AB} \qquad (157)$$

where $A, \tilde{A}, B$ are all exponentials of quadratic forms in the Clifford (or Fermi) algebra. We now use a variant of Wick's theorem in operator form which says the following. Let U and V be exponentials of quadratic forms in the Γ' s and O_1 and O_2 monomials in the Γ 's. Then

$$\frac{\mathrm{Tr}\ O_1 U O_2 V}{\mathrm{Tr}\ UV} = \text{Sum over products of contractions}$$

where by contractions we understand either $\mathrm{Tr}\,\Gamma_1 \Gamma_2 UV / \mathrm{Tr}\,UV$, $\mathrm{Tr}\,\Gamma_1 \Gamma_2 VU / \mathrm{Tr}\,UV$, or $\mathrm{Tr}\,\Gamma_1 U \Gamma_2 V / \mathrm{Tr}\,UV$ if Γ_1 and Γ_2 both belong to O_1 or to O_2, or one to O_1 and the other to O_2 respectively. A sign depending of the permutation of the Γ 's is understood on the r.h.s.. Clearly this property follows from a path integral representation. Apply this with $U = \tilde{A}$ and $V = B$ to obtain

$$\mathrm{Tr}\,\tilde{A} B \times \mathrm{Tr}\left(\Gamma_{x_1 - 1/2, x_1} \Gamma_{x_1, x_1 + 1/2} \tilde{A} \Gamma_{x_0 - 1/2, x_0} \Gamma_{x_0, x_0 + 1/2} B \right) =$$

$$= \mathrm{Tr}\left(\Gamma_{x_1 - 1/2, x_1} \Gamma_{x_1, x_1 + 1/2} \tilde{A} B \right) \mathrm{Tr}\left(\tilde{A} \Gamma_{x_0 - 1/2, x_0} \Gamma_{x_0, x_0 + 1/2} B \right)$$

$$- \mathrm{Tr}\left(\Gamma_{x_1 - 1/2, x_1} \tilde{A} \Gamma_{x_0 - 1/2, x_0} B \right) \mathrm{Tr}\left(\Gamma_{x_1, x_1 + 1/2} \tilde{A} \Gamma_{x_0, x_0 + 1/2} B \right)$$

$$+ \mathrm{Tr}\left(\Gamma_{x_1 - 1/2, x_1} \tilde{A} \Gamma_{x_0, x_0 + 1/2} B \right) \mathrm{Tr}\left(\Gamma_{x_1, x_1 + 1/2} \tilde{A} \Gamma_{x_0 - 1/2, x_0} B \right)$$

$$(158)$$

Let us divide both sides by $(\mathrm{Tr}\ AB)^2$ to turn all quantities into mean values, and interpret the six terms, using the Γ -algebra, the correlation functions of spins $\sigma^{(1)}_x$ called $\mathcal{C}$ and the correlation functions of disorder operators $\sigma^{(1)}_{x+1/2}$, or equivalently of spins at the dual temperature, called $\tilde{\mathcal{C}}$.

Thus, in thermodynamic limit $(L, M \to \infty)$

$$\frac{1}{\mathrm{Tr}\ AB}\,\mathrm{Tr}\,\tilde{A} B = \frac{1}{\mathrm{Tr}\ AB}\,\mathrm{Tr}\,\sigma^{(1)}_{x_1 - 1/2} A \sigma^{(1)}_{x_0 - 1/2} B = \tilde{\mathcal{C}}\left(x_1 - x_0, y_1 - y_0 \right).$$

$$\frac{1}{\mathrm{Tr}\,AB}\,\mathrm{Tr}\left(\Gamma_{x_1-1/2,x_1}\Gamma_{x_1,x_1+1/2}\tilde{A}B\right) = \frac{i}{\mathrm{Tr}\,AB}\,\mathrm{Tr}\left(\sigma^{(1)}_{x_1+1/2}A\sigma^{(1)}_{x_0-1/2}B\right)$$

$$= i\tilde{C}\left(x_1 - x_0 + 1, y_1 - y_0\right)$$

$$\frac{1}{\mathrm{Tr}\,AB}\,\mathrm{Tr}\left(\tilde{A}\Gamma_{x_0-1/2,x_0}\Gamma_{x_0,x_0+1/2}B\right) = i\tilde{C}\left(x_1 - x_0 - 1, y_1 - y_0\right)$$

$$\frac{1}{\mathrm{Tr}\,AB}\,\mathrm{Tr}\left(\Gamma_{x_1-1/2,x_1}\tilde{A}\Gamma_{x_0-1/2,x_0}B\right) = C\left(x_1 - x_0, y_1 - y_0\right)$$

With the notations (74), we compute using the equation of motion, i.e. the commutation with the transfer matrix T,

$$C\left(x_1 - x_0, y_1 - y_0 + 1\right) = \frac{1}{\mathrm{Tr}\,AB}\,\mathrm{Tr}\left(T^{-1}\sigma^{(1)}_{x_1}TA\sigma^{(1)}_{x_0}B\right)$$

$$= \tilde{C}C\left(x_1 - x_0, y_1 - y_0\right) - i\tilde{S}\frac{1}{\mathrm{Tr}\,AB}\,\mathrm{Tr}\left(\Gamma_{x_1,x_1+1/2}\tilde{A}\Gamma_{x_0-1/2,x_0}B\right)$$

and

$$C\left(x_1 - x_0, y_1 - y_0\right) = \left\langle\sigma_{x_1,y_1+1}\sigma_{x_0,y_0+1}\right\rangle$$

$$= \tilde{C}^2\frac{1}{\mathrm{Tr}\,AB}\,\mathrm{Tr}\left(T_{x_1-1/2,x_1}\tilde{A}\Gamma_{x_0-1/2,x_0}B\right)$$

$$- i\tilde{C}\tilde{S}\frac{1}{\mathrm{Tr}\,AB}\,\mathrm{Tr}\left(\Gamma_{x_1,x_1+1/2}\tilde{A}\Gamma_{x_0-1/2,x_0}B\right)$$

$$- i\tilde{C}\tilde{S}\frac{1}{\mathrm{Tr}\,AB}\,\mathrm{Tr}\left(T_{x_1-1/2,x_1}\tilde{A}\Gamma_{x_0,x_0+1/2}B\right)$$

Therefore we get

$$C\left(x_1 - x_0, y_1 - y_0\right)^2 - C\left(x_1 - x_0, y_1 - y_0 - 1\right)C\left(x_1 - x_0, y_1 - y_0 + 1\right)$$

$$= -\tilde{S}^2\frac{1}{(\mathrm{Tr}\,AB)^2}\left[\mathrm{Tr}\left(\Gamma_{x_1-1/2,x_1}\tilde{A}\Gamma_{x_0-1/2,x_0}B\right)\mathrm{Tr}\left(\Gamma_{x_1,x_1+1/2}\tilde{A}\Gamma_{x_0,x_0+1/2}B\right)\right.$$

$$\left. - \mathrm{Tr}\left(\Gamma_{x_1-1/2,x_1}\tilde{A}\Gamma_{x_0,x_0+1/2}B\right)\mathrm{Tr}\left(\Gamma_{x_1,x_1+1/2}\tilde{A}\Gamma_{x_0-1/2,x_0}B\right)\right]$$

i.e. we have an interpretation of the last terms in (158). Putting everything together, we obtain the first quadratic difference equation

$$S\left[C^2(x,y) - C(x,y+1)C(x,y-1)\right]$$
$$+ \tilde{S}\left[\tilde{C}^2(x,y) - \tilde{C}(x+1,y)\tilde{C}(x-1,y)\right] = 0 \tag{159a}$$

The second relation is obtained by interchanging the roles of the coordinates x and y, a discrete symmetry of the lattice (or by exchanging β and $\tilde{\beta}$)

$$S\left[C^2(x,y) - C(x+1,y)C(x-1,y)\right]$$
$$+ \tilde{S}\left[\tilde{C}^2(x,y) - \tilde{C}(x,y+1)\tilde{C}(x,y-1)\right] = 0 \qquad (159b)$$

Applying again Wick's theorem to another combination of operators one derives in the same vein a third relation (symmetric in x and y)

$$S\left[C(x+1,y+1)C(x,y) - C(x+1,y)C(x,y+1)\right] =$$
$$\tilde{S}\left[\tilde{C}(x+1,y+1)\tilde{C}(x,y) - \tilde{C}(x+1,y)\tilde{C}(x,y+1)\right] \quad (159c)$$

These relations exhibit an interesting symmetry (or antisymmetry) under duality. Supplemented by appropriate boundary conditions, they determine fully the two-point function. They also generalize to higher correlations. We now assume that a continuum limit is meaningful in the vicinity of T_c, and we study their continuum approximation. To do so, we first absorb a factor $S^{1/2} \equiv \sinh 2\beta^{1/2}$ into the definition of C

$$S^{1/2}C \to C \qquad (160)$$

expand all terms in equations (159) to second order in the lattice spacing, and use complex coordinates $z = x + iy$, $\bar{z} = x - iy$, to recast the system into the form

$$\left(C\partial\bar{\partial}C - \partial C\bar{\partial}C\right) + \left(\tilde{C}\partial\bar{\partial}\tilde{C} - \partial\tilde{C}\bar{\partial}\tilde{C}\right) = 0 \qquad (161a)$$

$$C\partial^2 C - (\partial C)^2 = \tilde{C}\partial^2\tilde{C} - \left(\partial\tilde{C}\right)^2 \qquad (161b)$$

$$C\bar{\partial}^2 C - \left(\bar{\partial}C\right)^2 = \tilde{C}\bar{\partial}^2\tilde{C} - \left(\bar{\partial}\tilde{C}\right)^2 \qquad (161c)$$

Let us assume C (and $\tilde{C}$) isotropic i.e. function only of the distance $|z|$. Then the two equations (161b) and (161c) collapse into a single one. For a while, let us set

$$z\bar{z} = e^{2t} \qquad C = e^f \qquad \tilde{C} = e^{\tilde{f}} \qquad (162)$$

Then

$$e^{2f}f'' + e^{2\tilde{f}}\tilde{f}'' = 0 \qquad (163a)$$

$$e^{2f}(f'' - 2f') = e^{2\tilde{f}}\left(\tilde{f}'' - 2\tilde{f}'\right) \qquad (163b)$$

where primes denote derivatives with respect to t. These equations have to satisfy a compatibility condition. To obtain this, we rewrite the system in the form

$$e^{2f} f' = e^{2\tilde{f}} \left(\tilde{f}' - \tilde{f}'' \right) \tag{164a}$$

$$e^{2\tilde{f}} \tilde{f}' = e^{2f} \left(f' - f'' \right) \tag{164b}$$

and take a derivative of the second equation with respect to t

$$e^{2\tilde{f}} \left(\tilde{f}'' + 2\tilde{f}'^2 \right) = e^{2f} \left(f'' - f''' + 2f'^2 - 2f'f'' \right)$$

we use (163a) to replace $e^{2\tilde{f}} \tilde{f}''$ by $-e^{2f} f''$ and (164b) to replace $2e^{2\tilde{f}} \tilde{f}'^2$ by $2\tilde{f}' e^{2f} (f' - f'')$, thus getting

$$f''' - 2f'^2 - 2f'' + 2f'f'' + 2\tilde{f}'(f' - f'') = 0$$
$$\tilde{f}''' - 2\tilde{f}'^2 - 2\tilde{f}'' + 2\tilde{f}'\tilde{f}'' + 2f' \left(\tilde{f}' - \tilde{f}'' \right) = 0$$

where the second equality follows from duality. Adding the two equations yields

$$\frac{d}{dt} \left[f'' + f'^2 + \tilde{f}'' + \tilde{f}'^2 - 2f'\tilde{f}' \right] - 2 \left[f'' + f'^2 + \tilde{f}'' + \tilde{f}'^2 - 2f'\tilde{f}' \right] = 0$$

Hence by integration

$$f'' + f'^2 + \tilde{f}'' + \tilde{f}'^2 - 2f'\tilde{f}' = \text{cst } e^{2t} = \text{cst } z\bar{z} \tag{165}$$

Returning to the original function, this is seen to be equivalent to

$$\frac{1}{C\tilde{C}} \left[\tilde{C}\partial\bar{\partial}C + C\partial\bar{\partial}\tilde{C} - \partial C \partial\tilde{C} - \bar{\partial}C\partial\tilde{C} \right] = \text{cst} = \tfrac{1}{4}m^2 \tag{166}$$

The constant on the right hand side is identified as $m^2/4$ assuming the following behaviour at large separation for $T > T_c$

$$\tilde{C} \to \text{cst} \qquad C \to \frac{\text{cst}}{|z|^a} \exp -m|z| \tag{167}$$

We have now what might be called the Painlevé system

$$C\partial\bar{\partial}C + \tilde{C}\partial\bar{\partial}\tilde{C} - \partial C\bar{\partial}C - \partial\tilde{C}\bar{\partial}\tilde{C} = 0 \tag{168a}$$
$$\tilde{C}\partial\bar{\partial}C + C\partial\bar{\partial}\tilde{C} - \partial C\bar{\partial}\tilde{C} - \bar{\partial}C\partial\tilde{C} = \tfrac{1}{4}m^2 C\tilde{C} \tag{168b}$$

The second equation provides a scale in which to measure length. We change variables by defining

$$u = \tfrac{1}{2}mz \tag{169}$$

and now understand derivatives as carried with respect to u and $\bar{u}$, which amounts to replace $m^2/4$ by unity in (168b). By adding and subtracting the two equations (168), we obtain the following system

$$C + \tilde{C} = e^L \qquad\qquad C - \tilde{C} = e^M \qquad\qquad u = \frac{m}{2}z \qquad (170a)$$

$$4\partial\bar{\partial}(L - M) = e^{2(L-M)} - e^{-2(L-M)} \qquad\qquad (170b)$$

$$4\partial\bar{\partial}(L + M) = 2 - e^{2(L-M)} - e^{-2(L-M)} \qquad\qquad (170c)$$

Recalling that C and $\tilde{C}$, and therefore L and M, only depend on $|z|$, equation (170b) determines $(L - M)$, while (170c) yields $L + M$ provided we include some information on boundary conditions. The system (170) therefore determines C and $\tilde{C}$. Painlevé's equation proper is (170b), when we insist that $L - M$ depends on $|z|$ only. When this is taken into account, it becomes a second order nonlinear differential equation with the specific property that its solutions have only poles as moveable singularities (i.e. dependent on the boundary conditions). Specifically, if

$$\eta = e^{-(L-M)} = \frac{\tilde{C} - C}{\tilde{C} + C} \qquad r = |u| = \tfrac{1}{2}m\,|z| \qquad (171a)$$

$$\frac{1}{\eta}\frac{d^2\eta}{dr^2} - \left(\frac{1}{\eta}\frac{d\eta}{dr}\right)^2 + \frac{1}{\eta}\frac{1}{r}\frac{d\eta}{dr} = \eta^2 - \frac{1}{\eta^2} \qquad (171b)$$

Formally this is invariant under $\eta \to 1/\eta$. We can satisfy ourselves with the solution for T slightly above T_c, since determining C and $\tilde{C}$ yields then also C below T_c. Conditions (167) imply that η goes to 1 for large r, up to exponentially small terms. If we neglect $(\eta'/\eta)^2$ as compared to η''/η, this yields

$$\eta = 1 + \text{cst}\,\frac{e^{-2r}}{r^{1/2}}$$

thus the exponent in (167) is $a = 1/2$, as expected from equation (107). We let the reader convince himself that the leading asymptotic behaviour is

$$|z| \to \infty \begin{cases} \eta = 1 - \dfrac{2}{\pi}K_0(2r) + \ldots = 1 - \sqrt{\dfrac{2}{\pi m\,|z|}}e^{-m|z|} + \ldots \\[2ex] \tilde{C} \to \text{cst} \\[2ex] C/\tilde{C} \sim \dfrac{1}{\pi}K_0(2r) \sim \sqrt{\dfrac{1}{2\pi m|z|}}e^{-m|z|} \end{cases}$$

$$(172)$$

Close to the origin, $\mathcal{C}$ and $\tilde{\mathcal{C}}$ have the expected $|z|^{-1/4}$ scaling behaviour

$$|z| \to 0 \begin{cases} \mathcal{C} \sim & \dfrac{1}{(m\,|z|)^{1/4}}\left[1 + \tfrac{1}{2}m\,|z|\ln\tfrac{1}{2}m\,|z| + \cdots\right] \\[2mm] \tilde{\mathcal{C}} \sim & \dfrac{1}{(m\,|z|)^{1/4}}\left[1 - \tfrac{1}{2}m\,|z|\ln\tfrac{1}{2}m\,|z| + \cdots\right] \end{cases} \tag{173}$$

Notes

Grassmannian variables and their properties are discussed by F.A. Berezin in *The Method of Second Quantization*, Academic Press, New York (1966). They are now used in many contexts, including the treatment of disordered systems. Examples will be given in chapter 10.

It is not possible to give here a serious and complete bibliography on the Ising model introduced in fact by Lenz, *Z. Physik* **21**, 613 (1920) and first studied by E. Ising, *Z. Physik* **31**, 253 (1925). We list here some of the fundamental works pertaining to the two-dimensional solution, as well as those related to our presentation. A proof of the existence of a low temperature ordered phase, not discussed in the text is the one of R. Peierls, *Proc. Cambridge Phil. Soc.* **32**, 477 (1936). Transfer matrix methods and duality appear in H.A. Kramers and G.H. Wannier, *Phys. Rev.* **60**, 252, 263 (1941). The free energy was computed, using Clifford algebra methods, by L. Onsager, *Phys. Rev.* **65**, 117 (1944). The proof was simplified by B. Kaufman, *Phys. Rev.* **76**, 1232 (1949). Onsager also announced the expression for the spontaneous magnetization in *Nuovo Cimento* **6**, Supplement, 261 (1949). A complete derivation was given by C.N. Yang, *Phys. Rev.* **85**, 808 (1952). M. Kac and J.C. Ward developed a combinatorial proof in *Phys. Rev.* **88**, 1332 (1952). Other reviews, and alternative methods can be traced in G.F. Newell and E.W. Montroll, *Rev. Mod. Phys.* **25**, 353 (1953), E.W. Montroll, R.B. Potts, J.C. Ward, *J. Math. Phys.* **4**, 308 (1963), T.D. Shultz, D.C. Mattis, E.H. Lieb, *Rev. Mod. Phys.* **36**, 856 (1964), H.S. Green and C.A. Hurst *Order disorder phenomena*, Interscience, London (1964). A discussion on finite size effects is found in A.E. Ferdinand and M.E. Fisher, *Phys. Rev.* **185**, 832 (1969).

The lemma on Töplitz determinants is due to G. Szegö in *Communications du séminaire mathématique de l'université de Lund*, tome supplémentaire dédié à M. Riesz, 228 (1952) and to M. Kac, *Duke Math. J.* **21**, 501 (1954).

A general summary, including many references and a presentation of their early contributions, is to be found in B.M. McCoy and T.T. Wu, *The Two-Dimensional Ising Model*, Harvard University Press, Cambridge, Mass. (1973).

The concept of order and disorder variables is developed in L.P. Kadanoff, H. Ceva, *Phys. Rev.* **B3**, 3918 (1971).

The construction of the equilibrium shape of crystals from the interfacial tension is due to G. Wulff, *Z. Kristallogr. Mineral.* **34**, 449 (1901). See also C. Herring, *Phys. Rev.* **82**, 87 (1951). The computation for the two-dimensional Ising case is based on the work of H. Cheng and T.T. Wu, *Phys. Rev.* **164**, 719 (1967), P.G. Watson, volume 3 of the Domb and Green series, D.B. Abraham and P. Reed, *J. Phys.* **A10**, L121 (1977) and C. Rottman and M. Wortis, *Phys. Rev.* **B24**, 6274 (1981). The curves in the text are from J.E. Avron, H. van Beijeren, L.S. Schulman and R.K.P. Zia, *J. Phys* **A15**, L81 (1982). See also R.K.P. Zia and J. Avron, *Phys. Rev.* **B25**, 2041 (1982).

The representation of the Ising model in terms of Grassmannian integrals is patterned after the work of S. Samuel, *J. Math. Phys.* **21**, 2806, 2815, 2820 (1980), and of one of the authors in *Nucl. Phys.* **B210** (**FS6**), 448 (1982).

The calculation of correlation functions in the scaling limit is found in the work of T.T. Wu, B.M. McCoy, C.A. Tracy, E. Barouch, *Phys. Rev.* **B13**, 316 (1976). For a relation with continuous field theory, see A. Luther, I. Peschel, *Phys. Rev.* **B12**, 3908 (1975), the work of one of the authors in collaboration with M. Bander, *Phys. Rev.* **D15**, 463 (1977), and with J.-B. Zuber, *Phys. Rev.* **D15**, 2875 (1977), as well as M. Bander and J.L. Richardson, *Phys. Rev.* **B17**, 1464 (1978).

Quadratic difference equations are discussed in B.M. McCoy and T.T. Wu, *Phys. Rev. Lett.* **45**, 675 (1980), J.H.H. Perk, *Phys. Lett.* **79A**, 1,3 (1980), B.M. McCoy, J.H.H. Perk, and T.T. Wu, *Phys. Rev. Lett.* **46**, 757 (1981).

Finally the work of M. Sato, T. Miwa and M. Jimbo is to be found in a series of publications, *Proc. Jap. Acad.* **53A**, 147, 153, 183 (1977) and *Publications R.I.M.S.*, Kyoto University, **14**, 223 (1978), **15**, 201, 577, 871 (1979).

For the Painlevé property, see E. Ince *Ordinary differential equations*, Dover, New York (1945).

3

SPONTANEOUS SYMMETRY BREAKING, MEAN FIELD

In many statistical models (but not all) one can introduce a local order parameter associated with a finite, discrete or continuous group of symmetries. The higher the temperature, the more important the fluctuations. One expects therefore in general to find pure phases at low temperature, with a reduced symmetry group, as a consequence of a nonvanishing expectation value of some order parameter. This situation is referred to as spontaneous symmetry breaking. A typical example is the classical Heisenberg model describing short-range interactions of an n-vector field φ, with an orthogonal $O(n)$ symmetry group. For $n = 1$, 2, 3, this can account for a Curie transition from a ferromagnetic to a paramagnetic phase. In particle physics the σ-model of Gell-Mann and Levy involves a spontaneous symmetry breaking of chiral invariance, typical of massless spinor fields, accompanied by soft excitation modes, the so-called Goldstone modes, associated to a π-meson triplet and leading to a nonvanishing dynamical fermion mass. In a first and rather crude approximation, one can analyse the action itself or an effective one incorporating some fluctuation effects, and look for extrema as a function of field configurations, generally translationally invariant. The remaining fluctuations are then treated perturbatively. This mean field method is common to a great variety of domains, ranging from the Clausius–Mossotti formula for a polarizable medium, the Weiss molecular field in the theory of magnetism, Landau's effective action in various statistical contexts, the effective medium approximation in disordered systems, the Hartree–Fock method in atomic or many-body physics, to the semiclassical approximation in the study of quantum systems. A common mathematical tool is, at one stage or another, the steepest descent or saddle point evaluation of a functional integral, implying a stability analysis and a perturbative expansion of the fluctuation effects.

The physical motivation is to substitute for the dynamics or statistics of a very large number of coupled degrees of freedom, an effective one, pertaining to a small number of carefully chosen relevant observables, interacting via a self-consistent mean field, representing the overall effect of the remaining ones. The examples will show how these ideas are implemented in practice.

3.1   Mean field approximation

3.1.1   Dielectric constant of a polarizable medium

A classical instance of mean field theory is the derivation of an effective dielectric constant of a polarizable medium. The introduction of a test charge in such an environment induces a charge density $\rho_{ind}(\mathbf{x})$, with a vanishing integral (since the medium is charge-neutral). The latter can therefore be written as the divergence of a dipole density

$$\rho_{ind}(\mathbf{x}) = -\nabla \cdot \mathbf{P}(x) \tag{1}$$

Multiply both sides by $\mathbf{x}$ and integrate over space. If $\rho_{ind}(\mathbf{x})$ and $\mathbf{P}(\mathbf{x})$ are sufficiently localized so that an integration by parts is justified, we find for the total induced dipole moment

$$\mathbf{P}_{tot} = \int d^3\mathbf{x} \; \mathbf{x}\rho_{ind}(\mathbf{x}) = \int d^3\mathbf{x} \; \mathbf{P}(\mathbf{x}) \tag{2}$$

A shift $\mathbf{x} \rightarrow \mathbf{x}+\mathbf{x}_0$ does not affect this result since the total induced charge vanishes. As a consequence, Gauss's law $\mathrm{div}\mathbf{E} = \rho/\varepsilon_0$, giving the electric field produced by the total charge density $\rho = \rho_{ext} + \rho_{ind}$, can be reshuffled to read

$$\mathrm{div}\left(\mathbf{E} + \frac{\mathbf{P}}{\varepsilon_0}\right) \equiv \mathrm{div}\mathbf{D} = \frac{\rho_{ext}}{\varepsilon_0} \tag{3}$$

For sufficiently small fields and an isotropic medium, the induced moment is proportional to the electric field. The electric induction can thus be written as

$$\mathbf{D} = \mathbf{E} + \frac{\mathbf{P}}{\varepsilon_0} = \varepsilon\mathbf{E} \tag{4}$$

where ε is the effective dielectric constant. Let us now look at this expression from a microscopic point of view. The dipole moment $\mathbf{P}$ can be written as $n\mathbf{p}$ where $\mathbf{p}$ denotes the individual molecular dipole moment and n is the density of carriers. In turn each molecular dipole $\mathbf{p}$ is proportional to a local electric field $\mathbf{E}_{\text{loc}}$, with $\mathbf{p} = \alpha\mathbf{E}_{\text{loc}}$, and α the polarizability of the isolated molecule. The point is now to find the effective local field. Assuming weak fields as in (3), the standard analysis decomposes $\mathbf{E}_{\text{loc}}$ into three parts. The first one is the electric field $\mathbf{E}$ due to external sources, the second $\mathbf{E}_1$ being the result of far away dipoles outside a volume, large on the microscopic scale but small on a macroscopic one (a classical, poorly defined, but convenient concept). This volume is chosen spherical to respect isotropy. Finally $\mathbf{E}_2$ will be the field produced by the nearby molecules in this spherical cavity. The latter vanishes in a medium with a large symmetry (a disordered one, as a liquid, or even in a crystalline cubic environment for instance). The problem of evaluating $\mathbf{E}_1$ at the center of the cavity is of course a well-known elementary exercise. It is obtained as the one deriving from a surface density $\sigma = -\mathbf{n} \cdot \mathbf{P}$, $\mathbf{n}$ being the unit vector along the normal to the sphere pointing outwards. Choosing the z-axis along $\mathbf{P}$, R being the radius of the sphere, the corresponding potential along the z axis is

$$\phi(z) = -P\frac{R^2 2\pi}{4\pi\varepsilon_0}\int_0^\pi d\cos\theta\,\frac{\cos\theta}{\sqrt{R^2 + z^2 - 2Rz\cos\theta}} = -z\frac{P}{3\varepsilon_0}$$

The field $\mathbf{E}_1 = -\nabla\phi$ at the center is

$$\mathbf{E}_1 = \frac{\mathbf{P}}{3\varepsilon_0}$$

The local field is the sum $\mathbf{E} + \mathbf{E}_1 = \mathbf{E} + \mathbf{P}/3\varepsilon_0$, and the individual polarization is

$$\mathbf{p} = \alpha\mathbf{E}_{\text{loc}} = \alpha\left(\mathbf{E} + \frac{\mathbf{P}}{3\varepsilon_0}\right)$$

Multiplying both sides by the density n

$$\mathbf{P} = n\alpha\left(\mathbf{E} + \frac{\mathbf{P}}{3\varepsilon_0}\right)$$

From this and the definition (4) it follows that

$$\frac{\varepsilon - 1}{\varepsilon + 2} = \frac{1}{3}\frac{n\alpha}{\varepsilon_0} \tag{5}$$

which is the Clausius–Mossotti relation (or the Lorentz–Lorentz relation in an optical context, with ε replaced by the square of the refraction index) expressing the phenomenological dielectric constant in terms of the microscopic polarizability α. Note the self-consistent character of the computation of the local field.

The mean field approximation (5) does not take into account fluctuations and correlations. It is interesting to see how these come about using the standard Kirkwood–Yvon theory. Consider the case of a uniform nonpolar medium of mean density n, from a microscopic point of view. Let $\mathbf{x}_i$ denote the positions of the molecules or atoms, and $\mathbf{E}$ stand as before for the external field. The induced dipole $\mathbf{p}_i$ on molecule i is given by

$$\mathbf{p}_i = \alpha \left[\mathbf{E} - \sum_{j,\, j \neq i} K_{ij} \mathbf{p}_j \right]$$

with K_{ij} the matrix

$$K_{ij} = \frac{1}{4\pi\varepsilon_0} \, \boldsymbol{\nabla}_i \otimes \boldsymbol{\nabla}_j \, \frac{1}{|\mathbf{x}_i - \mathbf{x}_j|}$$

Set $\langle \mathbf{p} \rangle = \langle \mathbf{p}_i \rangle$ independent of i, where the bracket is the statistical average. We have

$$\mathbf{p}_i = \alpha \left[\mathbf{E} - \sum_{j,\, j \neq i} \langle K_{ij} \rangle \langle \mathbf{p}_j \rangle - \sum_{j,\, j \neq i} \left(K_{ij}\mathbf{p}_j - \langle K_{ij} \rangle \langle \mathbf{p}_j \rangle \right) \right]$$

A calculation similar to the one above yields

$$\mathbf{p}_0 = \alpha \left[\mathbf{E} - \sum_{j,\, j \neq i} \langle K_{ij} \rangle \langle \mathbf{p} \rangle \right] = \alpha \left[\mathbf{E} + \frac{\mathbf{P}}{3\varepsilon_0} \right] = \frac{\alpha}{3}(\varepsilon + 2)\mathbf{E}$$

If one neglects the fluctuating contribution to $\mathbf{p}_i$, thus identifying $n\mathbf{p}_0$ with $\mathbf{P}$, one recovers the mean field result (5). To obtain the correction, we have only to iterate

$$\mathbf{p}_i = \mathbf{p}_0 - \alpha \sum_{j,\, j \neq i} \left[K_{ij}\mathbf{p_j} - \langle K_{ij} \rangle \langle \mathbf{p}_j \rangle \right]$$

by considering α formally as a small parameter. Let us content ourselves with a first iteration and then take the average. This yields

$$\langle \mathbf{p} \rangle = \mathbf{p}_0 + \alpha^2 \sum_{\substack{j,k \\ j \neq i \; k \neq j}} \left[\langle K_{ij} K_{jk} \rangle - \langle K_{ij} \rangle \langle K_{jk} \rangle \right] \mathbf{p}_0 + \cdots$$

Since $\mathbf{p}_0 = \frac{1}{3}\alpha(\varepsilon+2)\mathbf{E}$, it is clear that $\mathbf{p}$ is along $\hat{\mathbf{E}}$, the unit vector in the direction of the external field. Multiplying both sides by n/ε_0, we readily find

$$\frac{\varepsilon-1}{\varepsilon+2} = \frac{n\alpha}{3\varepsilon_0}[1+S] \qquad (6a)$$

$$S = \alpha^2 \sum_{\substack{j,k \\ j\neq i,\, k\neq j}} \hat{\mathbf{E}}\left[\langle K_{ij}K_{jk}\rangle - \langle K_{ij}\rangle\langle K_{jk}\rangle\right]\hat{\mathbf{E}} + \mathcal{O}(\alpha^3)$$

We split the above sum into a contribution with $j \neq i \neq k$ and another one with $k = i$. If N and V stand respectively for the total number of molecules and the total volume, both tending to infinity, we define the molecular distribution functions as

$$n_2(\mathbf{x}_1 - \mathbf{x}_2) = \sum_{i,j}' \langle \delta(\mathbf{x}_1 - \mathbf{x}_i)\delta(\mathbf{x}_2 - \mathbf{x}_j)\rangle = n^2 g_2(|\mathbf{x}_1 - \mathbf{x}_2|)$$

$$\cdots$$

where the prime means the omission of the contribution $i = j$, so that for instance

$$\sum_{j,\,j\neq i} \langle f(\mathbf{x}_i - \mathbf{x}_j)\rangle = \frac{V}{N}\frac{1}{V}\int d^3\mathbf{x}_1 d^3\mathbf{x}_2 f(\mathbf{x}_1 - \mathbf{x}_2)$$

$$\times \sum_{i,j}' \langle \delta(\mathbf{x}_1 - \mathbf{x}_i)\delta(\mathbf{x}_2 - \mathbf{x}_j)\rangle$$

$$= \frac{1}{n}\int d^3\mathbf{x}\, n_2(\mathbf{x})f(\mathbf{x}) = n\int d^3\mathbf{x}\, g_2(\mathbf{x})f(\mathbf{x})$$

Inserting such expressions in the correction term S and performing the sums and angular integrals gives

$$S = \frac{\alpha^2}{(4\pi\varepsilon_0)^2}\left\{ 8\pi n \int_0^\infty dx \frac{g_2(x)}{x^4} \right.$$

$$+ n^2 \int d^3\mathbf{x}_1 d^3\mathbf{x}_2 \frac{g_3(\mathbf{x}_1,\mathbf{x}_2) - g_2(x_1)g_2(x_2)}{x_1^3 x_2^3}[3\hat{\mathbf{x}}_1.\hat{\mathbf{x}}_2 - 1]\Bigg\}$$

$$+ \mathcal{O}(\alpha^3) \qquad (6b)$$

predicting to leading order a parabolic shape for the correction as a function of density or pressure. Long-range correlations close to a continuous transition invalidate the preceding perturbative approach and lead to critical opalescence.

3.1.2   Classical spin model with a finite symmetry group

Consider an Ising model on a hypercubical d-dimensional lattice where the coordination number $q = 2d$ grows linearly with d. The Boltzmann weight of a configuration in an external field h is

$$p(\sigma) = Z^{-1} \exp \left(\beta \sum_{(ij)} \sigma_i \sigma_j + h \sum_i \sigma_i \right) \qquad (7)$$

Strictly, the true external field is proportional to h/β ; we keep however the above terminology. For $h = 0$, the weights of the configurations $\{\sigma_i\}$ and $\{-\sigma_i\}$ are identical, exhibiting $Z/2Z$, or for short Z_2, the group with two elements ± 1, as the symmetry group of the model. We saw in chapter 2 that at least for $d = 2$, the Ising model has a low temperature ordered phase, with pure states characterized by a spontaneous magnetization breaking the Z_2 symmetry. Let us carry over the discussion to higher dimension in the mean field context. The probability weight (7) suggests that we consider the random variable

$$H_i = \beta \sum_{j(i)} \sigma_i \qquad (8)$$

as a field acting on the spin σ_i, very much as we did with the other dipoles which contributed to the local electric field in the previous example. The difference is here that interactions are of short range. As d increases, so does the number of neighbours contributing in (8). It is therefore tempting for large d to neglect the fluctuations of H_i, and to replace it by its mean value, borrowing the idea from the central limit theorem. If a nonvanishing mean value of H_i persists in the limit $h \to 0$, there is a spontaneous symmetry breaking. This mean field has to be computed self-consistently. Let us present several derivations to illustrate the point.

When computing the partition function corresponding to (7), which normalizes the weights, it is possible to reverse any set of (dummy) variables σ, since they are summed over. For instance, Z is invariant if we change the sign of a single spin σ_i, before summation. Taking into account the definition of H_i, this leads to the identity

$$1 = \langle \exp\left(-2\sigma_i(H_i + h)\right) \rangle \qquad (9a)$$

or equivalently

$$\langle \cosh 2(H_i + h) \rangle = 1 + \langle \sigma_i \sinh 2(H_i + h) \rangle \qquad (9b)$$

The approximation is to replace the random variable H_i by its translational invariant mean H, while $\langle \sigma_i \rangle$ can be identified with the magnetization M. From (9), we obtain

$$M = \frac{\cosh 2(H + h) - 1}{\sinh 2(H + h)} = \tanh(H + h) \qquad (10a)$$

and for consistency with (8), we must have

$$H = 2\beta d M \qquad (10b)$$

What is remarkable is that, even for zero external field h, these equations can have a nonvanishing solution. When $h = 0$, the Z_2 symmetry implies that for every solution (H, M) we have the opposite one $(-H, -M)$ and H and M have obviously the same sign. Therefore we can assume $H > 0$, which has then to satisfy

$$\frac{H}{2d\beta} = \tanh H \qquad (10c)$$

In general, $2d$ should be replaced by the coordination number of the lattice, assumed to be large. If $2d\beta > 1$, i.e. at sufficiently small temperature, equation (10c) admits a nonzero solution, such that $M = H/2d\beta$ decreases from 1 at $\beta \to \infty$, down to zero at $\beta_c = (2d)^{-1}$. The graphical solution is shown in figure 1. At high temperature, $\beta < \beta_c$, the spontaneous magnetization M vanishes with the local field H. One expects therefore a disordered symmetric phase. Observe that as the disorder increases, so does the symmetry. In a similar sense, the crystalline patterns observed after solidification express a loss of (translational or rotational) symmetry as compared to the liquid or gaseous phase.

Close to the transition point, one finds a square root behaviour typical of the present approximation

$$M \sim H \sim \sqrt{3}(\beta/\beta_c - 1)^{1/2} \qquad (11)$$

This critical exponent is denoted β, following the convention introduced in the previous chapter, where we saw that its value for $d = 2$ is $1/8$, rather than $1/2$ as predicted by the mean field approximation.

For $h \neq 0$, equations (10a) and (10b) admit either a unique solution or several ones, which have to be distinguished on the

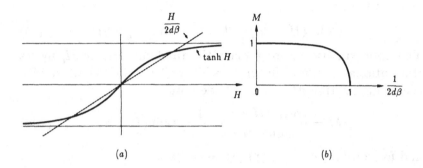

Fig. 1   (a) Graphical solution of the mean field equation. (b) Spontaneous magnetization.

basis of the maximization of $(\ln Z)/\Omega$, where Ω is the total volume. The interesting situation is the one near the critical point. At $\beta = \beta_c$, by expanding the hyperbolic tangent, we find $h \sim (M + h)^3/3$, thus

$$\beta = \beta_c \qquad\qquad M \sim (3h)^{1/3} \qquad\qquad (12a)$$

or

$$\beta = \beta_c \qquad\qquad h \sim M^\delta, \qquad\qquad \delta_{\text{mean field}} = 3 \qquad (12b)$$

Approaching criticality in the high temperature phase, M is linear in h, but as suggested by (12a) the corresponding coefficient, the susceptibility must diverge as $\beta \to \beta_c - 0$. Indeed we find

$$\beta \to \beta_c - 0 \qquad\qquad M \sim \chi h \qquad\qquad \chi \sim \frac{1}{1 - \beta/\beta_c}$$
$$(13a)$$

or

$$\chi \sim (1 - \beta/\beta_c)^{-\gamma} \qquad\qquad \gamma_{\text{mean field}} = 1 \qquad (13b)$$

Recall that $\gamma = 7/4$ in dimension two.

We return to the derivation of the mean field equations from another point of view, namely the minimization of the free energy $-\ln Z/\beta\Omega$. The latter is related to the convexity property of the exponential function, i.e.

$$\langle \exp A \rangle \geq \exp \langle A \rangle \qquad\qquad (14)$$

Let us apply this inequality by taking as a trial Boltzmann weight a factorized expression, the only one for which we can do an explicit computation. Alas, one has frequently no other choice but to search for one's keys under the street lamp! For N sites (the volume Ω being Na^d, with a the lattice spacing), write

$$Z = \frac{1}{2^N} \sum_{\sigma_i = \pm 1} \exp\left(\beta \sum_{(ij)} \sigma_i \sigma_j + h \sum_i \sigma_i\right) =$$
$$Z_H \left\{ Z_H^{-1} \frac{1}{2^N} \sum_{\sigma_i = \pm 1} \exp\left((H + h)\sum_i \sigma_i + A\right) \right\} \quad (15a)$$

with

$$Z_H = \frac{1}{2^N} \sum_{\sigma_i = \pm 1} \exp\left[(H + h)\sum_i \sigma_i\right] = [\cosh(H + h)]^N \quad (15b)$$

and

$$A = \beta \sum_{(ij)} \sigma_i \sigma_j - H \sum_i \sigma_i \quad (15c)$$

In other words, we have added and subtracted $H \sum_i \sigma_i$ in the action (or energy), and presented Z as the product of Z_H times a mean value taken on a factorized measure. The latter cannot be evaluated exactly, but we can apply the inequality (14), with the result that

$$F = \frac{1}{N} \ln Z \geq \underset{H}{\mathrm{Sup}} \left\{ \ln\cosh(h + H) + \frac{1}{N} \langle A \rangle_H \right\} \quad (16)$$

We have taken advantage of the arbitrariness on H to majorize the right-hand side over H, and of course $\langle \cdots \rangle_H$ means the average over the factorized measure. In the thermodynamic limit

$$M = \langle \sigma_i \rangle_H = \tanh(H + h)$$
$$\lim_{N \to \infty} \frac{1}{N} \langle A \rangle_H = \beta d \tanh^2(H + h) - H \tanh(H + h) \quad (17)$$

Thus

$$F \geq \underset{H}{\mathrm{Sup}} \, \mathcal{F}(H, h; \beta)$$

$$\mathcal{F}(H, h; \beta)$$
$$= \ln\cosh(h + H) + \beta d \tanh^2(H + h) - H \tanh(H + h) \quad (18)$$

Identifying $\tanh(H + h)$ with the magnetization M, the interpretation of (18) is straighforward. The piece

$$\beta d \tanh^2(H + h) + h \tanh(H + h) = \beta d M^2 + hM$$

is the energy contribution, while the remaining part

$$\ln \cosh(H + h) - (H + h) \tanh(H + h) =$$
$$= -\ln 2 - [\tfrac{1}{2}(1 + M)\ln \tfrac{1}{2}(1 + M) + \tfrac{1}{2}(1 - M)\ln \tfrac{1}{2}(1 - M)]$$

represents entropy. The $-\ln 2$ contribution arises from our normalization of Z. Thus

$$\mathcal{F}(H, h; \beta) = -V(M, \beta) + hM$$
$$V(M, \beta) = -\beta d M^2 + \tfrac{1}{2}(1 + M)\ln \tfrac{1}{2}(1 + M)$$
$$+ \tfrac{1}{2}(1 - M)\ln \tfrac{1}{2}(1 - M) + \ln 2 \qquad (19)$$

and we have to minimize $V(M, \beta) - hM$. The relation between the expressions (18) and (19) is known as a Legendre transformation. In either form, the condition for an extremum is the previous mean field equation

$$H = 2\beta d \tanh(H + h) \qquad (20)$$

The advantage of the present derivation is that it is based on a variational principle, thus allowing us to distinguish the various solutions. Figure 2 shows that for $h = 0$, we have a minimum for $V(M, \beta)$ at $M = 0$ for $\beta < \beta_c = 1/2d$, resulting from a dominance of entropy over energy, while in the low temperature regime $(\beta > \beta_c)$, two new equivalent minima emerge. An external field h, no matter how small, distorts the picture and leads to a unique absolute minimum with M of the same sign as h. For $\beta = \beta_c$, the curve $V(M)$ has a very flat minimum at the origin with a vanishing coefficient of the quadratic term. In the vicinity of the critical point, the expansion of $V(M, \beta)$ for small M reads

$$V(M, \beta) = \frac{1}{2}\left(1 - \frac{\beta}{\beta_c}\right)M^2 + \frac{1}{12}M^4 + \cdots \qquad (21)$$

The quartic term always has a positive coefficient, while the coefficient of the quadratic term vanishes linearly (and therefore changes sign). For $h = 0$, the extremum value of V is therefore 0 for $\beta < \beta_c$, and nonvanishing for $\beta > \beta_c$. In the immediate

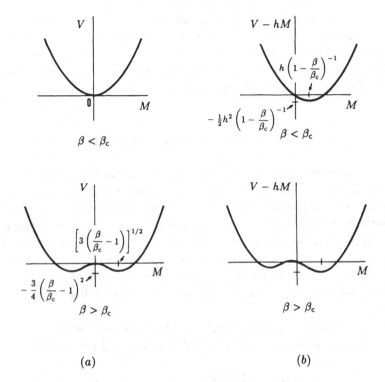

Fig. 2   Minimization of the trial free energy (a) in the absence of an external field and (b) in the presence of such a field.

vicinity, we have

$$\beta > \beta_c \qquad V_{\min} = -\frac{3}{4}\left(\frac{\beta}{\beta_c} - 1\right)^2 ; \qquad \beta < \beta_c \qquad V_{\min} = 0$$

$$(22)$$

Thus the specific heat c, the second derivative of $V_{\min}$ with respect to β, is discontinuous at β_c. If one insists on characterizing this behaviour as a power law in $|1 - \beta/\beta_c|^{-\alpha}$, with an exponent α, the only choice is to say that $\alpha = 0$. This same α was also zero for $d = 2$, but then the behaviour was logarithmic.

We have applied the mean field ideas in the case of the Ising model with a Z_2 symmetry group. The same reasoning generalizes when the symmetry group is finite. In general, such a group possesses nontrivial subgroups. According to the range of parameters involved in the effective action, the nature of the various ordered phases will depend on the residual symmetry group, as well as the transformation properties of the order parameter (there may be several ones). We shall not pursue this direction, except to say that the corresponding discussion generally requires some physical insight.

3.1.3   Continuous symmetry group

The spontaneous breaking of a continuous symmetry group leads to an interesting phenomenon in the low temperature phase, namely the existence of soft modes (Goldstone 1961). We shall investigate the simple case of an $O(n)$-invariant model, using as a dynamical variable an n−dimensional "vector"field φ with fixed length – say unity: $\varphi^2 = 1$. There exist many other possibilities – including variables transforming according to unitary groups, to be discussed at later stages – as well as tensor fields of various kinds.

We have already encountered the $O(n)$ vector model in chapter 1, where we introduced in (1.125) the integral ($H \equiv |\mathbf{H}|$)

$$\exp n u_n(H) = \int \frac{\mathrm{d}^n \varphi}{S_n} 2\delta\left(\varphi^2 - 1\right) \exp n\mathbf{H} \cdot \varphi$$

$$= \Gamma\left(\frac{n}{2}\right)\left(\frac{2}{nH}\right)^{n/2-1} I_{n/2-1}(nH) \qquad (23)$$

$$= \sum_{k=0}^{\infty} \left(\tfrac{1}{4}n^2 H^2\right)^k \frac{1}{k!} \frac{\Gamma(\tfrac{1}{2}n)}{\Gamma(\tfrac{1}{2}n + k)}$$

This is a generalization of $N^{-1} \ln Z_H$ appearing in equations (15). For $n = 1$, $u_1(H) = \ln \cosh H$. In general, for H small,

$$u_n(H) = \tfrac{1}{2}H^2 - \frac{n}{4(n+2)}H^4 + \cdots \qquad (24)$$

We write the partition function as

$$Z = \int \prod_i \left\{\frac{\mathrm{d}^n \varphi_i}{S_n} 2\delta\left(\varphi_i^2 - 1\right)\right\} \exp n \left\{\beta \sum_{(i,j)} \varphi_i \cdot \varphi_j + \sum_i \mathbf{h} \cdot \varphi_i\right\}$$

$$(25)$$

a direct generalization of the Ising case (corresponding to $n = 1$). The mean field approximation in variational form reads

$$F \geq \operatorname*{Sup}_{\mathbf{H}} n\mathcal{F}(\mathbf{H}, \mathbf{h}; \beta) \qquad (26)$$

$$\mathcal{F}(\mathbf{H}, \mathbf{h}; \beta) = u(\mathbf{H} + \mathbf{h}) + \beta d \left[\boldsymbol{\nabla}\, u(\mathbf{H} + \mathbf{h})\right]^2 - \mathbf{H} \cdot \boldsymbol{\nabla}\, u(\mathbf{H} + \mathbf{h})$$

Here the gradient operator $\boldsymbol{\nabla}$ refers to the argument of the function u and not of course to a spatial dependence. Both the external field $\mathbf{h}$, and the mean field $\mathbf{H}$, are assumed spatially homogeneous, as is the action in (25). It would seem at first that the maximization of $\mathcal{F}$ leads to results quite similar to those obtained in the case of a finite symmetry group. Set

$$\mathbf{M} = \boldsymbol{\nabla}\, u(\mathbf{H} + \mathbf{h}) = \langle \varphi \rangle \qquad (27)$$

then

$$\mathcal{F}(\mathbf{H}, \mathbf{h}; \beta) = -V(\mathbf{M}) + \mathbf{h} \cdot \mathbf{M}$$

$$V(\mathbf{M}) = \frac{1}{2}(1 - 2d\beta)\mathbf{M}^2 + \frac{1}{4}\frac{n}{n+2}(\mathbf{M}^2)^2 + \cdots \qquad (28)$$

With our choice of normalization, the critical value of β remains $\beta_c = 1/2d$, as it was in the case $n = 1$. A spontaneous magnetization appears for $\beta > \beta_c$, and the critical exponents are the same as those found before. In particular the specific heat is discontinuous at the transition, within this approximation.

There exists, however, a crucial difference from the finite case. In the ordered phase with $\mathbf{M} \neq 0$, there remains a continuous invariance group, $O(n-1)$, the little group which leaves the order parameter $\mathbf{M}$ invariant. As a consequence, the pure states are continuously degenerate, in other words pure phases are described by a point on a continuous manifold $O(n)/O(n-1)$, i.e. the unit sphere in n-dimensional space. This leads us to the consideration of the response to external fields with a slow spatial dependence. Let $\mathbf{h} = \mathbf{h}_0 + \mathbf{h}_1 \cos \mathbf{k} \cdot \mathbf{x}_i$, with $\mathbf{h}_0$ and $\mathbf{h}_1$ both very small, and assume $\beta > \beta_c$. The constant (or average) part $\mathbf{h}_0$ defines the mean direction of the magnetization. The susceptibility tensor, i.e. the coefficient of the additional piece of the magnetization proportional to $\mathbf{h}_1$, will have quite a different behaviour according to whether $\mathbf{h}_1$ is parallel or orthogonal to $\mathbf{h}_0$. If $\mathbf{h}_1$ is in these $n-1$ transverse directions, the corresponding variation in free energy vanishes in the limit of small wave numbers, $\mathbf{k} \to 0$, as it amounts simply to rotating in large spatial regions the global direction of the magnetization within the manifold of degenerate states. This is the Goldstone phenomenon, corresponding to the existence of $n - 1$ soft excitation modes in the ordered phase. Their possible entropic contribution will have to be considered at a later stage, since they will tend to provide a destabilizing mechanism. In the original σ-model the corresponding group is $O(4)$ (isomorphic to $SU(2) \times SU(2)/Z_2$, a direct product of two unitary unimodular two-dimensional groups, up to a global change in sign of both elements) and the three independent soft modes are associated to the three π-mesons. To be explicit we return to the convexity inequality in the presence of such an external field. Generalizing equations (26)–(27) and (28), we can write

$$
\frac{\ln Z}{n} \geq - \operatorname*{Inf}_{\{\mathbf{M}_i\}} \left[\sum_i V(\mathbf{M}_i) - \frac{\beta}{2} \sum_i \mathbf{M}_i \cdot \sum_{j(i)} (\mathbf{M}_j - \mathbf{M}_i) - \sum_i \mathbf{h}_i \cdot \mathbf{M}_i \right]
$$

$$(29)$$

The square of the spontaneous magnetization $\mathbf{M}_s$, in the absence of $\mathbf{h}$, and for β close to β_c, is approximately given by

$$
((n+2)/n)(\beta/\beta_c - 1),
$$

so that

$$
V(\mathbf{M}) = \frac{1}{4} \frac{n}{n+2} \left[(\mathbf{M}^2 - \mathbf{M}_s^2)^2 - (\mathbf{M}_s^2)^2 \right] + \cdots
$$

$$
\mathbf{M}_s^2 = \frac{n+2}{n} (\beta/\beta_c - 1)
$$

$$(30)$$

The spontaneous magnetization $\mathbf{M}_s$ is along $\mathbf{h}_0$. For $\mathbf{h}_1$ infinitesimal, we write

$$
\mathbf{M}_i = \mathbf{M}_s + \delta\mathbf{M}_i \tag{31}
$$

To first order in $\delta\mathbf{M}_i$ and $\mathbf{h}_1$, the above minimization condition yields

$$
\frac{2n}{n+2} \mathbf{M}_s (\mathbf{M}_s \cdot \delta\mathbf{M}_i) - \beta \sum_{j(i)} \left(\delta\mathbf{M}_j - \delta\mathbf{M}_i \right) = \mathbf{h}_1 \cos \mathbf{k} \cdot \mathbf{x}_i \tag{32}
$$

We look for a plane wave solution

$$
\delta\mathbf{M}_i = \operatorname{Re} \mathbf{A} \exp i \mathbf{k} \cdot \mathbf{x}_i \tag{33}
$$

such that

$$
\frac{2n}{n+2} \mathbf{M}_s (\mathbf{M}_s \cdot \mathbf{A}) + 2\beta \sum_{\mu=1}^{d} (1 - \cos a k_\mu) \mathbf{A} = \mathbf{h}_1 \tag{34a}
$$

Assuming ak_μ small (a is the lattice spacing), in the long wavelength limit, we can rewrite this amplitude equation as

$$\frac{2n}{n+2}\mathbf{M}_s(\mathbf{M}_s \cdot \mathbf{A}) + \beta a^2 \mathbf{k}^2 \mathbf{A} = \mathbf{h}_1 \tag{34b}$$

We now split $\mathbf{h}_1$ and $\mathbf{A}$ into longitudinal and transverse components with respect to $\mathbf{M}_s$, to find

$$\mathbf{A}^\perp = \frac{\mathbf{h}_1^\perp}{\beta a^2 \mathbf{k}^2} \qquad \mathbf{A}^\| = \frac{\mathbf{h}_1^\|}{\beta a^2 \left(\mathbf{k}^2 + \dfrac{2n}{n+2}\dfrac{\mathbf{M}_s^2}{\beta a^2} \right)} \tag{35}$$

where β can safely be replaced by β_c. In the limit $\mathbf{k} = 0$, we observe a typical infrared divergence in the transverse response (or susceptibility) corresponding to massless modes, while the longitudinal amplitude remains finite, the corresponding mass being proportional to the the spontaneous magnetization. Of course the number of soft modes is equal to the number of transverse components, i.e. $n - 1$. They are sometimes referred to as spin waves.

The conclusion of this analysis is that a low temperature phase in a system with a spontaneously broken continuous symmetry exhibits massless excitation modes. Their effective dynamics has to be carefully investigated. Potential infrared divergences become more and more severe as the dimension decreases. As a result, we shall see that would-be Goldstone modes are responsible for the absence of ordered phases in dimension two in the case of a noncommutative continuous symmetry group, where they play a pre-eminent role.

As schematic as it is, the mean field approximation reveals, using unsophisticated mathematical tools, a wealth of interesting behaviour. We shall study corrections at the end of this chapter. For the time being, let us look at an alternative point of view particularly useful in the case of discrete variables.

3.1.4 The Bethe approximation

We deal here with a variant of mean field theory, which will be exemplified again on the Ising case. Its main virtue is not that it allows to improve the computation of critical exponents, but that it offers a rather accurate means to locate the critical singularities.

The idea is to think of the system as a superposition of small clusters submitted to an effective field, while imposing a consistency condition by requiring the equality of magnetization, computed in alternative ways. To start with, let us imagine all two neighbouring spin clusters, their number being Nd, if N is the number of lattice sites, acted upon by an effective field H_2 which incorporates both the external field h, as well as the effect of the other interactions. If $\tau = \tanh\tilde{\beta} = e^{-2\beta}$, the corresponding two site partition function is proportional to

$$z_2 = 1 + 2\tau\rho_2 + \rho_2^2 \qquad \rho_2 = e^{-2H_2} \qquad (36a)$$

Up to a prefactor $\left[\frac{1}{2}\exp(\beta d + h)\right]^N$, we could think of replacing the partition function by $(z_2)^{Nd}$, but this would count each spin $2d$ times instead of one. We therefore correct for this overshooting effect by dividing by $z_1^{N(2d-1)}$, where z_1 is a one site partition function of a spin subject to another effective field H_1

$$z_1 = 1 + \rho_1 \qquad \rho_1 = e^{-2H_1} \qquad (36b)$$

The corresponding approximation is therefore

$$Z = \left[\tfrac{1}{2}\exp(\beta d + h)\right]^N \frac{z_2^{Nd}}{z_1^{N(2d-1)}} \equiv z_B^N \qquad (37)$$

We have two parameters at our disposal, H_1 and H_2. To determine them, we impose two conditions. Those are simply the equality of the magnetization per site in the complete system and in each of its subsystems. Taking into account the fact that z_2 includes two sites and z_1 only one, we find

$$M = \frac{\partial}{\partial h}\ln z_B(h,\beta)$$
$$= 1 - \rho_2\frac{\partial}{\partial\rho_2}\ln z_2(\rho_2,\tau) = 1 - 2\rho_1\frac{\partial}{\partial\rho_1}\ln z_1(\rho_1) \qquad (38a)$$

Thus from the last equation

$$z_2 = (1 + \rho_2\tau)z_1 \qquad \text{i.e.} \qquad \frac{\rho_1}{\rho_2} = \frac{\tau + \rho_2}{1 + \tau\rho_2} \qquad (38b)$$

while the first equation furnishes, through (37) and assuming ρ_2 and ρ_1 functions of h and τ, the relation

$$\rho_2 = e^{-2h}\left(\frac{\tau + \rho_2}{1 + \tau\rho_2}\right)^{2d-1} \qquad (38c)$$

Table I. Values of $\tanh \beta_c$ for the Ising model as a function of the dimension d, on a cubic lattice.

dimension	Bethe approximation	exact value
1	1	1
2	$1/3 = 0.33333...$	$\sqrt{2} - 1 = 0.41421...$
3	$1/5 = 0.2$	$0.218096(3)$
4	$1/7 = 0.14286...$	
$d \to \infty$	$1/(2d-1)$	$1/(2d-1)$

Therefore

$$z_B = \tfrac{1}{2}e^{h+d\beta}(1+\rho_2\tau)^{2d-1}(1+2\rho_2\tau+\rho_2^2)^{1-d} \tag{39}$$

Here ρ_2 is implicitly determined in terms of h and $\tau = e^{-2\beta}$ through equation (38c). When trying to solve for ρ_2, we find the critical singularity. To see this, use rather than ρ_2 the combination

$$p = \frac{\rho_2 + \tau}{1 + \rho_2\tau} \tag{40}$$

related to β and h by

$$\tau = e^{-2\beta} = \frac{p^{1-d}e^h - p^{d-1}e^{-h}}{p^{-d}e^h - p^d e^{-h}} \tag{41}$$

exhibiting the symmetry $h \leftrightarrow -h, p \leftrightarrow p^{-1}$. If the external field vanishes, $h = 0$, the branch $p(\beta) \sim e^{-2\beta}$ as $\beta \to \infty$ will become multivalued for a critical value β_c, given by

$$e^{-2\beta_c} = 1 - d^{-1}, \qquad 2d\beta_c = 1 + \frac{1}{2d} + \frac{1}{3d^2} + \frac{1}{4d^3} + \cdots \tag{42}$$

If we were to trust this approximation, it would imply a transition for any dimension $d > 1$, with $\beta_c(d) \to \infty$ as $d \to 1$. However, we expect that the scheme makes sense only for $d \to \infty$. We then observe an agreement with the mean field critical value, with an exact $1/2d$ correction to $2d\beta_c$. In the following table we compare the predictions of the Bethe approximation for $\tanh \beta_c$ with the exact value (in three dimensions the value is only known within a small error on the sixth digit).

The prediction for $\tanh \beta_c$ is remarkably accurate and could be improved using larger clusters. On any other lattice, one would have to replace $2d$ by the coordination number q.

For $\beta < \beta_c, h = 0$, the appropriate solution of the equation (38) is $p = \rho_2 = 1$, hence

$$\beta < \beta_c \qquad z_B = \tfrac{1}{2} e^{d\beta} \left(1 + e^{-2\beta}\right)^d 2^{1-d} = (\cosh \beta)^d \qquad (43)$$

as expected. At low temperature $\beta > \beta_c$, the magnetization is given by

$$M = \frac{1 - p^2}{1 - 2\tau p + p^2} \qquad (44)$$

As $\beta \to \beta_c + 0$

$$M^2 \sim \frac{2d}{2d - 1} 3 \left(\frac{\beta}{\beta_c} - 1\right) \qquad (45)$$

in agreement with the mean field prediction (11) for $d \to \infty$. The free energy is the logarithm of z_B. If we set $p = e^{-u}$, we have for $\beta > \beta_c$

$$z_B = \left[\cosh \beta \left(1 - e^{-2\beta}\right) \frac{\sinh du}{\sinh u}\right]^d (\cosh du)^{1-d}$$

$$e^{-2\beta} = \frac{\sinh(d-1)u}{\sinh du}, \qquad u^2 = \frac{12}{2d-1}(\beta - \beta_c) + \cdots \quad (46)$$

Close to criticality

$$\beta > \beta_c \qquad z_B/(\cosh \beta)^d = 1 + O((\beta - \beta_c)^2) \qquad (47)$$

meaning that Bethe's approximation predicts a second order transition with a discontinuous specific heat.

Compute the discontinuity of the specific heat, and compare it with the mean field prediction.

We observe within the same approximation an important property. Namely, if the external field is nonvanishing, for instance $h > 0$, all thermodynamic quantities become regular. In particular no singularity shows up when inverting equation (41) for p as a function of β (or $\tau = e^{-2\beta}$). In figure 3 we plot the inverse function $\tau(p)$, comparing the case $h = 0$ to $h \neq 0$ for $d = 3$.

It is legitimate to ask about the fate of the critical singularity as h takes a nonvanishing value. A bifurcation takes place and, for $\beta < \beta_c$, one finds two complex conjugate singularities on the imaginary h axis. More precisely, if we look for values where

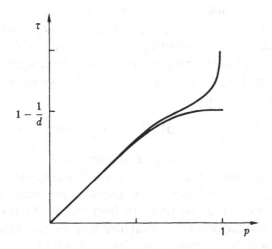

Fig. 3   Bethe approximation. Representation of τ as a function of p for h equal or different to zero, assuming $d = 3$.

the solution $p(\beta, h)$ of equation (41), becomes multivalued as a function of h for real $\beta < \beta_c$, we find that h has to be pure imaginary (and p a phase) with absolute value larger than h_c. As $\tau = e^{-2\beta} \to \tau_c$, one has

$$h_c \sim a \left(\frac{\tau}{\tau_c} - 1\right)^{3/2} \qquad a = \frac{2}{3}\left(\frac{\tau_c}{1 - \tau_c}\right)\left(\frac{2}{1 - \tau_c^2}\right)^{1/2} \qquad (48)$$

At any of the two points $\pm ih_c$, the free energy has a branch point of order 3/2, with a singular part

$$f_{\text{sing.}} \sim \text{cst}(h \pm ih_c)^{3/2} \qquad (49)$$

This phenomenon is not an artefact of the approximation, but is the reflection of a singularity discovered by Lee and Yang to be discussed below.

3.1.5   Critical exponents

Let us summarize what has been learnt from these elementary approximations. We are interested in the vicinity of critical points with long-range correlations for the dynamical variables. On this scale, the discontinuous lattice structure can be ignored, as was verified in the previous chapter in the case of the two-dimensional Ising model. Mean field ideas led us to minimize a trial free energy $V(\mathbf{M}) - \mathbf{h} \cdot \mathbf{M}$, a function of mean magnetization $\mathbf{M}$, which plays the role of an order parameter, of an external field $\mathbf{h}$ coupled linearly to $\mathbf{M}$, and other parameters such as the temperature. For instance, in an $O(n)$ symmetric vector model, we found a typical expression of the form

$$V(\mathbf{M}) = \frac{1}{2}(1 - 2d\beta)\mathbf{M}^2 + \frac{1}{4}\frac{n}{n+2}(\mathbf{M}^2)^2 + \cdots$$

Note that $\mathbf{M}$ is no longer a vector of unit length, as was the case for the original variables in the partition function. We recognize in $V(\mathbf{M})$ the potential part of a Lagrangian, with dynamical variables $\mathbf{M}$, in self-interaction. In field theory, in the classical limit, when one neglects the contributions of the kinetic term due to the possible spatial dependence of $\mathbf{M}(\mathbf{x})$, one similarly minimizes the potential term to find the ground state of the system. The coefficient of the quadratic term may be identified as a square mass, however not restricted to be positive in the present context. To take this into account, and recall that this square mass vanishes at criticality (here linearly), we denote this coefficient by θ. Also let us replace the coefficient $n/(n+2)$ of the $(\mathbf{M}^2)^2$ term by a positive coupling constant g to prepare for a more general situation (as was already envisioned in chapter 1). Finally, we neglect higher order terms and write

$$V(\mathbf{M}) = \tfrac{1}{2}\theta\mathbf{M}^2 + \tfrac{1}{4}g(\mathbf{M}^2)^2 \tag{50}$$

Let us now retrace the various steps which in one guise or another lead, within the mean field approximation, to the critical exponents. Minimizing $V(\mathbf{M}) - \mathbf{h} \cdot \mathbf{M}$ yields

$$\nabla_{\mathbf{M}} V(\mathbf{M}) \equiv \mathbf{M}(\theta + g\mathbf{M}^2) = \mathbf{h} \tag{51}$$

(i) In the absence of external field, $\mathbf{h} = 0$, the solution is

$$\begin{array}{lll} \theta > 0 & \mathbf{M} = 0 & V_{\min} = 0 \\ \theta < 0 & \mathbf{M}_s^2 = -\theta/g & V_{\min} = -\tfrac{1}{4}\theta^2/g \end{array} \tag{52}$$

The distinction $\theta > 0, \theta < 0$, corresponds to high and low temperature. The specific heat is proportional to $\partial^2 V_{\min}/\partial\theta^2$, and hence discontinuous. One writes generally

$$C(\theta) \underset{\theta \to \pm 0}{\sim} C_{\pm} |\theta|^{\alpha_{\pm}} \tag{53a}$$

To encompass the discontinuous mean field result, one agrees to set

$$\alpha = \alpha_{\pm} = 0 \tag{53b}$$

(ii) For $\theta < 0$, the spontaneous magnetization is given by $\mathbf{M}_s^2 = -\theta/g$ with

$$\theta < 0 \qquad\qquad |\mathbf{M}_s| \sim (-\theta)^{\beta} \qquad\qquad \beta = \tfrac{1}{2} \tag{54}$$

(iii) In the high temperature phase $\mathbf{M}(\mathbf{h})$ vanishes linearly with $\mathbf{h}$, $\mathbf{M}(\mathbf{h}) \sim \mathbf{h}/\theta$, the coefficient being the susceptibility, which diverges when $\theta \to 0$ as

$$\theta > 0 \qquad\qquad \chi \sim \chi_+ \theta^{-\gamma_+} \qquad\qquad \gamma_+ = 1 \tag{55}$$

In the low temperature phase we saw that the longitudinal and transverse response have to be distinguished. In tensor form

$$\mathrm{d}h_a = (\chi^{-1})_{ab}\mathrm{d}\mathbf{M}_b = \left[(\theta + 3g\mathbf{M}^2)\delta_{ab} + 2g(\mathbf{M}_a\mathbf{M}_b - \mathbf{M}^2\delta_{ab})\right]\mathrm{d}\mathbf{M}_b$$

For $\theta > 0$ we recover the previous result, but for $\theta < 0$ and a spontaneous magnetization $\mathbf{M}^2 = -\theta/g$, we have

$$(\chi^{-1})_{ab} \sim -2\theta\frac{\mathbf{M}_{sa}\mathbf{M}_{sb}}{\mathbf{M}_s^2}$$

In the transverse directions, χ^{-1} is strictly zero. This is Goldstone's phenomenon. In the parallel direction

$$
\begin{aligned}
&\qquad\qquad \chi^{\parallel} \sim \frac{1}{2}\frac{1}{(-\theta)}\\
&\theta < 0 \qquad\qquad \chi^{\parallel} \sim \chi_-(-\theta)^{-\gamma_-}, \qquad \gamma_- = 1\\
&\qquad\qquad \chi_+/\chi_- = 2
\end{aligned} \tag{56}
$$

(iv) At criticality, $\theta = 0$, the magnetization varies as the cubic root of the external field. In other words the response to an external field is much larger (as compared to the case when $\theta > 0$), signalling the transition to an ordered state. We write

$$|\mathbf{M}| \sim |\mathbf{h}|^{1/\delta} \qquad\qquad \delta = 3 \tag{57}$$

(v) Let us now assume that the external field varies from point to point, and reinstate the "kinetic term". Choose $n = 1$ for simplicity (although for $n > 1$ the following expressions will remain valid for the longitudinal components). We have to minimize the complete expression

$$S = \sum_i \left\{ -\tfrac{1}{2} M_i \sum_{j(i)} (M_j - M_i) + V(M_i) - h_i M_i \right\} \tag{58}$$

Thus

$$-\sum_{j(i)} (M_j - M_i) + (\theta + g M_i^2) M_i = h_i \tag{59}$$

For $\theta > 0$, this leads to

$$M_i = \sum_j \int \frac{d^d k}{(2\pi)^d} \frac{e^{i\mathbf{k} \cdot (\mathbf{x}_i - \mathbf{x}_j)}}{\theta + \sum_\mu 2(1 - \cos k_\mu)} h_j + O(gh^3) \tag{60}$$

As h goes to zero and at large distance $\left| \mathbf{x}_i - \mathbf{x}_j \right| \to \infty$, we find

$$\left\langle M_i M_j \right\rangle = \left. \frac{\delta M_i}{\delta h_j} \right|_{h=0} = \int \frac{d^d k}{(2\pi)^d} \frac{e^{i\mathbf{k} \cdot (\mathbf{x}_i - \mathbf{x}_j)}}{\theta + \sum_\mu 2(1 - \cos k_\mu)}$$

$$\underset{|\mathbf{x}_i - \mathbf{x}_j| \to \infty}{\sim} \frac{\theta^{(d-2)/2}}{2 \left(2\pi \theta^{1/2} \left| \mathbf{x}_i - \mathbf{x}_j \right| \right)^{(d-1)/2}} \exp \left(-\theta^{1/2} \left| \mathbf{x}_i - \mathbf{x}_j \right| \right)$$

$$\tag{61}$$

The exponential decrease as $\exp[-\left| \mathbf{x}_i - \mathbf{x}_j \right| / \xi]$ defines a correlation length ξ. The latter diverges as one approaches the critical point

$$\theta \to +0 \qquad \xi \sim \theta^{-\nu} \qquad \nu = \tfrac{1}{2} \tag{62}$$

For small negative θ, M_i fluctuates around the spontaneous magnetization M_s,

$$M_i = M_s + \sum_j \left. \frac{\delta M_i}{\delta h_j} \right|_{h=0} h_j + \cdots$$

$$\theta < 0, \qquad \left. \frac{\delta M_i}{\delta h_j} \right|_{h=0} = \int \frac{d^d k}{(2\pi)^d} \frac{e^{i\mathbf{k} \cdot (\mathbf{x}_i - \mathbf{x}_j)}}{-2\theta + \sum_\mu 2(1 - \cos k_\mu)} \tag{63}$$

Therefore for $\theta \to -0$, the correlation length behaves as $(-2\theta)^{-1/2}$ and the corresponding exponent remains $\nu = 1/2$ as above.

For an n-component model, transverse modes have an infinite correlation length below criticality.

(vi) At $\theta = 0$, both (61) and (63) reduce to

$$\left.\frac{\delta M_i}{\delta h_j}\right|_{h=0} = \int \frac{d^d k}{(2\pi)^d} \frac{e^{i\mathbf{k}\cdot(\mathbf{x}_i - \mathbf{x}_j)}}{\sum_\mu 2(1 - \cos k_\mu)}$$

$$\underset{|\mathbf{x}_i - \mathbf{x}_j| \to \infty}{\widetilde{}} \frac{\Gamma(d/2 - 1)}{4\pi^{d/2} \left|\mathbf{x}_i - \mathbf{x}_j\right|^{d-2}} \tag{64}$$

a generalization in d-dimensions of the Coulomb potential. Of course all this assumes d large enough. When comparing equations (61) and (64), we note that the inverse powers of $\left|x_i - x_j\right|$ differ (except when $d = 3$). One defines the exponent η through

$$\theta = 0 \qquad \left.\frac{\delta M_i}{\delta h_j}\right|_{h=0} = \left\langle M_i M_j \right\rangle \underset{|\mathbf{x}_i - \mathbf{x}_j| \to \infty}{\widetilde{}} \frac{\text{cst}}{\left|\mathbf{x}_i - \mathbf{x}_j\right|^{d-2+\eta}} \tag{65a}$$

and thus in the mean field case

$$\eta = 0 \tag{65b}$$

(vii) Assume now a uniform external field h, and consider the one component case only for simplicity. The minimization equation (51) becomes ambiguous when its derivative with respect to M vanishes, marking the onset of a multivalued solution for M. This corresponds to requiring simultaneously

$$\begin{aligned}\theta M + g M^3 &= h\\ \theta + 3g M^2 &= 0\end{aligned} \tag{66}$$

For a solution in M, the parameters θ, g and h have to be related by the vanishing of the discriminant

$$4\theta^3 + 27gh^2 = 0 \tag{67}$$

Real solutions with $\theta < 0$ can be left aside, since they ignore spontaneous symmetry breaking (recall that $g > 0$). For positive θ, one finds two complex conjugate values of h on the imaginary axis, $h = \pm i h_c$, with

$$h_c = \left(\frac{4}{27g}\right)^{1/2} \theta^{3/2} \tag{68}$$

This is the Lee–Yang edge singularity, derived in the previous subsection. If we set for small positive θ,

$$h_c \sim \theta^\Delta \tag{69a}$$

Table II. Critical exponents.

	Ising model $d = 2$	Mean field $d \to \infty$	Spherical model $(n \to \infty)$ $4 > d > 2$
Specific heat $C \sim C_\pm |\theta|^{-\alpha_\pm}$	$\alpha = 0^{(*)}$	$\alpha = 0^{(**)}$	$\alpha = \left(\dfrac{d-4}{d-2}\right)$
Spontaneous magnetization $\theta < 0 \qquad M_{sp} \sim (-\theta)^\beta$	$\beta = \frac{1}{8}$	$\beta = \frac{1}{2}$	$\beta = \frac{1}{2}$
Susceptibility $\chi \sim |\theta|^{-\gamma_\pm}$	$\gamma = \frac{7}{4}$	$\gamma = 1$	$\gamma = \dfrac{2}{d-2}$
Magnetization at $\theta = 0$ $M \sim h^{1/\delta}$	$\delta = 15$	$\delta = 3$	$\delta = \dfrac{d+2}{d-2}$
Correlation length $\xi \sim |\theta|^{-\nu_\pm}$	$\nu = 1$	$\nu = \frac{1}{2}$	$\nu = \dfrac{1}{d-2}$
Critical correlation $\langle M(\mathbf{x})M(\mathbf{y})\rangle \sim \dfrac{\text{cst}}{|\mathbf{x}-\mathbf{y}|^{d-2+\eta}}$	$\eta = \frac{1}{4}$	$\eta = 0$	$\eta = 0$
Lee-Yang edge singularity $h = \pm ih_c, \theta > 0$ $h_c \sim \theta^\Delta,\ M \sim (h \mp ih_c)^\sigma$	$\Delta = \frac{15}{8}$ $\sigma = -\frac{1}{6}$	$\Delta = \frac{3}{2}$ $\sigma = \frac{1}{2}$	

Notes: (*) up to logarithms. (**) discontinuous.

the mean field prediction is

$$\Delta = \tfrac{3}{2} \qquad\qquad (69b)$$

With $h_0 = \pm ih_c$, $M_0 = \pm iM_c$, $M_c = \frac{3}{2}h_c/\theta = \frac{3}{2}\theta^{1/2}$, the behaviour of the free energy in the vicinity of these points is

$$V(M) - V(M_0) - h(M - M_0) \approx \left[\frac{4}{27gM_0}(h - h_0)\right]^{3/2} - M_0(h - h_0)$$

$$(70)$$

with a singular part of the form

$$(h - h_0)^{1+\sigma} \qquad\qquad (71a)$$

such that the corresponding mean field value is

$$\sigma = \tfrac{1}{2} \qquad\qquad (71b)$$

To proceed further, we will have to study the corrections, due to fluctuations, which were neglected up to now. In large

enough dimension, these effects are expected not to modify in
an essential way the nature of the transition, in particular the
critical exponents. This will be discussed in the last section of this
chapter. The largest dimension in which the critical behaviour
is effectively modified, as indicated by the breakdown of the
perturbative estimate of fluctuations, is called the upper critical
dimension. From the preliminary discussion in chapter 1, we guess
that its value is $d = 4$ for the conventional transition in spin
models. It turns out to be $d = 6$ for the Lee–Yang (complex)
edge singularity. Other more powerful tools will be needed for
the analysis, which go under the name of the renormalization
group. This will be the subject of chapter 5. Before we undertake
a full-fledged discussion of these matters, we will take a closer
look at the meaning of the Lee–Yang singularity as a precursor
of symmetry breaking, and analyse an alternative (large n or
spherical model) approximation, which throws some light on the
upper critical dimension. For the time being, we summarize in
table II the mean field critical exponents and compare them with
those of the two-dimensional Ising model, as well as those of the
spherical model for $2 < d < 4$, where a departure with respect to
mean field predictions is observed, as discussed below.

3.2 Lee–Yang zeroes

In 1952 two famous papers appeared by Lee and Yang where
they took a different point of view towards the phenomenon of
spontaneous symmetry breaking by studying a lattice gas, or,
what amounts to the same, an Ising model, in an arbitrary external
field, in a finite (but arbitrarily large) domain. In their approach,
dimension or coordination does not play a pre-eminent role at first,
but the external field is allowed to take complex values in a search
of zeroes of the partition function. Since the latter describes a sum
over positive weights in the real (physical) domain, it is certainly
nonvanishing there, but, as parameters become complex, this in
not the case anymore. In a lattice gas interpretation, spin up or
down ($\sigma = +1$ or $\sigma = -1$) correspond to occupied or empty sites,
and the field h is conjugate to the density. Here we will take the
"magnetic" interpretation. The important point is that for real
β, up to a nonvanishing factor, the partition function in a finite

volume (i.e. for finitely many sites) is a polynomial in the activity $\exp(-2h)$.

3.2.1 The Lee–Yang theorem

The discovery of Lee and Yang is that the zeroes of the partition function are located on the unit circle of the complex activity plane, or equivalently on the imaginary h-axis. For any finite volume, or for that matter on an arbitrary finite graph, the number of zeroes is finite, as the partition function is a polynomial in the activity, and the degree of this polynomial increases with the size. In the infinite volume limit, the zeros will "condense" on a subset of the unit circle. As long as β is chosen in an appropriate range (high temperature), this singular set will leave a vicinity of unit activity (or zero field) free of zeroes. The free energy will remain analytic there. But, as β is increased, the analyticity gap will eventually close, forbidding analytic continuation from $\operatorname{Re} h > 0$ to $\operatorname{Re} h < 0$, a signal of a transition at $h = 0$. As seen along a path h real, h crossing zero, the transition is a first order one, with a jump in magnetization (the discontinuity) being twice the spontaneous magnetization, while the free energy at zero field might have a continuous derivative in β at β_c as is usual in an ordinary second order transition. The analysis is carried out in the activity plane because the Lee–Yang theorem controls the location of zeroes, but as long as the model allows identification of the partition function relative to a finite domain as a polynomial in some parameter, it can obviously be generalized. Such a study has been carried out for instance using the variable $\exp(-2\beta)$ for the Ising model, and gives an insight into the analytic structure of the free energy. Another extension is relative to models involving a continuous symmetry group. The subject has never been developed in as systematic a fashion as would be desirable, if only because in general these distributions of zeroes may be rather wild and lattice dependent. They enable one, however, to reach some interesting results, as will be shown later. Here we concentrate on the activity variable for a general Ising model on arbitrary graph of N sites (vertices of the graphs) with at most one link joining a pair of vertices, which are then said to be adjacent, or neighbours. The total number of links is L.

Let h_i denote the magnetic field at site i. It is convenient at first to have the field vary from site to site. Define

$$\rho_i = e^{-2h_i} \qquad \tau = e^{-2\beta} \qquad (72)$$

The partition function is of the form

$$Z_N = \frac{1}{2^N} \exp\left(\beta L + \sum_i h_i\right) P(\tau, \rho_i) \qquad (73)$$

with P a polynomial in τ and ρ_i, obtained as

$$P = \sum_{\sigma_i = \pm 1} \exp\left[\beta \sum_{(ij)} (\sigma_i \sigma_j - 1) + \sum_i h_i(\sigma_i - 1)\right] \qquad (74)$$

The summation $\sum_{(ij)}$ runs over all links. The polynomial P is of degree one in each ρ_i separately, and globally of degree N in all of them. For τ and ρ_i real and positive, P is obviously positive and hence cannot vanish. We assume in the sequel $0 < \tau < 1$, i.e. β real positive, which is the ferromagnetic regime. As examples, let us write the polynomials P corresponding to the simplest graphs

$$\overset{1}{\bullet}\!\!-\!\!\overset{2}{\bullet} \quad P_{12} = 1 + \tau(\rho_1 + \rho_2) + \rho_1\rho_2 \qquad (75)$$

$$\overset{1}{\bullet}\!\!-\!\!\bullet\!\!-\!\!\overset{3}{\bullet} \quad P_{123} = (1 + \rho_1\tau)(1 + \rho_3\tau) + \rho_2(\tau + \rho_1)(\tau + \rho_2)$$

Up to a regular function, the (total) free energy is the logarithm of P. Its only singularities are therefore the zeroes of the polynomial. If we have a growing sequence of graphs (as parts of a lattice say), in the limit $N \to \infty$, the free energy will remain analytic in any bounded region outside the accumulation points of zeroes.

We now look more closely at the way the polynomials can be generated by building the graph step by step. Notice first that on a disjoint union P factorizes, $P = P_1 P_2$, where P_1 and P_2 refer to the two disjoint subsets. Starting from such a situation, we can generate a new one, call it (12), by identifying a site a from subset 1 to a site b from subset 2. Let $P_{(12)}$ be the corresponding polynomial.

Now P_1 is linear in ρ_a, P_2 linear in ρ_b, in such a way that

$$P_1 = A_+ + \rho_a A_- \qquad P_2 = B_+ + \rho_b B_- \qquad (76)$$

Here $A_+(B_+)$ refers to the contribution with $\sigma_a = +1$ ($\sigma_b = +1$) while $A_-(B_-)$ refers to the one with $\sigma_a = -1$ ($\sigma_b = -1$). When a and b are identified, a new activity variable ρ_{ab} is attached to the

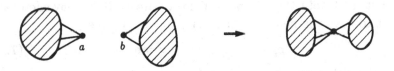

Fig. 4   The contraction process.

site, and we have the following contraction process (figure 4)

$$P_1 P_2 \equiv A_+ B_+ + \rho_a A_- B_+ + \rho_b A_+ B_- + \rho_a \rho_b A_- B_- \to$$
$$P_{12} \equiv A_+ B_+ + \rho_{ab} A_- B_- \tag{77}$$

Check the contraction process on the examples (75). Construct the polynomial P for a cyclic graph.

Nothing prevents the contraction process from being extended to a single connected part, where at first sites a and b are distinct, then identified as a single site ab with activity variable ρ_{ab}. Thus

$$P \equiv A_{++} + A_{-+}\rho_a + A_{+-}\rho_b + A_{--}\rho_a\rho_b \to P_{ab} \equiv A_{++} + \rho_{ab}A_{--} \tag{78}$$

It is assumed that in the contracted graph a pair of vertices is joined by at most one link.

The contraction process allows one to obtain the polynomial P for an arbitrary graph starting from the elementary result for the simplest graph of two sites joined by a link. In this case, from equation (75a), the polynomial vanishes for

$$\rho_1 = -\frac{1 + \tau\rho_2}{\tau + \rho_2} \tag{79}$$

This relation defines a global (conformal) one-to-one map from the ρ_2 complex plane to the ρ_1 plane. It obviously leaves the unit circle invariant (for real τ) and for $0 < \tau < 1$, exchanges the interior with the exterior of this circle. It follows therefore that if $|\rho_1| < 1$ and $|\rho_2| < 1$, or $|\rho_1| > 1$, and $|\rho_2| > 1$, the polynomial cannot vanish. This property generalizes in that, for an arbitrary graph, if all ρ_i lie inside, or all ρ_i lie outside the unit circle, P is different from zero. To prove that this is true, it is sufficient to check that this property survives the contraction process. Assume therefore that for a given graph $P(\rho_i) \neq 0$ when $|\rho_i| < 1$ for all i.

When the dependence on ρ_a and ρ_b is made explicit, P is given by (78), while P_{ab} is also a function of the set $\{\rho_i\} - \{\rho_a, \rho_b\}$, through A_{++} and A_{--}, and a new ρ_{ab}. Fix the ρ_i's distinct from ρ_a and ρ_b inside the unit circle. We want to show that $|A_{++}| > |A_{--}|$, in which case P_{ab} will be nonvanishing for $|\rho_{ab}| < 1$. But $P \neq 0$ for $|\rho_a| < 1, |\rho_b| < 1$. We can take $\rho_a = \rho_b = \rho$. Hence as a function of ρ the second degree polynomial $A_{++} + \rho(A_{+-} + A_{-+}) + \rho^2 A_{--}$ has its two roots of modulus equal or larger that unity, which indeed means that $|A_{++}| \geq |A_{--}|$. Thus P_{ab} is different from zero when all its ρ's are inside the unit circle, and this property holds for any graph. Now set all $\rho_i \equiv \rho$ and we have a uniform field. From the symmetry property under reversal of the field $h \to -h, \rho \to \rho^{-1}$ and $Z(h) = Z(-h)$, we have $e^{+Nh}P(\tau, \rho) = e^{-Nh}P(\tau, \rho^{-1})$ or

$$P(\tau, \rho) = \rho^N P(\tau, \rho^{-1}) \tag{80}$$

Hence if $P \neq 0$ for $|\rho| < 1$, one has also $P \neq 0$ for $|\rho| > 1$. The Lee–Yang theorem follows, stating that the partition function can (and does) only vanish on the unit circle $|\rho| = 1$.

We can check this in some limiting cases. At infinite temperature, $\tau = 1$, and P reduces to $P(1, \rho) = (1 + \rho)^N$, while at zero temperature, $\tau = 0$, $P(0, \rho) = 1 + \rho^N$. Thus decreasing τ, one starts from a degenerate zero with multiplicity N at $\rho = -1$ at infinite temperature, to a uniform distribution $\rho_k = e^{i\pi(2k+1)/N}$. As long as N is finite, no zero is on the real axis ($\rho = 1, h = 0$).

This remarkable result is so general that it does not allow us to find the transition in the infinite volume limit but only suggests the scenario that was described at the beginning of this section. To prove the existence of a transition (or the closure of the analyticity gap) requires more sophisticated means. As was also said before, the same result extends to a φ^4 self-interaction or to models with a continuous symmetry such as the $O(2)$ or rotator model. We have already seen that it shows up in mean field, as well as in the Bethe approximation; in each case it is therefore not an artefact.

3.2.2 The one-dimensional case

As $\tau \to \tau_c$ we saw that mean field ($d \to \infty$) predicted an analyticity gap shrinking as $(\tau - \tau_c)^{3/2}$ and a singular behaviour of the free energy as $(h \mp ih_c)^{3/2}$. At the other extreme, let us consider the case $d = 1$. For periodic boundary conditions,

using the transfer matrix formalism of section 1.2.5, the partition
function is given by

$$Z_N = \left(\tfrac{1}{2}e^{\beta+h}\right)^N \operatorname{Tr} T^N \qquad\qquad T = \begin{pmatrix} 1 & \tau\rho^{1/2} \\ \tau\rho^{1/2} & \rho \end{pmatrix} \qquad (81)$$

Thus

$$\operatorname{Tr} T^N = \lambda_+^N + \lambda_-^N \qquad (82)$$

where $\lambda_\pm$ are the two eigenvalues of T, given by

$$(1-\lambda)(\rho-\lambda) - \rho\tau^2 = 0$$

$$\lambda_\pm = \tfrac{1}{2}(1+\rho) \pm \sqrt{\tfrac{1}{4}(1-\rho)^2 + \rho\tau^2} \qquad (83)$$

In the thermodynamic limit $N \to \infty$, for $\rho > 0$ (h real), $0 < \tau < 1$,
only the largest eigenvalue λ_+ contributes

$$F = \lim_{N\to\infty} \frac{1}{N} \ln Z_N = \beta + h - \ln 2 + \ln\left\{\tfrac{1}{2}(1+\rho) + \sqrt{\tfrac{1}{4}(1-\rho)^2 + \rho\tau^2}\right\}$$
$$(84)$$

Singularities in ρ arise from the vanishing of the square root, i.e.

$$1 + 2\rho(2\tau^2 - 1) + \rho^2 = 0$$

$$\rho_c^{(\pm)} = e^{\pm i2h_c} \qquad (85)$$

$$\cos 2h_c = 1 - 2\tau^2 = 1 - 2e^{-4\beta}$$

We find a pair of singularities on the unit circle (in ρ) as expected.
As τ decreases from 1 to 0, the pair starts at -1 ($\cos 2h_c = -1$)
and ends at $+1$, zero temperature playing the role of critical
temperature in dimension one. As $\tau \to 0$, $h_c \sim \tau$ (compare with
$h_c \sim (\tau - \tau_c)^{3/2}$ for $d \to \infty$), while the singular part of the free
energy behaves as $(h \mp ih_c)^{1/2}$. Therefore the exponent σ defined
in equation (71a) is $-1/2$ for $d = 1$. The upper critical dimension
is six for the Lee–Yang edge singularity. As a result σ will vary
between $-1/2$ ($d = 1$) to $+1/2$ for $d = 6$, then stay constant above
six dimensions. It turns out that $\sigma = -1/6$ in dimension $d = 2$
(chapter 9), in agreement with the idea of smooth behaviour as a
function of dimensionality.

3.2.3   General properties

A point that needs emphasis is the claim that the Lee–Yang edge
singularity is independent of the temperature above T_c, or in

other words the exponent σ is temperature independent. The one-dimensional calculation supports this claim, as did the mean field approximation. Therefore the question may be studied for vanishingly small β and h close to $i\pi/2$ (i.e. $\rho = e^{-2h}$ close to -1). For h complex, one is not dealing anymore with a genuine statistical model, i.e. admitting a (classical) probabilistic interpretation. Rather we have reached a point where field theory proper is taken seriously. Along the imaginary h axis, the effective theory is described by a self-interacting scalar field φ, with a dominant symmetry breaking and imaginary interaction term $i\varphi^3$ for small fluctuations. This arises from a shift in the field φ designed so as to compensate the linear term induced by the coupling to the external field. This remark enables one to understand in a heuristic way the upper critical dimension, where, as we shall see later, the effective coupling constant is dimensionless. Since from chapter 1 a scalar field has dimension $(d-2)/2$, to insure that a term $\int d^d\mathbf{x}\varphi^3(\mathbf{x})$ is dimensionless we need $\frac{3}{2}(d-2) = d$, i.e. $d = 6$.

While for any finite N the Lee–Yang zeroes appear as singularities of the (total) free energy, this need not be the case in the infinite volume limit, as illustrated again by the one-dimensional calculation. Only the boundary points of the accumulation set survive certainly as singularities, hence the name "edge singularity". This phenomenon is familiar when a function has a branch point, but admits analytic continuation around the singularity, even though in some finite approximation it may turn out that there is a preferred locus for a branch cut. This is indeed the case if we return to the general discussion and approximate an infinite lattice of coordination number q by a finite volume of N sites, then take the infinite volume limit. The free energy per site is

$$F = \tfrac{1}{2}q\beta + h - \ln 2 + \lim_{N\to\infty} \frac{1}{N} \ln P_N(\tau,\rho) \qquad (86)$$

Since P_N is normalised to unity for $\rho = 0$, we have a factorization in terms of zeroes of unit modulus

$$P_N(\rho,\tau) = \prod_{a=1}^{N} \left(1 - \frac{\rho}{\rho_a(\tau)}\right) \qquad (87)$$

Let the zeroes accumulate for $N \to \infty$ on the unit circle $\rho = e^{i\varphi}$, with a limiting (τ-dependent) density $N\mu(\varphi)$, $\mu(\varphi) = \mu(-\varphi) \geq 0$.

The symmetry property is a reflection of the original symmetry in $h \leftrightarrow -h$. Normalization requires

$$\int_{-\pi}^{+\pi} d\varphi \, \mu(\varphi) = 1 \tag{88a}$$

This leads to a representation of F as

$$F = \tfrac{1}{2}q\beta + h - \ln 2 + \tfrac{1}{2} \int_{-\pi}^{+\pi} d\varphi \, \mu(\varphi) \ln(1 + \rho^2 - 2\rho \cos \varphi) \tag{88b}$$

where the contributions of conjugate zeroes have been grouped together. The representation (88) is valid in the whole range $0 < \tau < 1$. Below the critical temperature, where the support of $\mu(\varphi)$ is the full circle, it defines in general two distinct analytic functions, one for $|\rho| < 1$, the other for $|\rho| > 1$. At $\rho = 1$ ($h = 0$), F is continuous, but its derivative with respect to h, i.e. the magnetization

$$M(\rho) = 1 + 2 \int_{-\pi}^{+\pi} d\varphi \, \mu(\varphi) \frac{\rho}{e^{i\varphi} - \rho} = -1 + 2 \int_{-\pi}^{+\pi} d\varphi \, \mu(\varphi) \frac{1}{1 - \rho e^{-i\varphi}}$$

$$= \int_{-\pi}^{+\pi} d\varphi \, \mu(\varphi) \frac{1 - \rho^2}{1 - 2\rho \cos \varphi + \rho^2} \tag{89}$$

which vanishes as $h \to 0$ for $\tau > \tau_c$, becomes discontinuous for $\tau < \tau_c$, when the support of μ covers the circle. One has

$$\tau < \tau_c \qquad M_\pm = \lim_{h \to \pm 0} M = \pm 2\pi \mu(0) \tag{90}$$

So $\mu(0) = M_{sp}/2\pi$. From (88), the free energy and the magnetization have an expansion in terms of the (trigonometric) moments of $\mu(\varphi)$. Set

$$\mu_n = \int_{-\pi}^{\pi} d\varphi \, \mu(\varphi) \cos n\varphi \qquad \mu_0 = 1 \tag{91}$$

then

$$F = \tfrac{1}{2}q\beta + h - \ln 2 - \sum_{n=1}^{\infty} \mu_n \frac{\rho^n}{n}$$

$$M = 1 + 2 \sum_{n=1}^{\infty} \mu_n \rho^n \tag{92}$$

which appear as strong field expansions (small ρ) away from a fully ordered situation (all spins up say, hence $M = 1$). Conversely if such an expansion is known, from the standard moment problem, one can try to recover the τ-dependent distribution $\mu(\varphi)$. Unfortunately the best that can be done is to obtain finitely many terms in the series (92) and therefore additional information is needed (for instance a parametrization of the singular behaviour of μ near the edge of its support).

In the case $d = 1$, obtain the representation (88) in the form

$$F = \beta + h - \ln 2 + \frac{1}{2} \int_{-\pi}^{+\pi} d\varphi \frac{\theta(\cos\varphi_0 - \cos\varphi)}{2\pi}$$

$$\times \left[\frac{1 - \cos\varphi}{\cos\varphi_0 - \cos\varphi}\right]^{1/2} \ln\left[1 + \rho^2 - 2\rho\cos\varphi\right] \tag{93}$$

$$\cos\varphi_0 = 1 - 2\tau^2 = 1 - 2e^{-4\beta} \tag{94}$$

As $\tau \to 0$ the corresponding distribution $\mu(\varphi)$ tends to be uniform, while as $\tau \to 1$ it gets concentrated at $\varphi = \pm\pi$. For $0 < \tau < 1$, the edge singularity of $\mu(\varphi)$ is proportional to $(\cos\varphi_0 - \cos\varphi)^{-1/2}$. Compute the moments μ_n.

3.2.4 Zeroes in the temperature plane

No general theorems are known which constrain the zeroes in the complex temperature plane, but it is tempting to analyze their location either numerically or otherwise. We now assume the external field to vanish ($\rho = 1$). Fisher observed that for an Ising model on a square two-dimensional lattice, the zeroes are located in the complex τ plane on two circles, one of them being given by the equation $\bar\tau = (1-\tau)(1+\tau)^{-1}$, the other obtained by changing τ into τ^{-1}, i.e. inverting with respect to the unit circle. This holds in the thermodynamic limit. For a finite system, some zeroes might violate the above property, depending on the specific boundary conditions. The locus of zeroes is invariant under duality, and, as a consequence, goes through the transition point on the real axis. But more important is the fact that in this case

(i) the complex zeroes close to the critical point accumulate on curves,

(ii) for the two-dimensional model, these curves cross the real axis at a right angle.

Recover the above results from the expressions given in chapter 2.

There seems to be some evidence that, in higher dimension, the complex temperature zeroes accumulate also on curves close to the (real) critical temperature. If we assume this to be true, we can draw some interesting consequences regarding the nature of the singularity of the free energy. Because of the reality properties of the partition function, two curves of zeroes will be incident at T_c, one in the upper half plane and a reflected one in the lower half plane. Let $\pi - \varphi$ be the angle between the positive real T-axis and the upper line of zeroes, and let

$$f_{\text{sing}}^{(\pm)} \sim A_\pm \left(\frac{T - T_c}{T_c} \right)^{2-\alpha}$$

be the singular part of the free energy above (plus sign) and below (minus sign) on the real axis. Then both $f_{\text{sing}}^\pm$ extend as analytic functions up to the curves of zeroes where their real parts are equal. The curves of zeroes play the role of Stokes lines, where different asymptotic behaviours have to match (in modulus for the partition function). This is consistent only if the following relation holds between the above angle φ, the ratio A_-/A_+ of critical amplitudes and the exponent α

$$\tan \left[(2 - \alpha)\varphi \right] = \frac{\cos \pi \alpha - A_-/A_+}{\sin \pi \alpha} \tag{95}$$

as a consequence of the above matching condition. If we let $\alpha \to 0$ and $\varphi \to \pi/2$ as in $d = 2$, this relation is compatible only if $A_- = A_+$, which is the observed value. The mean field (or large d) approximation gives $\varphi = \pi/4$ (since $\alpha = 0$ and $A_+/A_- = 0$). When $d = 3$, inserting the value of $\alpha \sim 0.11$, and an observed angle φ of the order of 57 degrees, one predicts a ratio A_+/A_- close to 0.5. A more accurate measurement of the angle φ could, however, significantly modify this ratio.

3.3 Large n limit

In this section, we investigate the limit of an n-component vector model as $n \to \infty$. Of course $O(n)$ symmetry is assumed. It turns out that the solution agrees with the one of the spherical model, solved by Berlin and Kac in the fifties. The $1/n$ corrections are, however, different. Apart from its technical interest, this solution reveals, in as simple a setting as possible, the failure of mean field theory below the upper critical dimension $d = 4$, for just the reasons which were discussed in the first chapter. The spherical model involves scalar variables φ_i at each site, nearest neighbour interactions, and the constraint $\sum_i \varphi_i^2 = \text{cst}$. The clue to the equivalence between the two models lies in the saddle point treatment to which we turn now.

3.3.1 Saddle point method

As the number of field components n per site goes to infinity, we use a saddle point method to evaluate the path integral. This will be justified by showing that corrections are of order $1/n$. For N sites ($N \to \infty$) write the partition function as

$$Z = Z_0 \int \prod_j d^n \varphi_j \, \delta(\varphi_j^2 - 1) \exp n\beta \left\{ \sum_{(ij)} \varphi_i \cdot \varphi_j \right\}$$

$$= Z_0 \int \prod_j d^n \varphi_j \frac{n d\alpha_j}{2\pi}$$

$$\times \exp n \left\{ \sum_j (i\alpha_j + \lambda_j)(1 - \varphi_j^2) + \beta \sum_{(ij)} \varphi_i \cdot \varphi_j \right\}$$

$$Z_0 = \left(\frac{\Gamma(n/2)}{\pi^{n/2}} \right)^{Nd} \tag{96}$$

The normalizing factor Z_0 is such that $Z \to 1$ as $\beta \to 0$. In the second representation we have replaced the constraints $\varphi_j^2 = 1$ by a Fourier integral over a conjugate parameter α_j, while λ_j is arbitrary and will be chosen positive and sufficiently large to ensure convergence of later integrals. As a check, the solution should be independent of these values. The variables φ only occur in the exponential of a quadratic form. Set Δ for the lattice Laplacian, and define the matrix Q through

$$Q_{jk} = -\beta\Delta_{jk} + 2(\lambda_j + i\alpha_j - \beta d)\delta_{jk} \tag{97a}$$

Integrating over φ yields

$$Z = Z_0 \left(\frac{2\pi}{n}\right)^{\frac{1}{2}(n-2)N} \int \prod_j \mathrm{d}\alpha_j \, \exp \tfrac{1}{2} n \left\{ \sum_j 2(i\alpha_j + \lambda_j) - \mathrm{Tr} \ln Q \right\}$$
(97b)

Up to now, no approximation was involved. As $n \to \infty$ the prefactor in the exponential becomes infinite, while the term between brackets is n-independent. This is the basis for the saddle point method, which (unfortunately) amounts to interchanging the limits $N \to \infty$ (infinite volume) and $n \to \infty$ (infinitely many field components). The variational equation with respect to $2(\lambda_k + i\alpha_k)$, with k a space label, leads to the condition

$$1 = \langle k \,|\, Q^{-1} \,|\, k \rangle$$
(98)

We look for a translation invariant solution, hence such that $\lambda_k + i\alpha_k$ be k-independent. This in fact identifies the solution to the one pertaining to the spherical model, with scalar fields φ_k constrained to $\sum_k \varphi_k^2 = \mathrm{cst}$, up to the overall prefactor n. For convenience, define

$$2(\lambda + i\alpha - d\beta) = \beta \xi^{-2}$$
(99)

Averaging equation (98) over all sites, and taking the infinite volume limit, gives

$$1 = \lim_{N \to \infty} \frac{1}{N} \, \mathrm{Tr} \, Q^{-1} = \frac{1}{\beta} \int \frac{\mathrm{d}^d p}{(2\pi)^d} \; \frac{1}{\xi^{-2} + 2\sum_1^d (1 - \cos p_\mu)}$$
(100)

and to leading order

$$Z = Z_0 \left[\frac{1}{\beta} \left(\frac{2\pi}{\beta n}\right)^{(n-2)/2} \right]^N$$

$$\times \exp \frac{nN}{2} \left\{ \beta(2d + \xi^{-2}) - \int \frac{\mathrm{d}^d p}{(2\pi)^d} \ln \left[\xi^{-2} + 2 \sum_{\mu=1}^d (1 - \cos p_\mu) \right] \right\}$$
(101)

In both expressions the momentum p runs over a Brillouin zone, i.e. each component ranges from $-\pi$ to $+\pi$. The quantity ξ plays the role of a correlation length, and is determined self-consistently by equation (100) which is a stationarity condition of $\lim_{N \to \infty} N^{-1} \ln Z$ with respect to ξ^{-2}.

Let ξ vary from zero to infinity in equation (100). Correspondingly β runs from zero ($\xi^2 \sim \beta$ for $\xi \sim 0$) to some critical value $\beta_c(\xi = \infty)$ with

$$\beta_c = \int \frac{\mathrm{d}^d p}{(2\pi)^d} \ \frac{1}{2\sum_1^d(1 - \cos p_\mu)} = \frac{1}{2}\int_0^\infty \mathrm{d}\alpha \, e^{-\alpha d} I_0(\alpha)^d \quad (102)$$

This value is related to the probability (denoted $\Pi(0,1)$) that a Brownian curve returns to the origin, through

$$2d\beta_c = G(0,1) = \frac{1}{1 - \Pi(0,1)} \quad (103)$$

The behaviour of $2d\beta_c$ as a function of dimension d has therefore been discussed already, and for $d \to \infty$

$$2d\beta_c = 1 + \frac{1}{2d} + 3\left(\frac{1}{2d}\right)^2 + 12\left(\frac{1}{2d}\right)^3 + \cdots \quad (104)$$

The dominant term is the mean field prediction. What is new is the appearance of a *lower critical dimension*, here equal to two, below which no transition occurs. As $d \to 2$, $\beta_c \to \infty$. More precisely

$$2d\beta_c \underset{d \to 2}{\sim} \frac{2}{\pi(d-2)} \quad (105)$$

In two dimensions, there exists only a high temperature symmetric phase. The previous calculation, carried out in a phase with a positive parameter ξ^{-2}, is understood to be in an unbroken symmetry phase, that is in the range $0 \le \beta \le \beta_c$. The two-dimensional result is in agreement with the Mermin–Wagner theorem (to be discussed more fully in the next chapter) according to which, for a continuous symmetry group, short-range interactions, and dimension two, there cannot be an ordered phase characterized by a nonvanishing order parameter (the case of an Abelian or commutative symmetry group is set aside, transitions can occur, but again without macroscopic magnetization).

Let us estimate the divergence of the correlation length ξ as $\beta \to \beta_c - 0$, by computing from (100) and (102)

$$\xi^2(\beta_c - \beta) = \int \frac{\mathrm{d}^d p}{(2\pi)^d} \ \frac{1}{2\sum_1^d(1 - \cos p_\mu)\left[\xi^{-2} + 2\sum_1^d(1 - \cos p_\mu)\right]}$$

$$(106)$$

If $d > 4$, the right-hand side tends to a finite limit as $\xi^{-1} \to 0$, and $\beta \to \beta_c$. The sensitive region in the integral is the infrared one, where $\mathbf{p}$ goes to zero. The denominator behaves as $\mathbf{p}^2(\xi^{-2} + \mathbf{p}^2)$, and the integral converges for $d > 4$. Therefore

$$d > 4 \qquad\qquad \xi \sim (\beta_c - \beta)^{-1/2} \qquad\qquad \nu = \tfrac{1}{2} \qquad (107)$$

in agreement with mean field. However, for $d \leq 4$ a new phenomenon occurs, and the exponent ν departs from $\tfrac{1}{2}$. As a result, $d = 4$ is the *upper critical dimension*, an expected property which will remain true for finite n. At $d = 4$, an extra logarithm appears as $\xi^2(\beta_c - \beta) \sim \ln \xi$

$$d = 4 \qquad \xi \sim (\beta_c - \beta)^{-1/2} \left[\ln\left(\frac{1}{\beta_c - \beta}\right)\right]^{1/2} \qquad \nu = \tfrac{1}{2}$$
$$(108)$$

In the range $4 > d > 2$, one finds

$$4 > d > 2 \qquad\qquad \xi \sim (\beta_c - \beta)^{-1/(d-2)} \qquad\qquad \nu = 1/(d-2)$$
$$(109)$$

Finally for $d = 2$, ξ diverges with β exponentially

$$d = 2 \qquad\qquad \xi \underset{\beta \to \infty}{\sim} \exp 2\pi\beta \qquad\qquad (110)$$

In a sense, this means $\nu = \infty$. This behaviour is characterized as asymptotic freedom. If $\beta^{-1} \sim T$ is thought of as a coupling constant, then turning equation (110) around means that β^{-1} behaves like the inverse of the logarithm of the scale.

Returning to $d > 2$, for β larger than β_c, it would seem that the correlation length is infinite for all modes (indeed the massive longitudinal one is of relative weight $1/n \to 0$). A more careful analysis is required to extract the spontaneous magnetization, since the translational invariant solution is no longer an isolated saddle point. We shall use a slightly different approach below.

The variation of ν as a function of dimension is shown on figure 5, with $\nu = 1$ for $d = 3$. It might be remarked here that a natural extrapolation in d (see for instance equation (102)) is allowed as a result of the structure of the various integral representations, and that the upper critical dimension is signalled by the appearance of logarithms. These ideas will be formalized later on.

The specific heat is discontinuous above four dimensions as in mean field, so we might say that $\alpha = 0$. For $2 < d < 4$, the

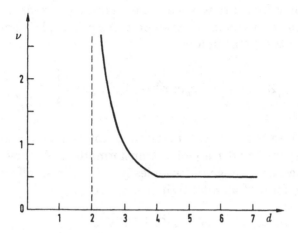

Fig. 5   The exponent ν as a function of dimension d, according to the spherical model.

singular part of the free energy is proportional to ξ^{-d}, hence the singular part of the specific heat is

$$4 > d > 2 \qquad C_{\text{sing}} \sim (\beta_c - \beta)^{-(d-4)/(d-2)} \qquad \alpha = \frac{d-4}{d-2}$$

$$(111)$$

Note that α is negative corresponding to a cusp in C. In particular for $d = 3$, $\alpha = -1$. When $d = 4$, there is a logarithmic singularity

$$d = 4 \qquad C_{\text{sing}} \sim \frac{1}{\ln[1/(\beta_c - \beta)]} \qquad (112)$$

3.3.2   Factorization

The study of the low temperature phase requires some care. As a matter of illustration, we shall use the factorization method of products of invariant observables in the large n limit, already encountered in chapter 1. For the computation of the spontaneous magnetization for instance, we look at the behaviour of the two point function at infinite separation. But first we derive a so-called Ward identity, as part of the equations of motion. This is an exact relation between correlation functions which follows from the structure of the partition function as an integral over an $O(n)$ invariant measure. Let γ, $\delta \ldots$ denote the indices of the field running from 1 to n, while i, j or coordinates $\mathbf{x}$ label the sites. Let us repeat an argument given previously, in slightly generalized form. The two point function is

$$\langle \varphi_0^\gamma \varphi_{\mathbf{x}}^\delta \rangle = Z^{-1} \int \prod_k \mathrm{d}^n \varphi_k \delta(\varphi_k^2 - 1) \varphi_0^\gamma \varphi_{\mathbf{x}}^\delta \exp \left\{ n\beta \sum_{(ij)} \varphi_i \cdot \varphi_j \right\}$$

(113)

In any finite volume, the measure is rotationally invariant (in internal space), under a global transformation. We now perform an infinitesimal change of variables in the integral, affecting only φ_i, in the form of an infinitesimal rotation

$$\varphi_i \to \varphi_i + \varepsilon(A\varphi)_i$$

(114)

with A an arbitrary antisymmetric matrix. This should not affect the integral. Hence the coefficient of ε is zero, leading to

$$0 = \delta_{i,0} \langle (A\varphi)_0^\gamma \varphi_{\mathbf{x}}^\delta \rangle + \delta_{i,\mathbf{x}} \langle \varphi_0^\gamma (A\varphi)_{\mathbf{x}}^\delta \rangle + n\beta \sum_{j(i)} \langle \varphi_0^\gamma \varphi_{\mathbf{x}}^\delta (\varphi_j \cdot (A\varphi)_i) \rangle$$

(115)

Since the antisymmetric matrix A is arbitrary, we identify the coefficient of $\frac{1}{2}(A^{\alpha\beta} - A^{\beta\alpha})$, and take the trace over indices, with the result that

$$(1 - n^{-1})(\delta_{i,\mathbf{x}} - \delta_{i,0}) \langle \varphi_0 \cdot \varphi_{\mathbf{x}} \rangle =$$
$$\beta \sum_{j(i)} \left\langle \left[\left(\varphi_0 \cdot \varphi_j \right) \left(\varphi_{\mathbf{x}} \cdot \varphi_i \right) - \left(\varphi_0 \cdot \varphi_i \right) \left(\varphi_{\mathbf{x}} \cdot \varphi_j \right) \right] \right\rangle \quad (116)$$

The sum over $j(i)$ runs over neighbours of the point i. In equation (116), we have already assumed an infinite volume limit, but as such it is an exact (Ward) identity. To make use of it as $n \to \infty$, we imagine evaluating all integrals by a saddle point method. Saddle points are in general degenerate owing to the $O(n)$ invariance, but rotational invariants are insensitive to this degeneracy, and their mean values are then expected to factorize. As a consequence, we may perform the replacement

$$\lim_{n \to \infty} \langle (\varphi_a \cdot \varphi_b) (\varphi_c \cdot \varphi_d) \rangle \quad \to \quad \lim_{n \to \infty} \langle \varphi_a \cdot \varphi_b \rangle \lim_{n \to \infty} \langle \varphi_c \cdot \varphi_d \rangle$$

(117)

In this limit we obtain, up to corrections of order n^{-1}, a quadratic equation for the two point function,

$$\left(\delta_{i,\mathbf{x}} - \delta_{i,0}\right) \langle \boldsymbol{\varphi}_0 \cdot \boldsymbol{\varphi}_{\mathbf{x}} \rangle$$
$$= \beta \sum_{j(i)} \left\{ \langle \boldsymbol{\varphi}_0 \cdot \boldsymbol{\varphi}_j \rangle \langle \boldsymbol{\varphi}_{\mathbf{x}} \cdot \boldsymbol{\varphi}_i \rangle - \langle \boldsymbol{\varphi}_0 \cdot \boldsymbol{\varphi}_i \rangle \langle \boldsymbol{\varphi}_{\mathbf{x}} \cdot \boldsymbol{\varphi}_j \rangle \right\} \quad (118)$$

Using a Fourier transform, with

$$\langle \boldsymbol{\varphi}_{\mathbf{x}} \cdot \boldsymbol{\varphi}_{\mathbf{y}} \rangle = \int \frac{d^d q}{(2\pi)^d} e^{i\mathbf{q}\cdot(\mathbf{x}-\mathbf{y})} G(\mathbf{q}) \quad (119)$$

the equation reads

$$G(\mathbf{q}) - G(\mathbf{k}) = 2\beta \sum_1^d (\cos q_\mu - \cos k_\mu) G(\mathbf{q}) G(\mathbf{k}) \quad (120)$$

For $\beta < \beta_c$, $\langle \boldsymbol{\varphi}_0 \cdot \boldsymbol{\varphi}_{\mathbf{x}} \rangle \to 0$ for $|\mathbf{x}| \to \infty$, hence $G(\mathbf{q})$ is regular at $\mathbf{q} = 0$, and we conclude that

$$G^{-1}(\mathbf{q}) = \beta \left\{ \xi^{-2} + 2 \sum_1^d (1 - \cos q_\mu) \right\} \quad (121)$$

is the required solution to (120), with ξ^{-2} a constant determined by $\langle \varphi_0^2 \rangle = 1$, which leads back to the saddle point equation (100). However, for $\beta > \beta_c$, a spontaneous magnetization appears, with $\lim_{|\mathbf{x}| \to \infty} \langle \boldsymbol{\varphi}_0 \cdot \boldsymbol{\varphi}_{\mathbf{x}} \rangle = M^2$, or equivalently a singular part in $G(\mathbf{q})$ of the form $M^2 (2\pi)^d \delta(\mathbf{q})$, where the δ function is assumed periodic on the torus $q_\mu = q'_\mu + 2\pi n_\mu$, n_μ integers. Subtracting M^2 from $\langle \boldsymbol{\varphi}_0 \cdot \boldsymbol{\varphi}_{\mathbf{x}} \rangle$ leads to the representation valid for $d > 2$

$$\beta > \beta_c, \; d > 2 \qquad \langle \boldsymbol{\varphi}_0 \cdot \boldsymbol{\varphi}_{\mathbf{x}} \rangle = M^2 + \frac{1}{\beta} \int \frac{d^d q}{(2\pi)^d} \frac{e^{-i\mathbf{q}\cdot\mathbf{x}}}{2 \sum_1^d (1 - \cos q_\mu)} \quad (122)$$

Now M^2 is determined by the condition $\langle \varphi_0^2 \rangle = 1$, i.e. using equation (102)

$$1 - M^2 = \frac{\beta_c}{\beta} \quad (123)$$

Therefore for $d > 2$, the singularity of M is $[1 - (T/T_c)]^{1/2}$, and the exponent β remains unchanged, $\beta = 1/2$. At criticality

$$\beta = \beta_c \qquad \langle \varphi_0 \cdot \varphi_x \rangle = \frac{1}{\beta_c} \int \frac{d^d q}{(2\pi)^d} \frac{e^{-iq \cdot x}}{2 \sum_1^d (1 - \cos q_\mu)}$$

$$\underset{|x| \to \infty}{\sim} \frac{\Gamma(d/2)}{\beta_c 2\pi^{d/2}(d-2)} \frac{1}{|x|^{d-2}} \qquad (124)$$

leading to $\eta = 0$ for $d > 2$. In the symmetric phase $\beta < \beta_c$, the susceptibility is isotropic in internal space, and reads

$$\chi = \sum_x \langle \varphi_0 \cdot \varphi_x \rangle = G(q = 0) = \xi^2/\beta \qquad (125)$$

It diverges as $(\beta_c - \beta)^{2\nu}$ and therefore

$$\gamma = 2\nu = 2/(d - 2) \qquad (126)$$

The analogous result in dimension two is that χ diverges as $\exp 4\pi\beta$ as $\beta \to \infty$.

Higher correlations functions in the large n limit easily follow from the factorization property. They vanish for an odd number of fields, while for instance in the symmetric phase

$$\left\langle \varphi_{x_1}^{\alpha_1} \varphi_{x_2}^{\alpha_2} \varphi_{x_3}^{\alpha_3} \varphi_{x_4}^{\alpha_4} \right\rangle = \frac{1}{n^2} \sum_{\text{pairings}} \delta^{\alpha_1 \alpha_2} \delta^{\alpha_3 \alpha_4} \left\langle \varphi_{x_1} \cdot \varphi_{x_2} \right\rangle \left\langle \varphi_{x_3} \cdot \varphi_{x_4} \right\rangle$$

$$(127)$$

where the sums includes three such terms.

3.3.3 Coupling to an external field

Let us introduce a coupling to an external field $n \sum_i h_i \cdot \varphi_i$. Applying the saddle point method leads to a generalization of equation (101) involving a function ξ_i^{-2} such that

$$Z = Z_0 \left[\frac{1}{\beta} \left(\frac{2\pi}{n\beta} \right)^{\frac{1}{2}(n-2)} \right]^N \exp \frac{n}{2} \left\{ \beta \sum_i (\xi_i^{-2} + 2d) \right.$$

$$\left. + \frac{1}{\beta} \sum_{i,j} h_i \cdot h_j \left(\frac{1}{\xi^{-2} - \Delta} \right)_{ij} - \text{Tr} \ln(\xi^{-2} - \Delta) \right\} \qquad (128a)$$

with the stationarity condition

$$\beta = \left(\frac{1}{\xi^{-2} - \Delta}\right)_{ii} + \frac{1}{\beta} \sum_{j,k} \mathbf{h}_j \cdot \mathbf{h}_k \left(\frac{1}{\xi^{-2} - \Delta}\right)_{ji} \left(\frac{1}{\xi^{-2} - \Delta}\right)_{ik} \tag{128b}$$

If $\mathbf{h}$ is uniform, so is ξ. Summing over sites in (128b) yields

$$\beta = \int \frac{d^d p}{(2\pi)^d} \frac{1}{\xi^{-2} + 2\sum_1^d (1 - \cos p_\mu)} + \frac{1}{\beta}\xi^4 \mathbf{h}^2 \tag{129a}$$

while, up to a β-independent constant F_0, the free energy per site is

$$F = F_0 + \frac{n}{2}\left\{-\ln\beta + \beta(\xi^{-2} + 2d) + \frac{\xi^2}{\beta}\mathbf{h}^2 \right.$$
$$\left. - \int \frac{d^d p}{(2\pi)^d} \ln\left[\xi^{-2} + 2\sum_1^d (1 - \cos p_\mu)\right]\right\} \tag{129b}$$

For any positive β, equation (129a) has a solution with ξ finite. In other words, no transition occurs in the presence of an external field. We can rewrite the relation between β, ξ and $\mathbf{h}$ as

$$\beta^2 - \beta\xi^2 a(\xi^2) - \mathbf{h}^2 \xi^4 = 0$$
$$a(\xi^2) = \int \frac{d^d p}{(2\pi)^d} \frac{1}{1 + \xi^2 2\sum_1^d (1 - \cos p_\mu)} \tag{129c}$$

with $a(\xi^2)$ positive, ranging from 1 to 0 as ξ^2 varies from zero to infinity ($\xi^2 a(\xi^2)$ goes to a finite limit). Hence β behaves as ξ^2 for small ξ and as $|\mathbf{h}|\xi^2$ for large ξ. The magnetization per site is given by

$$\mathbf{M} = \frac{1}{nN}\frac{\partial}{\partial \mathbf{h}} \ln Z = \frac{\xi^2}{\beta}\mathbf{h} \tag{130}$$

At β_c, we find from equation (129)

$$\xi^6 \mathbf{h}^2 = \beta_c \int \frac{d^d p}{(2\pi)^d} \frac{1}{2\sum_1^d (1 - \cos p_\mu)\left[\xi^{-2} + 2\sum_1^d (1 - \cos p_\mu)\right]} \tag{131}$$

which shows that, if $\mathbf{h} \to 0$, ξ diverges. More precisely if $d > 4$, the r.h.s. has a finite limit as $\xi \to \infty$, hence

$$d > 4 \qquad \beta = \beta_c \qquad M \sim h^{1/3} \qquad \delta = 3 \tag{132}$$

a classical mean field value. If $d = 4$, $\xi^6 h^2 \sim \text{cst} \ln \xi$ and

$$d = 4 \qquad \beta = \beta_c \qquad M \sim \left(h \ln \frac{1}{h}\right)^{1/3} \qquad \delta = 3$$
$$(133)$$

Finally for $d < 4$, one finds

$$d < 4 \qquad \beta = \beta_c \qquad M \sim h^{(d-2)/(d+2)} \qquad \delta = \frac{d+2}{d-2}$$
$$(134)$$

The spherical model therefore offers a simple laboratory to study departures from mean field theory due to fluctuations, between the two critical dimensions two and four. At the marginal upper critical dimension, the appearance of logarithms signals that new phenomena are taking place, and are naturally related to similar properties of Brownian motion. In two dimensions, the correlation length diverges exponentially as one approaches zero temperature, which is a point where a consistent continuum field theory can be reached (the nonlinear σ-model). The investigation of higher order corrections (in $1/n$) is a more tricky matter which relies on techniques similar to those developed in chapter 5.

(i) Obtain the behaviour of the free energy as a function of $\theta = \beta_c - \beta$, and h^2 in the vicinity of the critical point. Writing m^2 for ξ^{-2} one has from (129b)

$$F = F_0 + \frac{n}{2}\left\{\ln\frac{1}{\beta} + 2d\beta - \int \frac{d^d p}{(2\pi)^d} \ln\left[2\sum_1^d (1 - \cos p_\mu)\right] + f\right\}$$

$$f = -\theta m^2 + \frac{h^2}{\beta m^2}$$
$$(135a)$$

$$- \int \frac{d^d p}{(2\pi)^d}\left\{\ln\left[\frac{m^2 + 2\sum_1^d(1 - \cos p_\mu)}{2\sum_1^d(1 - \cos p_\mu)}\right] - \frac{m^2}{2\sum_1^d(1 - \cos p_\mu)}\right\}$$

with a small m^2 behaviour

$$f \underset{m^2 \to 0}{\sim} \theta m^2 + \frac{h^2}{\beta m^2} + \begin{cases} Km^4 & d > 4 \\ Km^4 \ln\frac{1}{m} & d = 4 \\ Km^d. & 2 < d < 4 \end{cases}$$
$$(135b)$$

Here K is a (d-dependent) positive constant, and equation (129a) expresses that F is stationary with respect to m^2, at fixed h and θ. Rederive the critical behaviour for small h, θ and m^2.

(ii) One can apply this same $n \to \infty$ limiting technique to the study of the Lee–Yang edge singularity in a complex external field. Set $\mathbf{h} = h\mathbf{u}$ where $\mathbf{u}$ is a fixed real unit vector. A singularity occurs when the solution $\xi(\beta, h)$ of (129a) becomes multivalued, i.e. when

$$\int \frac{d^d p}{(2\pi)^d} \; \frac{1}{[1 + \xi^2 2 \sum_1^d (1 - \cos p_\mu)]^2} + \frac{2h^2 \xi^2}{\beta} = 0 \qquad (136)$$

One indeed finds a pair of complex conjugate singularities at $\pm i h_c$. The singular part of the free energy is $(h \mp i h_c)^{3/2}$ in all dimensions greater than two. To leading order the critical exponent $\sigma = 1/2$, as in mean field, while for h_c one finds

$$
\begin{aligned}
d > 4 \quad & h_c \sim (\beta_c - \beta)^{3/2} \\
d = 4 \quad & h_c \sim \frac{(\beta_c - \beta)^{3/2}}{\ln\left[1/(\beta_c - \beta)\right]} \\
d < 4 \quad & h_c \sim (\beta_c - \beta)^{\frac{1}{2}(d+2)/(d-2)}
\end{aligned}
\qquad (137)
$$

This seems at first sight to contradict the idea that $d = 6$ is an upper critical dimension for the Lee–Yang edge singularity. However, a closer look at $1/n$ corrections reinstates $d = 6$ as a marginal dimension, where one observes departures from the value $\sigma = 1/2$. In essence, the Lee–Yang singularity has no particular sensitivity to the number of components as one (internal space) direction is singled out by the external field. Consequently, the limit $n \to \infty$ is not particularly suited to reveal the role of fluctuations.

3.4 Corrections to mean field

The mean field approximation has shown the existence of critical points where qualitative as well as quantitative properties are modified. We have encountered continuous transitions characterized by a divergence of the correlation length, where the free energy as well as its derivative, the internal energy (hence the entropy), are continuous. The transition is accompanied by a change of symmetry properties of pure phases. The exact solution of the two-dimensional Ising model, or the spherical model, have indicated that mean field predictions are not reliable in low dimensions, where fluctuations have to be taken into account. The spherical model confirms that, for the family of models under consideration, four is an upper critical dimension. This is the same dimension where the corresponding continuous φ^4 theory is (ultraviolet) renormalizable. These phenomena are related as shall become amply clear in the following.

One is therefore led to inquire about corrections to mean field (and similarly $1/n$ corrections to the large n limit). It is natural to look for a small parameter in which to develop a systematic expansion. The most naive choice is the inverse of the coordination number, or equivalently $1/d$, even though the dimension d only takes integer values. The result will be to take into account self-consistently correlations between second, third,... neighbours. In a similar spirit, one could study corrections to the Bethe approximation, although with different methods. This looks at first far from the desired goal, since it ignores collective long range excitations, which are presumed to be responsible for the deviations of critical exponents from their mean field values in low dimensions. Nevertheless, the method gives us some useful information. It enables one to locate accurately the critical points and gives quantitatively correct values outside the critical region. These data are useful for instance when compared to those of numerical simulations. It does not, however, modify the critical behaviour, which remains in the large distance limit the one of free fields (the Gaussian model). What is required to obtain a handle on long wavelength fluctuation effects, is a nonperturbative method, paradoxically emerging from a clever resummation of the perturbative series.

There are various ways to obtain $1/d$ corrections. Here we will illustrate one of them. In chapter 7, we shall present strong coupling expansions. Also we will limit ourselves to the Ising model, and derive the corrections from a study of fluctuations around a saddle point corresponding to mean field.

3.4.1   Laplace transform

If J_{ij} denotes a symmetric positive definite matrix, we can write the following Laplace transform formula

$$\exp\left(\tfrac{1}{2}\beta \sum_{i,j} \sigma_i J_{ij} \sigma_j\right) = (\det \beta J)^{-1/2} \int \prod_i \frac{dH_i}{(2\pi)^{1/2}}$$

$$\times \exp\left\{-\frac{1}{2\beta}\sum_{i,j} H_i (J^{-1})_{ij} H_j + \sum_i H_i \sigma_i\right\}$$

(138)

If the left-hand side is to be identified with a Boltzmann weight for the Ising model, we seem to encounter a difficulty. Choosing J_{ij} equal to 1 or 0 if sites are neighbours or not ($J_{ii} = 0$) leads to an operator with a symmetric spectrum of eigenvalues and the r.h.s. looks meaningless. This is reflected in the fact that the Gaussian integral is not convergent, nor is the square root of the determinant uniquely defined. This is however not a serious objection. Since $\sigma_i^2 = 1$, this can be corrected by shifting J_{ij} by a multiple of the identity at the price of multiplying all Boltzmann weights by a common factor. As we are only going to use the representation to perform an expansion around a saddle point, this shift would only complicate the expressions without producing any new information. We leave it to the careful reader to correct for this loose use of mathematics, and we proceed with (138) as it stands. Summing over $\sigma_i = \pm 1$ and dividing (conventionally) by 2^N, we obtain the Ising partition function in an external field h_i in the form

$$Z(\beta, h) = (\det \beta J)^{-1/2} \int \prod_i \frac{dH_i}{(2\pi)^{1/2}}$$

$$\times \exp\left\{-\frac{1}{2\beta}\sum_{i,j} H_i (J^{-1})_{ij} H_j + \sum_i \ln\cosh(H_i + h_i)\right\}$$

(139)

The r.h.s. describes an assembly of independent spins interacting with a random (displaced) field H_i with a Gaussian distribution. We note that the weight of this probability measure involves β^{-1} instead of β. The quantities H_i are continuous variables, and the saddle point method leads at once to the mean field equations

$$H_i = \beta \sum_j J_{ij} \tanh(H_j + h_j) \qquad (140)$$

We expand the action in equation (139) around a solution $\bar{H}_i$, and treat the higher order terms in $(H_i - \bar{H}_i)$ perturbatively. Terms in $\ln Z$ can be classified according to the number of loops in the corresponding Feynman diagrams. The perturbative technique will be elaborated in chapter 5. For simplicity, we assume a constant external field h, in which case the uniform solution $\bar{H}$ is given by

$$\bar{H} = 2\beta d \tanh(\bar{H} + h) \qquad (141)$$

To zeroth order (also called *tree level*), the free energy is given by the familiar expression

$$F_0 = \ln \cosh(\bar{H} + h) - \beta d \left[\tanh(\bar{H} + h)\right]^2 \qquad (142)$$

with $\bar{H}$ a function of β and h, according to equation (141). The magnetization is, to this order,

$$M_0 = \frac{d}{dh} F_0(\bar{H}(h,\beta), h, \beta) = \tanh(\bar{H} + h) \qquad (143)$$

Now perform in (139) the shift $H_i \rightarrow H_i + \bar{H}$ and expand the action to second order in H_i. With $Z_0 = e^{NF_0}$, we find

$$\frac{Z}{Z_0} = (\det \beta J)^{-1/2}$$

$$\int \prod_i \frac{dH_i}{(2\pi)^{1/2}} \exp -\tfrac{1}{2} H \times \left[(\beta J)^{-1} - \frac{1}{\cosh^2(\bar{H} + h)}\right] H$$

$$= \left\{ \det \left[1 - \frac{1}{\cosh^2(\bar{H} + h)} \beta J\right] \right\}^{-1/2} \qquad (144)$$

The singular (ill-defined) quantity $(\det \beta J)^{-1/2}$ has disappeared, and the one loop correction to the free energy, call it F_1, reads

$$F_1 = -\frac{1}{2} \int \frac{d^d q}{(2\pi)^d} \ln \left[1 - \frac{2\beta}{\cosh^2(\bar{H} + h)} \sum_1^d \cos q_\mu\right]$$

$$= \frac{\beta^2 d}{2 \cosh^4(\bar{H} + h)} + O\left(\frac{\beta^4 d^2}{\cosh^6(\bar{H} + h)}\right) \qquad (145)$$

The expansion coefficient, corresponding to the topological number of loops, is analogous to $\hbar$ in the WKB approximation of quantum mechanics.

We can also reorganize these corrections to obtain a $1/d$ expansion. The expressions become more and more cumbersome as we take into account higher order terms in $(H_i - \bar{H})$ in the action. Let us confine ourselves here to the above one loop correction. For simplicity, write H for $\bar{H} + h$. Then the leading term (142) is

$$F_0 = V_0(H) + h \tanh H \tag{146}$$
$$V_0(H) = \ln \cosh H - H \tanh H + \beta d \tanh^2 H$$

The variables H and h are related by the saddle point condition

$$\frac{\partial}{\partial H}(V_0(H) + h \tanh H) = 0 \tag{147a}$$

and to this order the magnetization is given by

$$M_0 = \frac{\partial}{\partial h}(V_0(H) + h \tanh H) \tag{147b}$$

The first order correction

$$V_1(H) = -\frac{1}{2} \int \frac{d^d q}{(2\pi)^d} \ln \left[1 - \frac{2\beta}{\cosh^2 H} \sum_1^d \cos q_\mu \right] \tag{148}$$

allows one to write the free energy as

$$F = V_0 + V_1 + h \tanh H$$
$$dF = \left(\tanh H + \frac{dV_1}{dH} \frac{dH}{dh} \right) dh \tag{149}$$

where we have taken into account the lowest order saddle point condition (147a). Therefore, up to first order, the magnetization is given by

$$M = \tanh H + \frac{dV_1}{dH} \frac{dH}{dh} \tag{150}$$

It is now natural to perform a *Legendre transformation*, which amounts to using M instead of H as a variable. Expanding H in terms of M up to first order again, we have

$$H = \text{arctanh}\, M + H_1 + \cdots$$
$$\tanh H = M - \frac{dV_1}{dH} \frac{dH}{dh} \bigg|_{H = \text{arctanh}\, M} + \cdots \tag{151}$$

Set also

$$V(M) = -F + hM \tag{152}$$

with $V(M)$ called the effective potential, and

$$dV = h\,dM \tag{153}$$

Thus to leading order in the corrections

$$
\begin{aligned}
V = & -V_0(H) - V_1(H) + h\frac{dV_1}{d\tanh H}\frac{d\tanh H}{dh} + \cdots \\
= & -[V_0(H) + V_1(H)]|_{H=\operatorname{arctanh} M} \\
& + \left(\frac{dV_0}{d\tanh H}\frac{dV_1}{d\tanh H} + h\frac{dV_1}{d\tanh H}\right)\frac{d\tanh H}{dh}
\end{aligned}
$$

By virtue of the fact that $dV_0/d\tanh H = -h$, the last term drops out, and we are left with

$$V(M) = -(V_0(H) + V_1(H))|_{H=\operatorname{arctanh} M} \tag{154}$$

Equation (153) expresses the fact that, in the absence of an external field h, the effective potential should be stationary. The zeroth order calculation suggests in fact that V should be at a minimum, if corrections are indeed sensible (that is small enough). This is the reason for the extra minus sign introduced in equations (152) and (154). One expects β_c to be defined by the condition that this minimum is for $M = 0$ as long as $\beta < \beta_c$, and that, beyond this point, $V(M)$ develops two new minima at $\pm M_{sp}$.

We can rewrite V as a function of M using

$$
\begin{aligned}
V_0(M) &= 2\beta d\frac{M^2}{2} - \frac{1}{2}\Sigma_\pm(1 \pm M)\ln(1 \pm M) \\
&= (2\beta d - 1)\frac{M^2}{2} - \frac{M^4}{12} + \cdots
\end{aligned}
$$

$$
\begin{aligned}
V_1(M) &= -\frac{1}{2}\int\frac{d^d q}{(2\pi)^d}\ln\left[1 - 2\beta(1 - M^2)\sum_1^d(1 - \cos q_\mu)\right] \\
&= -\frac{1}{2}\int\frac{d^d q}{(2\pi)^d}\ln\left[1 - 2\beta\sum_1^d(1 - \cos q_\mu)\right] \\
&\quad - M^2\int\frac{d^d q}{(2\pi)^d}\frac{\beta\sum_1^d(1 - \cos q_\mu)}{1 - 2\beta\sum_1^d(1 - \cos q_\mu)}
\end{aligned}
$$

$$+ M^4 \int \frac{\mathrm{d}^d q}{(2\pi)^d} \left[\frac{\beta \sum_1^d (1 - \cos q_\mu)}{1 - 2\beta \sum_1^d (1 - \cos q_\mu)} \right]^2 + \cdots \quad (155)$$

The expansion has coefficients which are singular at the original critical point $2\beta_c^{(0)} d = 1$, as is expected in a perturbative method. In any case, the origin is an extremum of $V(M)$, but new minima show up when the coefficient of the M^2 term vanishes, i.e. for a new value β_c given consistently to this order, by

$$2\beta_c d = \int \frac{\mathrm{d}^d q}{(2\pi)^d} \frac{1}{1 - 2d\beta_c \dfrac{1}{d} \displaystyle\sum_1^d (1 - \cos q_\mu)} = G(0, 2d\beta_c) \quad (156)$$

where G is the familiar Green's function for Brownian motion. We note the similarity with the large n result – equations (102) and (103) – which was $2d\beta_c = G(0, 1)$. From chapter 1, $G(0, 1)$ admits an expansion of the form $1 + 1/2d + \cdots$ hence the leading correction in both cases is

$$2d\beta_c = 1 + \frac{1}{2d} + \cdots \quad (157)$$

The sign is in agreement with the idea that the effect of fluctuations is to decrease the critical temperature, hence to increase β_c.

The perturbative expansion can be carried beyond quadratic terms. Show that for an $O(n)$ symmetric model the following d^{-1} expansion holds (Fisher and Gaunt, Abe)

$$\frac{1}{2d\beta_c} = 1 - \frac{1}{2d} - \frac{1}{(2d)^2} \left(2 - \frac{2}{n+2} \right) - \frac{1}{(2d)^3} \left(7 - \frac{8}{n+2} \right) + \cdots$$
$$(158)$$

For $n \to \infty$ we recover the result (102)–(104), namely $1/2d\beta_c = 1 - \Pi(0)$, where $\Pi(0)$ is the probability that a Brownian particle returns to the origin, which becomes 0 as $d \to 2$. One therefore expects the series (158) to become singular for $d = 2$, but this remains an open problem. In particular, $d = 4$ should not be a singularity. We also notice that the value $n = -2$ seems to play a special role. Such a value corresponds to a free complex fermionic model, provided the continuation in the parameter n is adequately defined.

Let us now compute the susceptibility in the high temperature phase

$$\beta < \beta_c \qquad\qquad \chi^{-1} = \frac{\partial^2 V}{\partial M^2}\bigg|_{M=0} \qquad\qquad (159)$$

The previous calculation yields the location of the zero β_c of χ^{-1}. By subtracting the two expressions, we get, by approximating β_c by its leading order value $1/2d$ in the denominator,

$$[2d(\beta_c - \beta)\chi]^{-1} = 1 - \int \frac{d^d q}{(2\pi)^d}$$

$$\times \frac{\frac{1}{d}\sum_1^d (1 - \cos q_\mu)}{\left[1 - \frac{1}{d}\sum_1^d (1 - \cos q_\mu)\right]\left[1 - 2\beta \sum_1^d (1 - \cos q_\mu)\right]} + \cdots \quad (160)$$

For $d > 4$, as $\beta \to \beta_c \sim 1/2d$, the integral converges, and the critical exponent γ is 1. But for $d \le 4$, the (infrared) effect of fluctuations is catastrophic in this perturbative scheme. The integral develops logarithmic singularities at $d = 4$, and the higher order terms have an even more divergent behaviour. Thus one finds a criterion of upper critical dimensionality (Landau–Ginzburg) telling us how the mean field approximation breaks down under the effect of long wavelength fluctuations. These divergences will be resummed by the renormalization group method to produce the expected finite deviations to critical exponents.

One could proceed to compute various $1/d$ corrections for other quantities above $d = 4$, and one can also generalize the method to other models characterized by one or several order parameters. Other strong coupling expansions are also available.

We will also have to return to the implementation of mean field theory when dealing with systems which do not admit a local order parameter, the typical cases being gauge field models.

Effect of an inhomogeneous external field. Let F denote now the total free energy. Up to one loop order

$$F = V_0 + V_1 + \sum_i h_i \tanh H_i$$

$$V_0 = \sum_i \ln \cosh H_i - H_i \tanh H_i + \tfrac{1}{2}\beta \sum_{i,j} (\tanh H_i) J_{ij} (\tanh H_j)$$

$$V_1 = -\tfrac{1}{2} \operatorname{Tr} \ln \left[\delta_{ij} - \beta J_{ij}(1 - \tanh^2 H_j)\right] \qquad (161)$$

The relation between H and h is given by

$$\frac{\partial}{\partial H_i}\left\{V_0 + \sum_i h_i \tanh H_i\right\} = 0 \tag{162}$$

Let M_i denote the derivative of F with respect to h_i

$$M_i = \frac{d}{dh_i}F(H_i(h), h_i) \tag{163}$$

then the Legendre transform Γ as a functional of M_i is defined through

$$\Gamma(M) + F(h) = \sum_i h_i M_i \tag{164}$$

and it generalizes the effective potential (which appears as the density of Γ if M is homogeneous). Up to one loop,

$$\begin{aligned}
\Gamma(M) &= -\left.(V_0 + V_1)\right|_{M_I = \tanh H_i} \\
&= -\tfrac{1}{2}\beta \sum_{ij} M_i J_{ij} M_j + \tfrac{1}{2}\sum_{i,\pm}(1 \pm M_i)\ln(1 \pm M_i) \\
&\quad + \tfrac{1}{2}\operatorname{Tr}\ln\left[\delta_{ij} - \beta J_{ij}(1 - M_j^2)\right]
\end{aligned} \tag{165}$$

and h_i is then obtained as

$$\frac{\partial \Gamma}{\partial M_i} = h_i \tag{166}$$

Generalize the Legendre transformation to all orders.

Before presenting the general analysis of the perturbation series, we will devote the next chapter to general considerations on scaling and models of renormalization.

Notes

Mean field is almost as old as statistical mechanics. A standard reference is W.L. Bragg, E.J. Williams, *Proc. Roy. Soc.* **A145**, 699 (1934). Corrections to the Clausius Mossotti formula for dense systems discussed in J. De Boer, F. Van der Maesen and C.A. Ten Seldam, *Physica* **19**, 265 (1953) and G. Stell and G.S. Rushbrooke, *Chem. Phys. Lett.* **24**, 531 (1974) are based on the formalism developed by J.G. Kirkwood, *J. Chem. Phys.* **4**, 592 (1936) and J. Yvon, *Actualités scientifiques et industrielles*, **543**,

Hermann, Paris (1937). The case of polar media was investigated by L. Onsager, *J. Am. Chem. Soc.* **58**, 1496 (1936).

For an overall view on critical phenomena as well as their physical context, see H.E. Stanley *Introduction to Phase Transitions and Critical Phenomena*, Clarendon Press, Oxford (1971). The Landau approach is also discussed extensively in L.D. Landau and E.M. Lifshitz *Statistical Physics*, third edition revised by E.M. Lifshitz and L.P. Pitaevskii, Pergamon Press, Oxford (1980). The Bethe approximation is introduced in H. Bethe, *Proc. Roy. Soc.* **A-216**, 45 (1935). See also G.S. Rushbrooke and H.I. Scoins, *Proc. Roy. Soc.* **A-230**, 74 (1953). Zero modes arising in a model with a spontaneous broken continuous symmetry are discussed in J. Goldstone, *Nuovo Cimento*, **19**, 154 (1961). Analogies between the mechanism of superconductivity and particle mass generation, a subject not covered here, is found in Y. Nambu and G. Jona-Lasinio, *Phys. Rev.* **122**, 345 (1961) and **124**, 246 (1961). An influential paper in the context of particle physics, introducing the σ-model, is the one by M. Gell-Mann and M. Lévy, *Nuovo Cimento* **16**, 705 (1960). Various aspects of symmetry breaking, and much more, are discussed in depth in the collected Erice lectures of S. Coleman *Aspects of Symmetry*, Cambridge University Press, Cambridge (1985).

The investigation of complex zeroes is the work of C.N. Yang and T.D. Lee, *Phys. Rev.* **87**, 404,410 (1952). Our presentation is inspired by D. Ruelle, *Phys. Rev. Lett.* **26**, 303 (1971), *Comm. Math. Phys.* **31**, 265 (1973). For further developments see J.D. Bessis, J.-M. Drouffe, P. Moussa, *J. Phys.* **A9**, 2105 (1976) and D.A. Kurtze, M.E.Fisher, *Phys. Rev.* **B20**, 2785 (1979). The zeroes in the complex temperature plane are but one minor item in the Boulder Lectures by M.E. Fisher, in *Lectures in Theoretical Physics*, **VII-C**, edited by W.E. Brittin, University of Colorado Press, Boulder (1964). See also C. Itzykson, R.B. Pearson, J.-B. Zuber, *Nucl. Phys.* **B220 (FS8)** 415 (1983), and numerical studies by E. Marinari, Nucl. Phys. **B235[FS11]**, 123 (1984), and G. Bhanot, R. Salvador, S. Black, P. Carter and R. Toral, *Phys. Rev. Lett.* **59**, 803 (1987).

The spherical model was solved by T.H. Berlin and M. Kac, *Phys. Rev.* **86**, 821 (1952). For the connection with the large n-expansion, see H.E. Stanley, *Phys. Rev.* **176**, 718 (1968). The Coleman lectures quoted above contain a review as well as further developments initiated by G. 't Hooft for models involving matrix valued fields, to be discussed in chapter 10.

The large d expansions quoted in the last section are from M.E. Fisher and D.S. Gaunt, *Phys. Rev.* **133A**, 224 (1964) and R. Abe, *Prog. Theor. Phys.* **47**, 62 (1972).

We have not treated the question of defects in ordered media, a legitimate part of the physics of phase transitions. Numerous aspects of this fascinating subject are covered in the book of M. Kléman, *Points, Lines and Walls in Liquid Crystals, Magnetic Systems and Various Ordered Media*, John Wiley and Sons, New York (1983), including references to many original works. For a review of the topology of defects, see N.D. Mermin, *Rev. Mod. Phys.* **51**, 591 (1979) and L. Michel, *Rev. Mod. Phys.* **52**, 617 (1980).

4

SCALING
TRANSFORMATIONS
AND THE XY-MODEL

Condensed media show a large variety of critical phenomena, ranging from critical opalescence at the end point of the liquid–gas coexistence curve, the Curie transition of ferromagnetic materials, to the superfluid transition of helium, the behaviour of solutions of polymers, the conductivity of random media, ... To this list should be added systems with local symmetries, such as those suggested by particle physics in order to understand quark confinement. Surprisingly all these phenomena can be classified into a few universality classes, characterized by specific large distance behaviour with the same critical exponents. In this chapter, we sketch the methods and ideas of the renormalization procedure. We illustrate the concepts with simple approximations in the language of classical spin models in the first section, treating in more detail the XY-model as an example in the second section.

4.1 Scaling laws. Real space renormalization

4.1.1 Homogeneity and scale invariance

The discussion given in previous chapters suggests that, close to a continuous transition, critical systems exhibit universal properties. Correlations at large distance are not sensitive to the details of microscopic interactions. Their behaviour is described by a specific dimensional analysis governed by some essential characteristics of the system, such as the dimension of space, the nature of the order parameter and the underlying symmetries.

The mean field approximation gave a first idea of a simple critical behaviour. Fluctuations have only a quantitative effect in dimension greater than the upper critical one. This dimension is generally equal to four for spin systems, and the role of fluctuations becomes nontrivial in four and lower dimensions. With T_c being the critical temperature, we use as a control parameter the reduced variable

$$\theta = \left(\frac{T}{T_c} - 1\right) \tag{1}$$

In the language of field theory, the bare mass squared is proportional to θ. The domain $\theta > 0$ corresponds to a symmetric disordered phase, and, for $\theta < 0$, the symmetry is spontaneously broken. In this case, any uniform boundary condition or any infinitesimal uniform external field allows one to isolate a pure phase in which a given direction for the spontaneous magnetization has been singled out among all those equivalent with respect to the symmetry group. A continuous transition is characterized by a spontaneous magnetization vanishing continuously at $\theta = 0$. In this phase, there generally remains a reduced subgroup of symmetry, the isotropy group around this privileged direction. The size of the domain near $\theta = 0$ in which the spacing a, or more generally the interaction range, can be neglected with respect to the correlation length ξ depends on the microscopic parameters, and can be therefore difficult to estimate in a concrete case. Our hypothesis is $\xi \gg a$. This correlation length can be visualized (for a discrete symmetry group) by considering a typical configuration near $\theta = 0$, and estimating the size of clusters of spins with the same orientation. At the critical point, ξ is infinite, and this length scale itself disappears. We also assume that correlations become isotropic near the critical point.

In order to simplify the reasoning, let us consider a scalar field and let us measure lengths in terms of a microscopic spacing a. Ignoring anisotropies, a correlation function

$$G(\mathbf{x}) = \langle \varphi(\mathbf{x})\varphi(\mathbf{0}) \rangle \tag{2}$$

satisfies *a priori*

$$\frac{G(r_2)}{G(r_1)} = \gamma\left(\frac{r_2}{r_1}, \frac{r_1}{a}\right)$$

with $r = |\mathbf{x}|$. The aim of the theory is to justify the existence of the critical limit $a \to 0$ for fixed r_1 and r_2, in which case

$$\frac{G(r_2)}{G(r_1)} = \gamma\left(\frac{r_2}{r_1}\right) \qquad r_1 \gg a \quad r_2 \gg a \tag{3}$$

The group law

$$\frac{G(r_3)}{G(r_1)} = \frac{G(r_3)}{G(r_2)} \times \frac{G(r_2)}{G(r_1)}$$

requires that $\gamma(\rho)$ be a homogeneous function equal to unity for $\rho = 1$. In other words, one should have

$$G(r_2) = G(r_1) \left(\frac{r_1}{r_2}\right)^{d-2+\eta} \qquad \text{at } \theta = 0 \qquad (4)$$

with the conventional definition of the critical exponent η. This exponent vanishes in the mean field approximation, and thus measures the deviation from the Gaussian model. Hence $\eta \neq 0$ corresponds to a nontrivial scale invariant field theory. In the disordered phase $\theta > 0$, very close to the critical point (so that $\xi \gg a$), we can distinguish three domains of lengths r, large with respect to the lattice spacing or to any short-distance characteristic scale

$$\begin{cases} (i) & a \ll r \ll \xi \\ (ii) & a \ll r \sim \xi \\ (iii) & a \ll \xi \ll r \end{cases}$$

In the first domain, the preceding considerations remain valid since the correlation length can be considered as infinite. This region is equivalent to the ultraviolet region of a renormalizable field theory, as will be shown in the next chapter. The second region is a delicate zone of transition, where one begins to deviate from a homogeneous behaviour. Finally, correlation functions decrease exponentially in the third region, diverging more and more from the scale invariant regime. The temperature, characterized by the parameter θ, is an essential or relevant parameter for the large-distance (infrared) behaviour. This behaviour shows an instability as θ departs from 0. In other words, as the distance increases, the system becomes more and more sensitive to a deviation from the critical temperature.

These considerations are summarized by the expression

$$G(r, \theta) \sim \frac{1}{r^{d-2+\eta}} g\left(\frac{r}{\xi(\theta)}\right) \qquad (5)$$

The function g is regular at the origin, up to logarithmic terms in marginal cases, while it decreases exponentially at infinity.

As θ vanishes, the mean field approximation suggests a divergence of the correlation length according to a power law

$$\xi(\theta) \sim \xi_+ \theta^{-\nu} \qquad \text{as } \theta \to 0 \qquad (6)$$

which introduces a second exponent ν. The quantity ξ_+ is of course dependent on the microscopic structure of the interactions. The hypothesis (6) allows one to relate the length (or mass) scale to the temperature (in field theoretic language, the physical mass to the bare mass). Landau's theory predicts $\nu = \frac{1}{2}$, while $\nu = 1$ for the two-dimensional Ising model.

Generalizing these ideas, one is naturally led to a series of scaling laws for the various physical quantities, stemming from the *scaling hypothesis* that ξ is the only relevant intrinsic scale length. Let us consider the free energy per unit volume. It is expected that fluctuations contribute to its singular part as the ratio of the unit volume to the volume $\xi(\theta)^d$

$$F_{\text{sing}}(\theta) \quad \sim \quad \frac{1}{\xi(\theta)^d} \quad \sim \quad \theta^{\nu d} \qquad (7)$$

This is the so-called *hyperscaling hypothesis*. As a consequence, the exponent α of the singular part of the specific heat C, proportional to the second derivative of F, is given by

$$\begin{aligned} C(\theta) \quad &\sim \quad \theta^{-\alpha} \\ \alpha &= 2 - \nu d \end{aligned} \qquad (8)$$

A positive value of α implies a divergence of the specific heat, whereas a negative value corresponds to a cusp in its representative curve as a function of temperature.

In order to justify equation (7), Pippard and Ginsberg gave the following argument. Suppose that the free energy per unit volume has a singular part behaving as $\theta^{2-\alpha}$. At a temperature greater than the critical temperature ($\theta > 0$), a fluctuation corresponding to a departure from an ordered state in a range ξ produces an increase $\Delta F \sim \theta^{2-\alpha} \xi^d$ of the free energy. The probability $e^{-\Delta F}$ of such a fluctuation becomes negligible when $\Delta F \simeq 1$, and one gets the estimate $\xi \sim \theta^{-(2-\alpha)/d}$.

The logarithmic singularity of the specific heat encountered in the Ising model is interpreted as $\alpha = 0$. This is in agreement with (7), according to which $\nu = 1$. Similarly, the spherical model verifies this relation for $2 < d < 4$, with

$$\alpha = \frac{d-4}{d-2} \qquad \nu = \frac{1}{d-2} \tag{9}$$

Relation (7) is violated by the mean field approximation where we found $\alpha = 0$ and $\nu = \frac{1}{2}$, independently of d. The above reasonning is indeed incompatible with the one leading to Landau's approximation, where fluctuations are ignored. To estimate the upper critical dimension d_c, we can nevertheless use these mean field values to predict that fluctuations, which are negligible in large dimensions, should become important as the dimension d decreases below d_c given by $\alpha_{mf} = 2 - \nu_{mf}d_c$, i.e. $d_c = 4$. We recover here the Landau–Ginzburg criterion for the upper critical dimension.

The susceptibility, which is the second derivative of the free energy with respect to the external field when this external field vanishes, is proportional to the integral of the two-point correlation function. Using (5), one deduces that

$$\chi(\theta) = \int_{|\mathbf{x}|<\xi(\theta)} d^d\mathbf{x} \frac{1}{|\mathbf{x}|^{d-2+\eta}} \sim \xi(\theta)^{2-\eta} \tag{10}$$

The corresponding critical exponent denoted γ is therefore given by

$$\chi(\theta) \sim \theta^{-\gamma}$$
$$\gamma = \nu(2-\eta) \tag{11}$$

In the presence of an external field H, the transition disappears. The magnetization $M(H,\theta)$ vanishes linearly with H for $\theta > 0$, with a slope $\chi(\theta)$. For $\theta < 0, M(H,\theta)$ tends towards the spontaneous magnetization $M(\theta)$ when $H \rightarrow +0$. As one approaches the critical temperature from below, this spontaneous magnetization vanishes as

$$M(\theta) \sim (-\theta)^\beta \quad \text{as } \theta \rightarrow -0 \tag{12}$$

At the critical temperature, the magnetization is singular when H vanishes, with an exponent δ^{-1}

$$M(H,\theta = 0) \sim H^{1/\delta} \quad \text{as } H \rightarrow 0 \tag{13}$$

Relations (12) and (13) suggest that the quantities H, M^δ and $(-\theta)^{\beta\delta}$ are related. The *equation of state* to be derived in the next subsection expresses $HM^{-\delta}$ as a regular function of $\theta M^{-1/\beta}$

$$H = M^\delta f_H(\theta M^{-1/\beta}) \qquad (14)$$

The first negative zero of f_H corresponds to the spontaneous magnetization, and its value at zero yields the proportionality coefficient in equation (13). For $\theta > 0, M$ and H are linearly related when they vanish simultaneously, $M = \chi(\theta)H$. Hence the function $f_H(z)$ behaves as $z^{\beta(\delta-1)}$ for large z. This yields $H \sim M\theta^{\beta(\delta-1)}$. Comparing with (11), we obtain the relation

$$\gamma = \beta(\delta - 1) \qquad (15)$$

This illustrates the general method for relating singular quantities through regular relations. If the specific heat singularity is the same above and below the critical temperature, the equation of state (14) can be reinterpreted as a relation for the singular part of the free energy expressed in terms of H and θ. For $\theta > 0$, one writes

$$F_{\text{sing}}(\theta, H) \quad \sim \quad \theta^{2-\alpha} f_F(H\theta^{-\beta\delta}) \qquad (16)$$

the second derivative with respect to H gives the susceptibility, and hence

$$\gamma = 2\beta\delta + \alpha - 2 \qquad (17)$$

The relations (15) and (17) allow the determination of β and δ.

To summarize, the two critical exponents, η for the behaviour of the Green function at the critical point, and ν for the correlation length in the critical domain yield the four other exponents (using the scaling hypothesis) which are

$$
\boxed{
\begin{aligned}
\alpha &= 2 - \nu d \\
\beta &= \tfrac{1}{2}\nu(d - 2 + \eta) \\
\gamma &= \nu(2 - \eta) \\
\delta &= \frac{d + 2 - \eta}{d - 2 + \eta}
\end{aligned}
}
\qquad (18)
$$

The goal of the renormalization group is to justify the above relations and to provide a tool for computing critical exponents and other universal quantities, which can also be obtained from perturbative expansions at low or high temperature. It seems that these two different approaches, including other numerical methods (such as finite size scaling, Monte-Carlo renormalization,...) are all in reasonable agreement. Delicate experiments confirm the theoretical predictions.

Let us add a remark concerning the Lee–Yang singularity. The magnetization is a regular function of the temperature in the real domain $H \neq 0$, However, its analytic continuation for complex H is singular (for $\theta > 0$) on the imaginary axis, starting at $\pm iH_c$. The value H_c can be estimated from equation (16) in the critical region

$$H_c \sim \theta^{\beta\delta} = \theta^{\beta+\gamma} \qquad (19)$$

For d large, $\beta = \frac{1}{2}, \delta = 3$ and $\gamma = 1$ and one recovers $H_c \sim \theta^{3/2}$. On the other hand, the nature of the singularity cannot be obtained from the behaviour in the critical domain. Another length scale is involved, which becomes infinite at the Lee–Yang point. Relation (19) only expresses a compatibility relation between the two singularities.

4.1.2 Recurrence relations in real space

The behaviour of systems near the critical point can be analyzed by studying the effect of a change in scale (dilatation). This is the essential point elaborated by the various methods of renormalization, which will lead us, in the framework of field theory, to the Callan–Symanzik equations.

We have just established relations between critical exponents, using directly the observables and assuming only homogeneity. To study a scaling transformation, we will proceed at first in configuration space, although the sophisticated methods of the continuous theory use momentum space. We start from the elementary spacing (characterizing the finite range of the forces). In order to implement a change in scale and to erase progressively the microscopic details, there exist several techniques (decimation, block variables, ...). None of these methods is completely satisfactory, although they offer the simplest approach conceptually.

On the example of the Ising model, we present first the block spin technique introduced by Kadanoff. Let us consider a new lattice with spacing λa, λ being an integer, grouping the spins in clusters. To each of these blocks, we associate a new dichotomic variable σ_b For instance, one can choose the majority rule

$$\sigma_b = \text{Sign} \sum_{i \in b} \sigma_i \qquad (20)$$

with an additional convention in the case where this sum vanishes (when this can happen). If the correlation length is large with respect to the elementary spacing, one expects that fluctuations within a given block are inessential, as long as λa is much smaller than ξ, In the original variables, the statistical weight e^S is given in terms of the action

$$S_a = \beta \sum_{(i,j)} \sigma_i \sigma_j + H \sum_i \sigma_i \qquad (21)$$

with H small and β near the critical point. One wants to define a new action $S_{\lambda a}$ for the block variables, leading of course to the same partition function. This action is formally simple to construct by performing the constrained sum

$$e^{S_{\lambda a}} = \sum_{\{\sigma_i = \pm 1\}} \left[\prod_b \delta \left(\sigma_b - \text{Sign} \sum_{i \in b} \sigma_i \right) \right] e^{S_a} \qquad (22)$$

It is clear that the partition function is invariant

$$Z = \sum_{\{\sigma_b = \pm 1\}} e^{S_{\lambda a}} = \sum_{\{\sigma_i = \pm 1\}} e^{S_a} \qquad (23)$$

In contradistinction with the decimation method (where some initial variables are held fixed while others are summed over), the block variables cannot be identified to some of the original ones. Hence, correlations are not *a priori* invariant in a change of scale $a \to \lambda a$.

The equation (22) leads to an intricate form for $S_{\lambda a}$, and the constrained summation is almost as difficult to perform as the original one. Nevertheless, $S_{\lambda a}$ has the same global symmetry as the initial system. Moreover, if the parameter λ could vary continuously, one would expect that $S_{\lambda a}$ is similar to S_a in the vicinity of $\lambda = 1$. On the lattice, the transformation $S_a \to S_{\lambda a}$ is difficult to study without additional hypotheses or approximations. Simplifying drastically and using the idea of universality, we assume that, up to a constant, $S_{\lambda a}$ is well approximated by nearest neighbour interactions and a linear symmetry breaking term

$$S_{\lambda a} \cong \beta_\lambda \sum_{(b,b')} \sigma_b \sigma_{b'} + H_\lambda \sum_b \sigma_b + S_0 \qquad (24)$$

Therefore, interactions between next-to-nearest neighbours or farther away sites, as well as interactions between more than two sites (including odd symmetry breaking terms) are neglected. This will be justified later, showing that these interactions become inessential after several iterations of the rescaling transformation. In a general reasoning, they should however be included, and thus the initial action might as well be assumed to depend on an infinite number of parameters. For our present purposes, we restrict ourselves to the approximation (24). Even more, in a first stage, we suppress the external field H, so that H_λ also vanishes. Hence, up to the constant term S_0, the block variables lead to a new Ising model, with nearest neighbour interactions on a new length scale. The transformation reduces to

$$\begin{cases} \mathbf{x} \to \mathbf{x}_\lambda = \mathbf{x}/\lambda \\ \theta \to \theta_\lambda = \varphi(\lambda, \theta) \end{cases} \qquad (25)$$

which expresses that the physical quantities of the initial model relative to distances x and temperature θ are related to the transformed model with distances x_λ and temperature θ_λ. The additive quantity S_0 and the transformation law $\theta_\lambda = \varphi(\lambda, \theta)$ are regular functions of the initial parameters. This is an essential point, the renormalization transformation does not deal with singular functions.

   Relation (23) expresses the conservation of the partition function and allows the determination of the free energy per site

$$\lambda^d F(\theta) = F(\theta_\lambda) + F_0(\theta) \qquad (26)$$

where $F_0(\theta)$ is a regular function of θ depending smoothly on λ. Note that the same function F appears on both sides. This is due to the fact that renormalization preserves the structure of the model (with modified parameters). Here this relation is only approximate since the number of parameters has been limited; it becomes exact, but almost useless, in the infinite dimensional space of parameters describing all short–range interactions.

   The effect of iterations of the transformation can be analyzed qualitatively. When the temperature differs from the critical tem-

Fig. 1   Instability of the critical point with respect to the temperature.

perature, the block size increases and approaches the correlation length. Hence, in units of block lattice spacing, the new corre- lation length decreases, so that the deviation from the critical temperature increases. In particular, if $\beta < \beta_c$, β_λ decreases to zero, the high temperature attractive fixed point, as λ increases indefinitely. A similar scheme is valid for $\beta > \beta_c$, where renor- malization drives the effective temperature to zero. Recall that, in the ordered phase, ξ describes the range of fluctuations around the spontaneous magnetization. In other words, renormalization amplifies the deviations from the critical temperature, which ap- pears as a nontrivial, unstable infrared fixed point (figure 1). This property allows one to identify with a great sensitivity the critical point.

In the critical domain, assuming that it is possible to consider λ as a continuous parameter, one expects a behaviour near $\lambda = 1$ of the form

$$\theta_\lambda = \lambda^{y_\theta} \theta \tag{27}$$

with $y_\theta > 0$. The correlation length transforms as

$$\xi(\theta_\lambda) = \xi(\theta)/\lambda$$

One can iterate the transformation until $\xi(\theta_\lambda)$ and θ_λ are of order unity which yields $\lambda \sim \theta^{-1/y_\theta}$. Hence

$$\xi \sim |\theta|^{-1/y_\theta} \qquad \longrightarrow \qquad \nu = \frac{1}{y_\theta} \tag{28a}$$

The exponent is given by

$$y_\theta = \frac{1}{\nu} = \frac{\lambda}{\theta_\lambda} \frac{\partial}{\partial \lambda} \theta_\lambda \Big|_{\lambda=1} \tag{28b}$$

again assuming that λ is a continuous variable.

The hypothesis that the renormalization transformation is reg- ular, even in the neighborhood of the critical point, leads to the same exponent ν above and below the critical temperature. Equa- tion (28) is only meaningful if one can perform renormalization operations depending continuously on the dilatation parameter λ.

On a lattice, λ can only take discrete values, and, writing near the critical point $\theta_\lambda = \mu\theta + o(\theta)$, an estimate of y_θ is given by $\ln\mu/\ln\lambda$ with μ the derivative of the transformation at the critical point.

Asuuming in equation (26) that the dependence of $F_0(\theta)$ on λ is not a serious problem, even when λ gets as large as θ^{-1/y_θ} (admittedly a weak point of the argument), the singular part of the free energy satisfies the homogeneous equation

$$\lambda^d F_{\text{sing}}(\theta) = F_{\text{sing}}(\theta_\lambda) \tag{29}$$

and hence, according to (27), if we pick $\lambda \sim \theta^{-1/y_\theta}$,

$$F_{\text{sing}}(\theta) = f_\pm |\theta|^{2-\alpha} \tag{30}$$

which introduces the exponent α, given as before by

$$2 - \alpha = \nu d \tag{31}$$

The exponent α is the same above and below the critical temperature, although the amplitudes f_+ and f_- may differ.

This reasoning can be generalized in the presence of an external field H. Now $\beta_\lambda(\beta, H)$ and $H_\lambda(\beta, H)$ are both functions of β and H, the last one vanishing with H for reasons of symmetry. The critical point is characterized by the vanishing of both θ and H. We have in general to diagonalize the renormalization flow in the vicinity of the fixed point. We call again these variables θ and H, and hence we get in the critical domain

$$\theta_\lambda = \lambda^{y_\theta} \theta$$
$$H_\lambda = \lambda^{y_H} H \tag{32}$$
$$\lambda^d F_{\text{sing}}(\theta, H) = F_{\text{sing}}(\theta_\lambda, H_\lambda)$$

Previous hypotheses and definitions yield

$$F_{\text{sing}}(\theta, H) = |\theta|^{\nu d} F_{\text{sing}}(\pm 1, H/|\theta|^{\nu y_H}) \tag{33}$$

Differentiating with respect to H, one obtains the magnetization and the susceptibility

$$\begin{array}{ll} M \sim (-\theta)^{\nu(d-y_H)} & \beta = \nu(d - y_H) \\ \chi \sim \theta^{-\nu(2-y_H-d)} & \gamma = \nu(2y_H - d) \end{array} \tag{34}$$

In the present scheme, $y_\theta = 1/\nu$ and $y_H = \frac{1}{2}(d + \gamma/\nu)$ are the fundamental exponents which enable us to derive all the others. Combining relations (31) and (34), one gets

$$\alpha + 2\beta + \gamma = 2 \tag{35}$$

By differentiating equation (33) with respect to H, the magnetization can be re-expressed in terms of the field and the temperature as

$$H = M^{\nu y_H/\beta} f(\theta M^{-1/\beta}) \tag{36}$$

Comparing with (14), one gets

$$\delta = \frac{\nu y_H}{\beta} = \frac{y_H}{d - y_H} \tag{37}$$

$$2 - \alpha = \nu d = \beta(1 + \delta)$$

Scaling transformations can also be applied to correlation functions, taking into account the invariance of the partition function. For example, in the domain $\theta \geq 0$, one finds

$$Z^{-1}\delta^2 Z = \sum_{\text{blocks}} \delta H_\lambda(\mathbf{x}_1')\delta H_\lambda(\mathbf{x}_2')G(\mathbf{x}_1' - \mathbf{x}_2', \theta_\lambda, H_\lambda)$$

$$= \sum_{\text{sites}} \delta H(\mathbf{x}_1)\delta H(\mathbf{x}_2)G(\mathbf{x}_1 - \mathbf{x}_2, \theta, H) \tag{38}$$

$$G(\mathbf{x}, \theta, H) = \lambda^{2(y_H - d)}G(\mathbf{x}/\lambda, \theta_\lambda, H_\lambda) \tag{39}$$

The prefactor arises from condition (38) and expresses the change of the unit of length as well as of the scale of the magnetic field. It is interpreted in the framework of field theory as a "wavefunction" renormalization, corresponding to a nontrivial transformation on the spin variables.

When iterating (39), $\mathbf{x}$ decreases and θ and H increase, so that one leaves the critical region. Assuming that correlations are isotropic in this domain, we have

$$G(\mathbf{x}, \theta, H) = \frac{1}{|\mathbf{x}|^{2(d - y_H)}} g(\theta |\mathbf{x}|^{1/\nu}, H |\mathbf{x}|^{y_H}) \tag{40}$$

In the absence of an external field, we recover the interpretation of $\xi \sim \theta^{-\nu}$ as a correlation length. At the critical temperature $\theta = 0$, one has rather $\xi \sim H^{-1/y_H}$. At the critical point $\theta = H = 0$, the correlation function is homogeneous in $|\mathbf{x}|$ and the critical exponent η verifies

$$d - 2 + \eta = 2(d - y_H) \tag{41}$$

or equivalently

$$\nu(2 - \eta) = \nu(2y_H - d) = \gamma \tag{42}$$

Summarizing, the renormalization method in real space introduces two fundamental exponents $y_\theta = 1/\nu$ and y_H which characterize the instability in even (θ) and odd (H) directions in coupling space. All other exponents are related to the previous ones

$$
\begin{aligned}
\alpha &= 2 - \nu d \\
\beta &= \nu(d - y_H) \\
\gamma &= \nu(2y_H - d) \\
\delta &= \frac{y_H}{d - y_H} \\
\eta &= 2 + d - 2y_H
\end{aligned}
\tag{43}
$$

More generally, let us discuss, following Wegner, the behaviour of a set of couplings $\{K_i\}$ under a scale transformation λ which can be taken arbitrarily close to unity. The critical point is assumed to be at $K_i = 0$. Linearizing the transformation near this point, one writes

$$
\lambda \frac{\mathrm{d}}{\mathrm{d}\lambda} K_i(\lambda) \Big|_{\lambda=1} = y_{ij} K_j
\tag{44}
$$

The eigenvectors of the matrix y_{ij} define linear combinations of the interactions, called *scaling* (or *operator*) *fields*. Their coefficient in the action have a simple power law behaviour under the renormalization transformation. They are classified in three categories

 i) *relevant* if the corresponding eigenvalue is positive,
 ii) *irrelevant* if it is negative,
 iii) *marginal* if it vanishes.

In particular, in our example, the deviation from the critical temperature and the magnetic field are *relevant*. Irrelevant couplings tend toward zero by renormalization transformation and can thus be neglected in the critical domain, where they only generate corrections to dominant terms. The case of marginal operators is more interesting and requires a study beyond the linear approximation.

This global approach allows one to sharpen the hypotheses of scale invariance theory when one describes the effects of renormalization in the space of coupling constants (figure 2). The critical manifold corresponds to the vanishing of relevant couplings. The

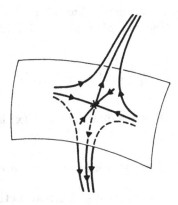

Fig. 2   A pictorial view of the renormalization group trajectories near the critical point.

codimension of this manifold is equal to the number of positive eigenvalues of the matrix y_{ij}. On this manifold, the renormalization transformations result in a contraction. Conversely, starting from a point out of this manifold, renormalization trajectories will finally run away. In other words, the concept of renormalization allows us to concentrate on a few relevant or marginal couplings, thus being intimately tied with the idea of universality.

Generalizing (39) to the correlations of n spins, show that, near the critical point,

$$G(\mathbf{x}_1,\ldots,\mathbf{x}_n;\theta,H) = Z(\lambda)^{n/2} G(\mathbf{x}_1/\lambda,\ldots,\mathbf{x}_n/\lambda;\theta_\lambda,H_\lambda)$$
$$Z(\lambda)^{1/2} = \lambda^{-(d-y_H)} = \lambda^{-(d-2+\eta)/2} \tag{45}$$

Here, $Z(\lambda)$ is the wave renormalization factor and should be distinguished from the partition function. This is again an example of conflicting traditions in what used to be distinct theoretical areas! In the case of a free field, we showed in chapter 1 that the continuous limit allows one to assign the dimension $(d-2)/2$ to the field in terms of wavenumbers (inverse length). The interpretation of the relation (46) is that the corresponding continuous field $\varphi(\mathbf{x})$ has an effective dimension

$$d_\varphi = d - y_H = \frac{d-2}{2} + \frac{\eta}{2} \tag{46}$$

The deviation from the canonical dimension $(d - 2)/2$ is $\eta/2$, the so-called *abnormal dimension of the field* φ. The first equation (45) expresses that

$$\sum_i H_i \sigma_i \rightarrow \int H(\mathbf{x}) \varphi(\mathbf{x}) \mathrm{d}^d \mathbf{x}$$

is dimensionless, i.e. that the dimension of $H(\mathbf{x})$ is y_H.

4.1.3 Examples and approximations

We shall return later to effective numerical methods, and present here some rather drastic approximations designed to illustrate the renormalization transformations. We begin by a trivial case in which the decimation method (integration over a subset of variables) is possible and leads to the exact result.

One-dimensional decimation   We take this opportunity to introduce the Potts model which generalizes the Ising model. Dynamical variables describe now q distinct states at each site. The nearest neighbour interaction energy assumes two possible values, distinguishing the case when the two states are identical (favoured case) from the one when they differ. The Ising model corresponds to $q = 2$. For a one-dimensional lattice of size L, the partition function reads

$$Z_L = \frac{1}{q^L} \sum_{\{\sigma\}} \exp\left(\beta \sum_{n=1}^{L} (2\delta_{\sigma_n, \sigma_{n+1}} - 1)\right) \qquad (47)$$

where the Boltzmann weight is e^β if the neighbouring spins are equal, $e^{-\beta}$ if they differ. The partition function has been normalized to unity for $\beta = 0$. The global symmetry of the system is the permutation group π_q of q objects (with $\pi_2 = Z_2$), and the set of pure states can be identified with the elements of the cyclic group Z_q.

Each factor $\exp \beta(2\delta_{\sigma\sigma'} - 1)$ is expanded as

$$e^{\beta(2\delta_{\sigma\sigma'} - 1)} = \left(\frac{e^\beta + (q - 1)e^{-\beta}}{q}\right) \times 1 + \left(\frac{e^\beta - e^{-\beta}}{q}\right) \times (q\delta_{\sigma\sigma'} - 1)$$

$$(48)$$

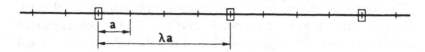

Fig. 3   Decimation on a one-dimensional lattice.

in the basis 1 and $q\delta_{\sigma\sigma'} - 1$. These matrices (with indices σ and σ') are orthonormal, in the following sense

$$\frac{1}{q}\sum_\mu [a_1 + b_1(q\delta_{\sigma\mu} - 1)][a_2 + b_2(q\delta_{\mu\sigma'} - 1)] = a_1 a_2 + b_1 b_2(q\delta_{\sigma\sigma'} - 1)$$

(49)

The partition function appears as the trace of the matrix (48) to the power L, and the free energy is thus given by

$$F = \lim_{L\to\infty} \frac{\ln Z_L}{L} = \ln \frac{e^\beta + (q-1)e^{-\beta}}{q}$$

(50)

Of course, there is no transition. The known result $F = \ln \cosh \beta$ is recovered for $q = 2$.

In the decimation method (figure 3), one considers blocks of size $\lambda a, a$ being the lattice spacing and λ an integer. The spin variables located at sites with a coordinate which is a multiple of λa are kept fixed, while we sum over intermediate ones using formula (49). Generalizing the notation $t = \tanh \beta$ of the Ising model through

$$t = \frac{1 - e^{-2\beta}}{1 + (q-1)e^{-2\beta}}$$

(51)

one immediately obtains

$$t_\lambda \equiv t(\beta_\lambda) = t(\beta)^\lambda$$

(52)

This is the *exact* renormalization equation for this very simple case in which the decimation does not modify the form of the interaction. To be more precise, equation (52) is complemented by the transformation

$$\frac{Z_L(\beta)}{\left(\dfrac{e^\beta + (q-1)e^{-\beta}}{q}\right)^L} = \frac{Z_{L/\lambda}(\beta_\lambda)}{\left(\dfrac{e^{\beta_\lambda} + (q-1)e^{-\beta_\lambda}}{q}\right)^{L/\lambda}}$$

(53)

The fixed points of the transformation (52) are $t = 0$ ($\beta = 0$), attractive high temperature point, and $t = 1$ ($\beta = \infty$), repulsive low temperature fixed point which plays the role of the critical point. In the present case, we have a natural continuation in the variable λ. Hence, one can obtain the behaviour of the correlation length at low temperature. Indeed, near $t = 1$, $1 - t$ plays the role of θ in the general discussion, with

$$1 - t_\lambda \sim \lambda(1 - t) \tag{54}$$

and one deduces the expected result

$$\xi \sim \frac{1}{1 - t} \sim \frac{1}{q} e^{2\beta} \tag{55}$$

In higher dimension, the decimation procedure generates interactions beyond nearest neighbours. Truncations are then necessary, and it is rather difficult to assert their validity. Moreover, there is some difficulty in interpreting this type of transformation due to the absence of wavefunction renormalization.

It is then interesting to develop other approximation schemes in which the calculations are also rather simple. One of those has been proposed by Migdal and interpreted by Kadanoff.

Link displacement   Consider the initial partition function

$$Z = \sum_{\{\sigma\}} e^{S(\sigma)} \tag{56}$$

Let $\Delta(\sigma)$ be a functional of the spins, with a vanishing mean value

$$\langle \Delta(\sigma) \rangle = Z^{-1} \sum_{\{\sigma\}} \Delta(\sigma) e^{S(\sigma)} = 0 \tag{57}$$

Due to the convexity inequality

$$Z_\Delta = \sum_{\{\sigma\}} e^{S(\sigma) + \Delta(\sigma)} = Z \left\langle e^{\Delta(\sigma)} \right\rangle \geq Z e^{\langle \Delta(\sigma) \rangle} = Z \tag{58}$$

The approximation replaces the action $S(\sigma)$ by a modified one $S(\sigma) + \Delta(\sigma)$, with a choice of $\Delta(\sigma)$ so as to generate recurrence relations. We analyze again the Potts model, but now in arbitrary dimension. The lattice is hypercubic. We introduce a transformation which treats separately the different directions, and therefore we allow distinct couplings $\beta_1, ..., \beta_d$, for the interactions between

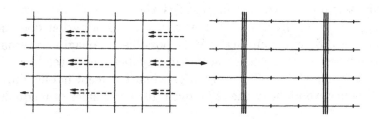

Fig. 4 Link displacement.

nearest neighbours on a link parallel to the direction $1,..., d$. The action reads

$$S(\sigma) = \sum_{\mathbf{x},\mu} \beta_\mu (2\delta_{\sigma_\mathbf{x},\sigma_{\mathbf{x}+\hat{\mu}}} - 1) \qquad (59)$$

Let us now look for a perturbation Δ satisfying the condition (57) and such that the new action $S + \Delta$ can be interpreted as an action similar to S on a lattice with larger spacing. The approximation will then replace the r.h.s. of (58) by the l.h.s.. Due to translational invariance, the variables attached to two parallel links verify

$$\Big\langle (2\delta_{\sigma_\mathbf{x},\sigma_{\mathbf{x}+\hat{\mu}}} - 1) - (2\delta_{\sigma_\mathbf{y},\sigma_{\mathbf{y}+\hat{\mu}}} - 1) \Big\rangle = 0 \qquad (60)$$

This equality allows one to displace parallel links. Choosing a direction μ, we displace links parallel to all directions $\nu \neq \mu$ in order to change the lattice spacing a_μ in direction μ to λa_μ (see figure 4). Links in any direction different from μ are replicated λ times. In the direction μ, it is now possible to use the technique of one-dimensional decimation previously described. With the same notation as in equations (51)–(52), the result of this transformation reads

$$\begin{cases} t(\beta'_\mu) = t(\beta_\mu)^\lambda \\ \beta'_\nu = \lambda\beta_\nu \qquad \nu \neq \mu \end{cases} \qquad (61)$$

This operation is performed consecutively for the different directions $\mu = 1, 2,..., d$. After this process, the lattice is again cubic with spacing λa, and the couplings β_μ are replaced by $\beta_\mu(\lambda)$ according to the Migdal-Kadanoff recurrence formulae

$$t\left(\lambda^{\mu-d}\beta_\mu(\lambda)\right) = \left[t\left(\lambda^{\mu-1}\beta_\mu\right)\right]^\lambda \qquad 1 \leq \mu \leq d \qquad (62)$$

This asymmetric treatment justifies the choice of distinct couplings in each direction. However, the transformation does not mix the different couplings. It is possible to imagine alternate procedures which restore the symmetry between directions. In any case, equation (62) is only approximate, except in one dimension, since we identify the left- and right-hand sides of inequality (58).

Write, in the same approximation, the multiplicative factor of the partition function.

Let $\left\{\beta_\mu^*\right\}$ be a nontrivial fixed point. The first component satisfies

$$t\left(\lambda^{1-d}\beta_1^*\right) = \left[t\left(\beta_1^*\right)\right]^\lambda \tag{63}$$

and from equation (62), one gets

$$\beta_\mu^* = \lambda^{-(\mu-1)}\beta_1^* \tag{64}$$

These critical values coincide when λ is analytically continued up to 1. Duality affords an interesting interpretation in two dimensions. We observe that the Potts model is self-dual, as is the Ising model. In the anisotropic case, the duality transformation reads

$$(\beta_1, \beta_2) \leftrightarrow (\tilde{\beta}_1, \tilde{\beta}_2) \tag{65}$$
$$e^{-2\tilde{\beta}_1} = t(\beta_2)$$
$$e^{-2\tilde{\beta}_2} = t(\beta_1) \tag{66}$$

The invariant curve corresponds to $\exp(-2\beta_1) = t(\beta_2)$ and is given by

$$(q-1)e^{-2\beta_1}e^{-2\beta_2} + e^{-2\beta_1} + e^{-2\beta_2} - 1 = 0 \tag{67}$$

Thus in two dimensions, for any choice of λ, the Migdal–Kadanoff fixed point lies on the self-dual line. The expressions (66)–(67) generalize those obtained in chapter 2. In chapter 6, we shall present a general study of duality. In particular, the critical isotropic point $\beta_1 = \beta_2 = \beta$ satisfies

$$e^{-2\beta} = t(\beta) = \frac{\sqrt{q}-1}{q-1} \tag{68}$$

Prove the self-duality of the 2-d Potts model and check that the fixed point given by equations (63)–(64) lies on the curve (67).

In terms of the variables y_1 and y_2 defined as

$$y_1 = e^{-2\beta_1}$$

$$y_2 = t(\beta_2) = \frac{1 - e^{-2\beta_2}}{1 + (q-1)e^{-2\beta_2}} \qquad (69)$$

the Migdal–Kadanoff transformation reads

$$y_\mu(\lambda) = F_\lambda(y_\mu) \qquad (\mu = 1, 2)$$

$$F_\lambda(y) = \left\{ \frac{[1 + (q-1)y]^\lambda - [1-y]^\lambda}{[1 + (q-1)y]^\lambda + (q-1)[1-y]^\lambda} \right\}^\lambda \qquad (70)$$

The function $F_\lambda(y)$ is rational when λ is an integer, with $F_1(y) = y$. In the Ising case ($q = 2$) and for $\lambda = 2$, it reads

$$F_2(y) = \left(\frac{2y}{1 + y^2} \right)^2 \qquad (71)$$

We illustrate in figure 5 the typical instability characteristic of the critical point given by

$$y_1 = y_2 = y^* = F_\lambda(y^*) \qquad (72)$$

This point depends on λ, but lies on the self-dual curve $y_1 = y_2$, as does the exact critical point. The transformation is at first meaningful only when λ is an integer. If, however, we perform a natural continuation for $\lambda \to 1$, we find the isotropic critical point

$$y_1 = y_2 = \frac{1 - y_1}{1 + (q-1)y_1} = \frac{\sqrt{q} - 1}{q - 1} \qquad (73)$$

The critical exponents are rather unrealistic, and depend on the choice of λ. Near the fixed point, one can linearize the transformation as

$$(y_\lambda - y^*) \simeq F'_\lambda(y^*)(y - y^*) \qquad (74)$$

According to the definitions (27-28), we can relate the coefficient in this relation to the exponent ν through

$$1 < F'_\lambda(y^*) = \lambda^{1/\nu}$$

$$\nu = \frac{\ln \lambda}{\ln F'_\lambda(y^*)} \qquad (75)$$

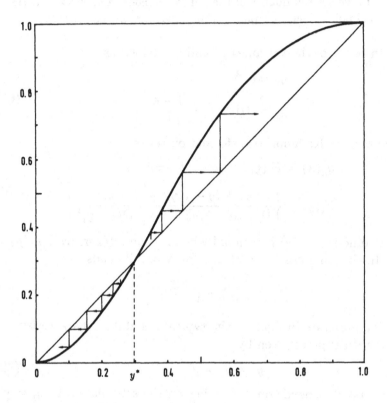

Fig. 5   The transformation $y \to F_2(y) = [2y/(1+y^2)]^2$ and its unstable fixed point $y^* \simeq 0.2956$, $F_2'(y^*) \simeq 1.6786$.

Let us compare this value with the one pertaining to the two-dimensional Ising model for which $\nu = 1$. For $q = 2$ and $\lambda = 2$, one finds

$$y^* = 0.2956 \qquad F_2'(y^*) = 1.6786 \qquad \nu = 1.3383 \quad (76)$$

This rather poor result does not improve in the limit $\lambda \to 1$. In this case, y^* tends to the familiar value

$$y^* = \sqrt{2} - 1 = 0.4142 \tag{77}$$

and one has

$$\nu^{-1} = \frac{d}{d\beta}(\beta + \sinh\beta\cosh\beta\ln\tanh\beta)\Big|_{\beta_*} = 2 + \cosh 2\beta^*\ln\tanh\beta^*$$

$$= 2 + \sqrt{2}\ln\left(\sqrt{2}-1\right) \qquad\rightarrow\qquad \nu = 1.327$$

(78)

In all these decimation methods, the block variables are among the original ones and hence $G(\mathbf{x},\beta) = G(\mathbf{x}/\lambda,\beta_\lambda)$ for the correlations. This implies

$$d - y_H = \beta/\nu = 0$$

(79)

In spite of these shortcomings, the Migdal–Kadanoff approximation is nevertheless interesting. One may hope to construct systematic improvements due to its variational character. It is exact at the lower critical dimension ($d = 1$) and the general idea can be applied to more complex systems such as gauge models.

Majority rule As an illustration of the majority rule, we consider the Ising model on a two-dimensional triangular lattice, following Niemeijer and Van Leeuwen. Each site has six neighbours. Using the methods developed in chapter 2, the exact free energy per site, normalized to zero at low temperature, is found to be

$$F(\beta) = \frac{1}{2}\int_{-\pi}^{\pi}\frac{d^2\theta}{(2\pi)^2}\ln[(\cosh 2\beta)^3 + (\sinh 2\beta)^3$$

$$- \sinh 2\beta(\cos\theta_1 + \cos\theta_2 + \cos(\theta_1 + \theta_2))]$$

(80)

The critical point is

$$\beta_c = \tfrac{1}{4}\ln 3 = 0.2747$$

(81)

and the critical exponents are the same as those on a square lattice.

    Prove these results.

    The triangular lattice is not self-dual. The duality transformation leads to a similar model on a honeycomb lattice where each site has three neighbours, with a new coupling β' such that

$$e^{-2\beta'} = \tanh\beta$$

(82)

    An exact decimation procedure called the *star–triangle* transformation, allows one to recover a triangular lattice. Let us split

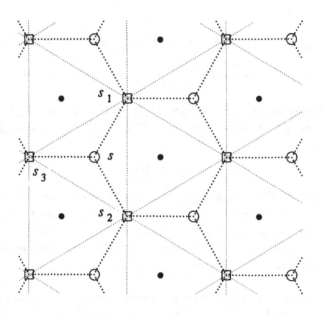

Fig. 6 The star–triangle transformation. The original sites appear as black dots. The dual sites on the honeycomb lattice are split into two families, represented by white circles and squares.

the dual honeycomb lattice, with twice as many sites as the original one, into two triangular sublattices. Now we can sum over the values of the spins s located on the first sublattice, which only interact with three neighbouring spins s_1, s_2, s_3 located on the second (figure 6)

$$
\begin{cases}
\frac{1}{2} \sum_{s=\pm 1} \exp[\beta' s(s_1 + s_2 + s_3)] = A \exp[\beta''(s_1 s_2 + s_2 s_3 + s_3 s_1)] \\
A^4 = \cosh 3\beta' (\cosh \beta')^3 \\
e^{4\beta''} = \dfrac{\cosh 3\beta'}{\cosh \beta'} = 2 \cosh 2\beta' - 1 = 2 \dfrac{\cosh 2\beta}{\sinh 2\beta} - 1
\end{cases}
$$

$$(83)$$

The last relation may be rewritten in a more symmetric way

$$(e^{4\beta''} - 1)(e^{4\beta} - 1) = 4 \tag{84}$$

We have then a correspondence between two isomorphic models with respective couplings β and β''. The critical point (if it

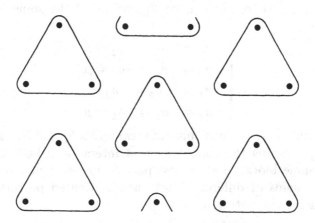

Fig. 7 Blocks of three spins.

exists and if it is unique, as is the case) is the fixed point of the transformation, and agrees with (81).

(1) Write the corresponding relation between the free energies and check that it is satisfied by the expression (80).

(2) Can the star–triangle transformation be generalized to the Potts model?

Let us now apply a renormalization transformation, according to the majority rule, with blocks containing three sites (an odd number which avoids ambiguities), as shown in figure 7. Larger blocks could of course also be considered. In this case, as the number of sites is divided by 3, the scaling factor is $\lambda = \sqrt{3}$. To each block is assigned the new dichotomic variable

$$\mu = \text{Sign}(\sigma_1 + \sigma_2 + \sigma_3) \tag{85}$$

The transformation is defined by assuming that the partition function is invariant. Upon inserting a product of factors $1 = \sum_\mu \delta_{\mu,\text{Sign}(\Sigma\sigma)}$, we write

$$Z = \sum_{\{\mu_a = \pm 1\}} e^{S'(\mu)} = \sum_{\{\mu_a = \pm 1\}} \sum_{\{\sigma_i = \pm 1\}} \left(\prod_a \delta_{\mu_a, \text{Sign}(\sigma_1^a + \sigma_2^a + \sigma_3^a)} \right) e^{S(\sigma)}$$

$$\tag{86}$$

For fixed μ's there are four configurations of the spins σ_i inside each block

$$\begin{cases} \sigma_1 = \sigma_2 = \sigma_3 = \mu \\ \sigma_1 = \sigma_2 = -\sigma_3 = \mu \\ \sigma_1 = -\sigma_2 = \sigma_3 = \mu \\ -\sigma_1 = \sigma_2 = \sigma_3 = \mu \end{cases} \tag{87}$$

The initial action is then split into two terms S_0 and S_1. The first part S_0 contains the contribution of interactions between spins of the same block. The second part S_1 arises from interactions between spins of different blocks and is treated perturbatively. Introducing the notation

$$Z_0 = \sum_\sigma \prod_a \delta_{\mu_a, \mathrm{Sign}(\sigma_1^a + \sigma_2^a + \sigma_3^a)} e^{S_0(\sigma)} = z^{N/3}$$
$$z = e^{3\beta} + 3e^{-\beta} \tag{88}$$

and

$$\langle A(\sigma) \rangle = Z_0^{-1} \sum_\sigma \prod_a \delta_{\mu_a, \mathrm{Sign}(\sigma_1^a + \sigma_2^a + \sigma_3^a)} e^{S_0(\sigma)} A(\sigma) \tag{89}$$

one wants to compute

$$e^{S'(\mu)} = Z_0 \langle \exp S_1 \rangle \tag{90}$$

In the cumulant expansion of this mean value

$$\ln \langle \exp S_1 \rangle = \langle S_1 \rangle + \tfrac{1}{2} \left(\langle S_1^2 \rangle - \langle S_1 \rangle^2 \right) + \cdots \tag{91}$$

we only retain the first term. In this approximation, the block variables have only nearest neighbour interactions, as do the initial spins. For a given pair (a, b), one finds the term

$$\beta \langle (\sigma_1^a + \sigma_2^a) \sigma_3^b \rangle = \beta \left(\langle \sigma_1^a \rangle + \langle \sigma_2^a \rangle \right) \langle \sigma_3^b \rangle$$

Since

$$\langle \sigma_i^a \rangle = \mu_a \frac{e^{3\beta} + e^{-\beta}}{e^{3\beta} + 3e^{-\beta}}$$

the renormalized coupling reads

$$\beta_\lambda = 2\beta \left(\frac{e^{3\beta} + e^{-\beta}}{e^{3\beta} + 3e^{-\beta}} \right)^2 \qquad (\lambda = \sqrt{3}) \tag{92}$$

Higher order terms in (91) would generate longer-range interactions. Within the present approximation, the method reduces to

a coupling constant renormalization (92). Its fixed point is given
by

$$\beta^* = \tfrac{1}{4}\ln(1 + 2\sqrt{2}) = 0.3356 \qquad (93)$$

to be compared to (81). Near this fixed point, the expansion

$$(\beta_\lambda - \beta^*) = (\beta - \beta^*)[1 + 2\beta^*(8 - 5\sqrt{2})] + \cdots$$

yields the critical exponent

$$\left(\sqrt{3}\right)^{1/\nu} = 1 + 2\beta^*(8 - 5\sqrt{2}) = 1.6235 \qquad \rightarrow \qquad \nu = 1.1335 \tag{94}$$

This value is not very far from the exact one $\nu = 1$. One may
hope that the approximation improves when further terms are
kept in the expansion (91). Of course, one should then start
from an action involving longer-range interactions with additional
couplings. In any case, it is rather difficult to find the appropriate
truncation without a thorough numerical study.

Sticking to this simple framework, let us analyze the effect of
an external field H. A term $H\sum_i s_i$ is added to the action which
is again split into S_0 and S_1. Within each block, the contribution
to the partition function is given by

$$z_H e^{H_\lambda^{(0)}\mu} = \sum_{\{\sigma_i = \pm 1\}} e^{\beta(\sigma_1\sigma_2 + \sigma_2\sigma_3 + \sigma_3\sigma_1) + H(\sigma_1 + \sigma_2 + \sigma_3)} \delta_{\mu,\,\mathrm{Sign}(\sigma_1 + \sigma_2 + \sigma_3)}$$

$$\rightarrow \begin{cases} z_H^2 = \left(e^{3H+3\beta} + 3e^{H-\beta}\right)\left(e^{-3H+3\beta} + 3e^{-H-\beta}\right) \\ e^{2H_\lambda^{(0)}} = \left(e^{3H+3\beta} + 3e^{H-\beta}\right) \big/ \left(e^{-3H+3\beta} + 3e^{-H-\beta}\right) \end{cases} \tag{95a}$$

with

$$\langle\sigma\rangle = \alpha_1\mu + \alpha_2$$

$$\alpha_1 \pm \alpha_2 = \frac{e^{\pm 3H+3\beta} + e^{\pm H-\beta}}{e^{\pm 3H+3\beta} + 3e^{\pm H-\beta}} \tag{95b}$$

The relations (88) and (90) are replaced by

$$e^{S'(\mu)} \cong (z_H)^{N/3} \exp\left\{\sum_a H_\lambda^{(0)}\mu_a + 2\beta\sum_{(ab)}(\alpha_1\mu_a + \alpha_2)(\alpha_1\mu_b + \alpha_2)\right\} \tag{96}$$

and the new couplings are

$$\begin{cases} \beta_\lambda = 2\beta\alpha_1^2 \\ H_\lambda = H_\lambda^{(0)} + H_\lambda^{(1)} \end{cases} \tag{97}$$

with $H_\lambda^{(0)}$ given by equation (95) and

$$H_\lambda^{(1)} = 12\beta\alpha_1\alpha_2 \tag{98}$$

The critical point corresponds to $\beta = \beta^*$ given by (93), and $H = 0$.
Linearizing the transformation (which is already diagonal) in the
vicinity of this point, leads to

$$\begin{cases} \beta_\lambda - \beta^* = \lambda^{1/\nu}(\beta - \beta^*) + \cdots \\ H_\lambda = \lambda^{y_H} H + \cdots \end{cases} \tag{99}$$

$$\begin{cases} \lambda^{1/\nu} = 1 + 2\beta^*(8 - 5\sqrt{2}) = 1.6235 \\ \lambda^{y_H} = \dfrac{3}{\sqrt{2}} + 3\beta^*(8 - 5\sqrt{2}) = 2.123 + 0.9353 = 3.0566 \end{cases} \tag{100}$$

We have distinguished in the last expression two contributions of
respective orders 0 and 1. Using the value $\lambda = \sqrt{3}$, this yields the
exponents

$$\begin{cases} y_H^{(0)} = 1.369 \\ y_H^{(1)} = 2.034 \end{cases} \tag{101}$$

In the present case $\nu = 1, \gamma = \frac{7}{4}$ correspond to the exact value
$y_H = \frac{15}{8} = 1.875$. The result (101) is thus not so bad. Niemeijer
and van Leeuwen, Braathen and Hemmer have performed the
computation up to second order, including interactions between
second and third neighbours. The coupling space is now three-
dimensional, and $\lambda^{1/\nu}$ is identified with the largest eigenvalue of
the linearized approximation near the fixed point. We only display
the results

order	$y_\theta = 1/\nu$	y_H	β^*
0		1.369	
1	0.822	2.034	0.3356
2	1.053	1.847	0.2514
exact	1.000	1.875	0.2747

$$\tag{102}$$

which show a spectacular improvement. Of course, the preceding
discussion should only be considered as a short introduction to
numerous works which cannot be listed here.

Critical amplitudes Up to now we focused our attention on the calculation of the critical exponents. It is also interesting to compute the amplitude of singular terms. We limit ourselves to the simple example of a recurrence relation, with a fixed dilatation coefficient λ, of the type (26). This is rewritten as

$$F(\theta) = f(\theta) + \lambda^{-d} F(\theta_\lambda) \tag{103}$$

Near the critical point, θ measures the deviation from the critical temperature and θ_λ is its transform. The function $f(\theta)$ is assumed to be regular at $\theta = 0$. We work in the high temperature phase ($\theta > 0$) and normalize the free energy to $F(\infty) = f(\infty) = 0$. Near the origin, the fixed point of the map $\theta \to \theta_\lambda$, the linearized transformation is

$$\theta_\lambda = \mu_\lambda \theta + \mathcal{O}(\theta^2) \tag{104}$$

and we know that the critical exponent α is given by $\mu_\lambda^{2-\alpha} = \lambda^d$. The factor μ_λ is larger than unity, expressing the instability of the fixed point. Let n be an integer such that $n < 2 - \alpha < n + 1$ (for the time being, the inequalities are assumed to be strict). Let us iterate the transformation $\theta \to \theta_\lambda \to \theta_{\lambda^2} \to \cdots \to \theta_{\lambda^p} \to \cdots$. One obtains

$$F(\theta) = \sum_{k=0}^{p-1} \lambda^{-dk} f(\theta_{\lambda^k}) + \lambda^{-dp} F(\theta_{\lambda^p}) \tag{105}$$

As $p \to \infty, \theta_{\lambda^p}$ tends towards the high temperature stable point. The series converges and hence

$$F(\theta) = \sum_{k=0}^{\infty} \lambda^{-dk} f(\theta_{\lambda^k}) \tag{106}$$

The knowledge of an exact renormalization relation of the type (103) would allow the complete determination of the free energy. The same reasoning can also be applied in the low temperature phase (changing the normalization of F in order to ensure convergence). These hypotheses are of course very strong since we assume an exact renormalization formula. Let us nevertheless proceed within this framework. The estimation of (106) is rather difficult without numerical methods. As an approximation, one can replace the transformation $\theta \to \theta_\lambda$ by its linearized form. Better, we substitute a new variable $\tilde{\theta}(\theta)$ such that the renormalization transformation reads $\tilde{\theta}_\lambda \equiv \tilde{\theta}(\theta_\lambda) = \mu_\lambda \tilde{\theta}$. This change of variable is always possible locally. We shall not discuss here the delicate question of its global domain of validity. In order to simplify the notation, we suppress the tilde on this variable and write

$$F(\theta) = \sum_{p=0}^{\infty} \lambda^{-dp} f(\mu_\lambda^p \theta) \qquad (107)$$

In particular, we get $F(0) = f(0)/(1 - \lambda^{-d})$. As f is regular at $\theta = 0$, it can be expanded in a Taylor series as $f(\theta) = \sum_{k=0}^{\infty} f_k \theta^k$. Naively, one could think of inserting this expansion in (107) and inverting the order of the two summations. However, this is incorrect, since the summation over p does not converge when $k > n = \lfloor 2 - \alpha \rfloor$ ($\mu_\lambda^k > \lambda^d$). Nevertheless, let us introduce the regular function

$$\varphi(\theta) = \sum_{k=0}^{\infty} f_k \frac{\theta^k}{1 - \mu_\lambda^k/\lambda^d} \qquad (108)$$

Each term in this series is well-defined, since we exclude the case where $2 - \alpha$ is an integer. Moreover, if the radius of convergence of $f(\theta)$ is R, the expansion of φ converges in a disk of radius $\mu_\lambda R > R$. Let us now split f into a subtraction term $\sum_0^n f_k \theta^k$ and a remainder $f_1(\theta)$, so that the subtraction term leads to a convergent summation over p. Hence

$$F(\theta) = \sum_{k=0}^{n} f_k \frac{\theta^k}{1 - \mu_\lambda^k/\lambda^d} + \sum_{p=0}^{\infty} \lambda^{-dp} f_1(\mu_\lambda^p \theta) \qquad (109)$$

In the last sum, each term is of order θ^{n+1} although we know that the sum has a singular behaviour in $\theta^{2-\alpha}$. In order to show this, we add and subtract in (109) the absolutely convergent term $\sum_{-\infty}^{-1} \lambda^{-dp} f_1(\mu_\lambda^p \theta)$. Substituting for $f_1(\theta)$ its Taylor expansion $\sum_{n+1}^{\infty} f_k \theta^k$, one obtains

$$\sum_{p=-\infty}^{-1} \lambda^{-dp} \sum_{k=n+1}^{\infty} f_k \mu_\lambda^{kp} \theta^k = -\sum_{k=n+1}^{\infty} f_k \frac{\theta^k}{1 - \mu_\lambda^k/\lambda^d}$$

the convergence being ensured by the fact that $\mu_\lambda^k > \lambda^d$ for $k \geq n + 1$. Under these conditions,

$$F(\theta) = \varphi(\theta) + \sum_{p=-\infty}^{\infty} \lambda^{-dp} f_1(\mu_\lambda^p \theta) \qquad (110)$$

This formula gives an explicit representation of F as a sum of a regular part φ, and a singular one which satisfies the homogeneous equation

$$F_{\text{sing}}(\theta) = \sum_{p=-\infty}^{\infty} \lambda^{-dp} f_1(\mu_\lambda^p \theta) = \lambda^{-d} F_{\text{sing}}(\mu_\lambda \theta) \qquad (111)$$

and which behaves as $\theta^{2-\alpha}$. Hence

$$F_{\text{sing}}(\theta) = \theta^{2-\alpha} A(\theta)$$

$$A(\theta) = A(\mu_\lambda \theta) = \theta^{\alpha-2} \sum_{p=-\infty}^{\infty} \lambda^{-dp} f_1(\mu_\lambda^p \theta) \qquad (112)$$

For a fixed renormalization factor λ, the amplitude A depends on θ and is in fact a periodic, bounded function of $\ln\theta$ with period $\ln\mu_\lambda$. Writing

$$A(\theta) = \sum_{s=-\infty}^{+\infty} A_s \exp\left(2i\pi s \frac{\ln\theta}{\ln\mu_\lambda}\right) \qquad (113)$$

we have for the coefficients

$$\begin{aligned} A_s &= \frac{1}{\ln\mu_\lambda} \int_1^{\mu_\lambda} \frac{d\theta}{\theta^{3-\alpha+2i\pi s/\ln\mu_\lambda}} \sum_{p=-\infty}^{\infty} \lambda^{-dp} f_1(\mu_\lambda^p \theta) \\ &= \frac{1}{\ln\mu_\lambda} \int_0^{\infty} d\theta\; \theta^{\alpha-3-2i\pi s/\ln\mu_\lambda} f_1(\theta) \end{aligned} \qquad (114)$$

The last expression has been derived using the relation $\mu_\lambda^{2-\alpha} = \lambda^d$, so that each term of the sum is written as an integral over the interval $\left[\mu_\lambda^p, \mu_\lambda^{p+1}\right]$. Finally, the last integral in (114) can be integrated by parts until f_1 is replaced by the $(n+1)$-th derivative of f. In general, (113) and (114) are only approximate, and one sees, on numerical examples, that the amplitude of the oscillations is very small, so that it is possible to replace $A(\theta)$ by A_0. Of course, similar expressions can be deduced for the low temperature region.

When $2-\alpha$ is an integer, the singularity does not disappear, but yields a logarithmic term. By continuity in $\varepsilon = n-2+\alpha \to 0$, one of the terms in φ becomes singular and one gets

$$\begin{aligned} F(\theta) &= F_{\text{reg}}(\theta) - \frac{f_n \theta^n}{\varepsilon \ln\mu_\lambda} + \frac{\theta^n(1-\varepsilon\ln\theta)}{\ln\mu_\lambda} \int_0^{\infty} \frac{d\theta'\, f_1(\theta')}{\theta'^{1+n-\alpha}} \\ &\simeq F_{\text{reg}}(\theta) - \frac{f_n \theta^n}{\varepsilon \ln\mu_\lambda} + \frac{\theta^n(1-\varepsilon\ln\theta)}{\ln\mu_\lambda} f_n \int_0^{\theta_0} \frac{d\theta'}{\theta'^{1-\varepsilon}} \end{aligned} \qquad (115)$$

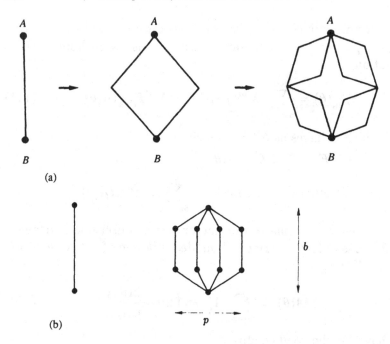

Fig. 8   (*a*) Recursive construction of a hierarchical lattice. (*b*) Generalization with p strings of b links. A dimension can be defined by comparing the increase in "length" $L' = bL$ between the two points AB to the increase in "volume", i.e. the number of bonds $N' = pbN$. If N scales like L^d, $d = \ln(pb)/\ln b$. Figure (*a*) corresponds to $d = 2$.

$$\simeq F_{\text{reg}}(\theta) - \frac{f_n}{\ln \mu_\lambda} \theta^n \ln |\theta|$$

The same reasoning can be applied for $\theta < 0$. Hence, while the amplitudes A_+ and A_- above and below the critical point seem to be unrelated when $2 - \alpha$ is not an integer, the logarithmic singularity has the same amplitude on both sides when $2 - \alpha$ is an integer. This is what has been observed in the two-dimensional Ising model.

Consider the *hierarchical* lattice obtained by the following recursive process. Starting from an elementary graph H_2 with two sites joined by one link, H_n is deduced from H_{n-1} by substituting four links for each one, according to the scheme depicted on figure 8a.

i) Show the existence of a thermodynamical limit $n \to \infty$ for a q-states Potts model on H_n with a partition function

$$Z_n = \sum_{\sigma_i=1}^{q} \exp 2\beta \sum_{ij} \delta_{\sigma_i \sigma_j} \qquad (116)$$

ii) Obtain by decimation a recurrence relation between Z_n and Z_{n-1} generalizing the Migdal–Kadanoff relation (71). Using $x = e^{2\beta}$ instead of $y = e^{-2\beta}$,

$$Z_n(x) = (2x + q - 2)^{2 \cdot 4^{n-2}} Z_{n-1}(x')$$

$$x' = T(x) = \left(\frac{x^2 + q - 1}{2x + q - 2} \right)^2 \qquad (117)$$

iii) Study the critical exponents and amplitudes.

iv) To represent the behaviour of the partition function, one can plot the limiting set of its zeroes in the complex x-plane as $n \to \infty$. One obtains remarkable Julia sets indicating a very interesting structure as shown in figure 9. The linearization of the corresponding renormalization map near the fixed points, as described in the text above, yields very inaccurate amplitudes of the singular part of the free energy. The matter is discussed in the references.

4.2   The XY-model

Real space renormalization transformations meet spectacular success when applied to the XY-model in two dimensions. This is why we present it in this chapter. This model deals with two component classical spins $\left(S_x, S_y \right)$ normalized to unit length $S_x^2 + S_y^2 = 1$, and interacting with their nearest neighbours. The corresponding global symmetry is $O(2)$. After some transformations, this model also describes a classical Coulomb gas, a set of fluctuating surfaces, or superfluid films. It undergoes a remarkable transition, such that the free energy and all its derivatives remain continuous (Kosterlitz and Thouless). There is no macroscopic spontaneous magnetization (Mermin and Wagner). The renormalization group arguments reveal the nature of the singularity, as can be checked by comparing with the exact solution of a related model.

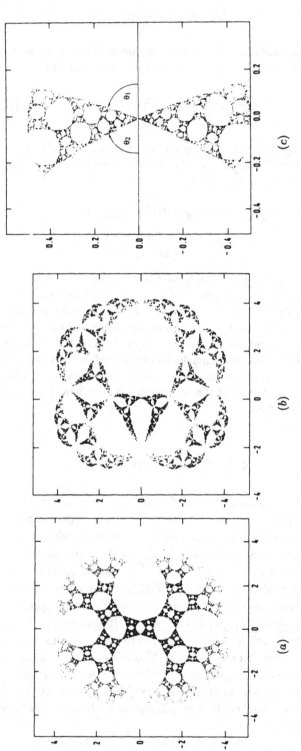

Fig. 9 (a) Julia set of zeroes of the partition function in the $n \to \infty$ limit for the Ising case $q = 2$. The points shown in the figure are the images under the map T^{-1} of the unique zero of Z_1. (b) Same as before for $q = 3$ with a bias designed to obtain a sharper view. (c) An enlargement of the Julia set for $q = 2$ near the ferromagnetic fixed point x_c showing two (equal) limiting angles θ_1 and θ_2 for the support of the zeroes.

Spin variables can also be described by angles, defined modulo 2π, with the partition function written as

$$Z = \int \prod_i \frac{d\theta_i}{2\pi} \exp\left(\beta \sum_{(ij)} \cos(\theta_i - \theta_j)\right) \qquad (118)$$

The interaction is ferromagnetic, in the sense that the configurations with equal θ_i's are favored. The global rotational symmetry appears here as a translation invariance $\theta_i \to \theta_i + \alpha$. We study this model on a two-dimensional square lattice.

4.2.1 High temperature behaviour

We have already considered this model in one dimension and shown that it is in a disordered phase at any temperature. One expects a similar disordered phase in two dimensions for sufficiently high temperature. This is confirmed by a strong coupling (high temperature) expansion (β small). Using a general procedure discussed in chapter 7, each factor is expanded on the basis of irreducible characters of the group $O(2)$, which amounts, in simpler terms, to the usual Fourier decomposition

$$e^{\beta \cos \theta} = I_0(\beta) + \sum_{n=1}^{\infty} I_n(\beta) \left(e^{in\theta} + e^{-in\theta}\right)$$

$$= I_0(\beta) \left(1 + \sum_{n \neq 0} t_n(\beta) e^{in\theta}\right) \qquad (119a)$$

where the coefficients are expressed in terms of modified Bessel functions

$$t_n(\beta) = t_{-n}(\beta) = \frac{I_n(\beta)}{I_0(\beta)} \qquad (119b)$$

with the following behaviour for small or large β

$$t_n(\beta) \sim \begin{cases} \dfrac{\beta^n}{2^n n!} & \beta \to 0 \\ e^{-n^2/2\beta} & \beta \to \infty \end{cases} \qquad (119c)$$

These coefficients generalize the quantity $\tanh \beta$ of the Ising model. They lie between 0 and 1 in the domain $0 \le \beta < \infty$.

For a lattice with N sites and $2N$ links, one has

$$\frac{Z}{I_0^{2N}(\beta)} = \int \prod_i \frac{d\theta_i}{2\pi} \prod_{(ij)} \left(1 + \sum_{n_{ij} \neq 0} t_{n_{ij}}(\beta) e^{in_{ij}(\theta_i - \theta_j)}\right) \qquad (120)$$

The interest of the expansion (119) is clear for small β, since the coefficients t_n decrease very rapidly with n and since integrations can be performed explicitly over each term in the expansion of the product (120). The result is

$$\frac{Z}{I_0^{2N}(\beta)} = \sum_{\substack{\{n_{ij}\} \\ (\partial n)_i = 0}} \prod_{(ij)} t_{n_{ij}}(\beta) \qquad (121)$$

One sums over all configurations of relative integers n_{ij}, with the notations $t_0 \equiv 1, n_{ji} \equiv -n_{ij}$. These configurations satisfy the zero divergence condition $(\partial n)_i \equiv \sum_j n_{ij} = 0$. At high temperature, only finitely many links carry nonzero n_{ij}'s.

Similarly, one can compute the correlation function

$$\langle \mathbf{S}_1 \cdot \mathbf{S}_2 \rangle = \text{Re}\left\langle e^{i(\theta_1 - \theta_2)} \right\rangle \qquad (122)$$

and more generally

$$\left\langle e^{im(\theta_1 - \theta_2)} \right\rangle = \frac{\displaystyle\sum_{\substack{\{n_{ij}\} \\ (\partial n)_i = m(\delta_{i1} - \delta_{i2})}} \prod_{(ij)} t_{n_{ij}}(\beta)}{\displaystyle\sum_{\substack{\{n_{ij}\} \\ (\partial n)_i = 0}} \prod_{(ij)} t_{n_{ij}}(\beta)} \qquad (123)$$

The point 1 acts as a source and the point 2 as a sink of intensity m for the "field" n_{ij}. Analogous expressions are easily written for correlation functions involving more points.

For $m = 1$, the dominant term in the r.h.s. of equation (123) at high temperature is $N_{12} t_1(\beta)^{d_{12}}$, where d_{12} is the minimal distance on the lattice between points 1 and 2, and N_{12} is the number of corresponding paths of minimal length. Hence correlations decrease exponentially at high temperature to lowest order, and this extends to all orders because the high temperature series has a finite radius of convergence. The question arises whether this situation remains true at low temperature. In appendix A, we present the proof that a system with a continuous symmetry cannot have a nonvanishing mean value for the order parameter in dimension two, in the absence of a symmetry breaking external field. This is to be contrasted with the case of a discrete symmetry. This phenomenon is due in essence to an excess of entropy over energy.

4.2.2 Low temperature expansion. Vortices.

At low enough temperature, the energetic term favours the alignment of spins, and one is tempted to expand $\cos(\theta_i - \theta_j)$ near $\theta_i - \theta_j = 0$. In this treatment, the periodic character of the angles is neglected, but the existence of an order, at least at short range, partly justifies this method. This point will be further studied later. The model is replaced by a so-called *spin wave* approximation. The partition function is the one of a pure Gaussian model

$$Z_{sw} = \int \prod_i \frac{d\theta_i}{(2\pi)^{1/2}} \exp -\frac{1}{2}\beta \sum_{(ij)} (\theta_i - \theta_j)^2 \qquad (124)$$

where we have omitted a constant additive term in the action. The inverse propagator is the kernel of the quadratic form, i.e.

$$G^{-1}(\mathbf{x}, \mathbf{x}') = \int_{-\pi}^{\pi} \frac{d^2 q}{(2\pi)^2} e^{i\mathbf{q}\cdot(\mathbf{x}-\mathbf{x}')} [4 - 2(\cos q_1 + \cos q_2)] \qquad (125)$$

The rotational invariance of the spins, replaced here by the translation invariance of the angles, is associated with a zero mode. The corresponding infinite factor does not play a significant role in the thermodynamic limit; one can omit, for instance, the integration over one of the variables θ. The free energy per site associated to (124) is thus well-defined

$$F = \lim_{N \to \infty} \frac{1}{N} \ln Z_{sw} = -\frac{1}{2} \int_{-\pi}^{\pi} \frac{d^2 q}{(2\pi)^2} \ln\{\beta[4 - 2(\cos q_1 + \cos q_2)]\} \qquad (126)$$

However, low energy fluctuation modes have a sufficiently high entropy to disorder the spins at large distances. This is the content of a theorem due to Hohenberg, Mermin and Wagner, and Coleman. Consider the expression of the correlation function in the spin wave approximation

$$G_{sw}^{(p)}(\mathbf{x}_1, \mathbf{x}_2) = \left\langle \exp[ip(\theta_{\mathbf{x}_1} - \theta_{\mathbf{x}_2})] \right\rangle \tag{127}$$

$$= Z_{sw}^{-1} \int \prod \frac{d\theta_{\mathbf{x}}}{2\pi} \exp\left(-\tfrac{1}{2}\beta \sum_{\mathbf{x}, \mathbf{x}'} \theta_{\mathbf{x}} G_{\mathbf{x},\mathbf{x}'}^{-1} \theta_{\mathbf{x}'} + ip(\theta_{\mathbf{x}_1} - \theta_{\mathbf{x}_2}) \right)$$

which leads, after integration, to the expression

$$G_{sw}^{(p)}(\mathbf{x}_1, \mathbf{x}_2) = \exp\left(\frac{-p^2}{2\pi\beta} \Gamma(\mathbf{x}_1 - \mathbf{x}_2) \right) \tag{128}$$

The notation $\Gamma(\mathbf{x})$ denotes the integral

$$\Gamma(\mathbf{x}) = \int_{-\pi}^{\pi} \frac{d^2\mathbf{q}}{2\pi} \frac{1 - e^{i\mathbf{q}\cdot\mathbf{x}}}{4 - 2(\cos q_1 + \cos q_2)} \tag{129}$$

Assuming an infrared regularization for the free field Green function $G(\mathbf{x})$, the well-defined quantity $\Gamma(\mathbf{x})$ is equal to the subtracted form $2\pi(G(\mathbf{0}) - G(\mathbf{x}))$. Observe that Γ appears in an exponential, since we compute mean values of exponentials.

The function $\Gamma(\mathbf{x})$ is well-defined for all $\mathbf{x}$ on a two-dimensional square lattice, vanishes for $\mathbf{x} = 0$, is equal to $\pi/2$ for $|\mathbf{x}| = 1$ and has an isotropic behaviour at large distances

$$\Gamma(\mathbf{x}) \underset{|\mathbf{x}|}{\sim} \ln(2\sqrt{2}e^{\gamma}|\mathbf{x}|) + \mathcal{O}(1/|\mathbf{x}|) \tag{130}$$

where γ is Euler's constant. The asymptotic behaviour given in equation (130) is an excellent approximation, since it gives $\Gamma(\mathbf{x}) \simeq 1.6169$ for $|\mathbf{x}| = 1$, to be compared with the exact value $\tfrac{1}{2}\pi \simeq 1.5708$.

To prove the estimate (130), it is sufficient to evaluate Γ for $\mathbf{x} = (2p, 0)$. Hence

$$\Gamma(2p, 0) = \int_{-\pi}^{\pi} \frac{dq_1}{2\pi}(1 - \cos 2pq_1) \int_{-\pi}^{\pi} dq_2 \frac{1}{4 - 2(\cos q_1 + \cos q_2)}$$

$$= \frac{1}{2} \int_{-\pi}^{\pi} dq_1 \frac{1 - \cos 2pq_1}{\sqrt{(1 - \cos q_1)(3 - \cos q_1)}}$$

This can be rewritten, with $q_1 = 2q$,

$$\Gamma(2p, 0) = 2 \int_0^{\frac{1}{2}\pi} dq \frac{\sin^2 2pq}{\sin q \sqrt{1 + \sin^2 q}} = \Gamma_0 + \Gamma_1$$

$$\begin{cases} \Gamma_0 = 2 \int_0^{\frac{1}{2}\pi} dq \, \dfrac{\sin^2 2pq}{\sin q} \\[3mm] \Gamma_1 = 2 \int_0^{\frac{1}{2}\pi} dq \, \dfrac{\sin^2 2pq}{\sin q} \left(\dfrac{1}{\sqrt{1 + \sin^2 q}} - 1 \right) \end{cases}$$

With the decomposition $\sin 2pq / \sin q = 2 \sum_{s=1}^{p} \cos(2s - 1)q$, the first integral reads

$$\Gamma_0 = 4 \int_0^{\frac{1}{2}\pi} dq \sin 2pq \sum_{s=1}^{p} \cos(2s - 1)q$$

$$= 2 \sum_{s=0}^{2p-1} \frac{1}{2s + 1} = 2 \sum_{n=1}^{4p-1} \frac{1}{n} - \sum_{n=1}^{2p-1} \frac{1}{n}$$

$$\sim 2 \ln(4p - 1) e^{\gamma} - \ln(2p - 1) e^{\gamma} + \mathcal{O}\left(\frac{1}{p}\right)$$

$$\sim \ln 2p + 2 \ln 2 + \gamma + \mathcal{O}\left(\frac{1}{p}\right)$$

To estimate Γ_1, we can replace it by its limit when $p \to \infty$, trading $\sin^2 2pq$ for its average $\frac{1}{2}$, so that

$$\Gamma_1 \sim \int_0^{\frac{1}{2}\pi} \frac{dq}{\sin q} \left(\frac{1}{\sqrt{1 + \sin^2 q}} - 1 \right)$$

Expanding now

$$(1 + \sin^2 q)^{-1/2} - 1 = \sum_{s=1}^{\infty} \frac{\sin^{2s} q}{s!} \, \frac{\Gamma(\frac{1}{2})}{\Gamma(\frac{1}{2} - s)}$$

one obtains

$$\Gamma_1 \sim \sum_{s=1}^{\infty} \frac{\Gamma(\frac{1}{2})}{s! \Gamma(\frac{1}{2} - s)} \int_0^{\frac{\pi}{2}} dq \sin^{2s-1} q = \sum_{s=1}^{\infty} \frac{\pi}{2s \Gamma(\frac{1}{2} + s) \Gamma(\frac{1}{2} - s)}$$

$$= \frac{1}{2} \sum_{s=1}^{\infty} \frac{(-1)^s}{s} = -\frac{1}{2} \ln 2$$

Gathering these results, we find Γ up to order $1/p$ and obtain equation (130).

In terms of the lattice spacing, we define

$$r_0^{-1} = 2\sqrt{2} e^{\gamma} \tag{131}$$

so that $\Gamma(\mathbf{x})$ is isotropic to a very good approximation

$$G_{sw}^{(p)}(\mathbf{x}) \underset{|\mathbf{x}|\gg 1}{\sim} \exp\left[-(p^2/2\pi\beta)\ln(|\mathbf{x}|/r_0)\right] = (r_0/|\mathbf{x}|)^{p^2/2\pi\beta} \quad (132)$$

The correlation function corresponds to the case $p = 1$. The spin wave approximation predicts a decrease according to a power law behaviour, qualitatively different from the high temperature exponential decrease. This suggests the existence of a transition at an intermediate temperature. The exponent characterizing the decrease of $G(\mathbf{x})$ varies with temperature and vanishes linearly at zero temperature. An isotropic power law behaviour is characteristic of a critical theory. We have thus here a whole continuous zone of critical temperatures. The underlying continuous field theory describes a free massless scalar field, with the subtractions which are implied by formulae (129)–(130) and which automatically arise when one computes invariant correlation functions. The result (132) shows the importance of large wavelength fluctuations which prevent ordering.

Fluctuations of the angles increase logarithmically at large distance. Hence the approximation neglecting their periodic character might become unjustified. Indeed, as Berezinskii, Kosterlitz and Thouless have shown, the topological effects resulting from this periodicity play an essential role. They combine with the spin wave fluctuations and increase disorder, so that correlation functions will acquire an exponential decrease beyond some critical temperature.

At a sufficiently large scale with respect to the lattice spacing, the microscopic structure of the latter can be neglected. Let us study the variation of the angle between the spin and a fixed direction as one follows a simple closed curve. Implicitly, fluctuations are supposed to be weak, so that this angle varies continuously along the path. If the spins are almost aligned, the variation of course vanishes. However, one can imagine configurations such that this angle varies by an integral multiple of 2π. This is a topological invariant, in the sense that it does not change when the curve or the configuration changes continuously. The simplest case corresponds to $n = \pm 1$ and describes a *vortex* of intensity $n = \pm 1$. For a continuous field of unit length, such a configuration should present at least one singular point. On a lattice, this concept is meaningless on a microscopic scale.

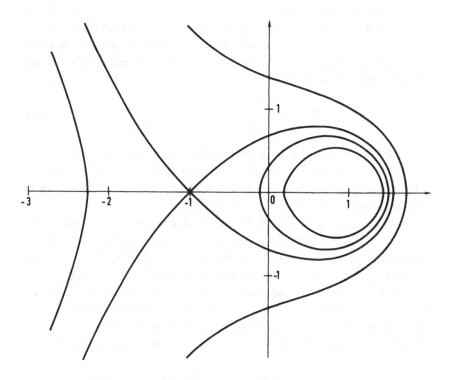

Fig. 10   A vortex pair of intensities ±1.

We assume now that all spins are aligned on the boundary of the system, so that the total vorticity (i.e. the sum of the n_i's assigned to singular points) vanishes.

Figure 10 displays an example of two vortices with intensity ±1. Identifying the (x, y) plane with the complex plane $z = x + iy$, we choose $S_x + iS_y = e^{i\theta} = \exp(i\mathrm{Arg}[i(z - z_0)/(z + z_0)])$. Lines are tangent to the vector field $\mathbf{S}(x, y)$, with $z_0 = 1$, and their equation is given by $y^2 = Ae^{-(x+1)} - (x - 1)^2$, with $0 < A < \infty$. This particular example of a function $\theta(x, y)$, harmonic everywhere except at the points $\pm z_0$ (singularities of $\nabla \times \mathbf{S}$), is not arbitrary. It corresponds to an extremum of the classical action which is the continuous counterpart of (124)

$$S_{\mathrm{class}} = \tfrac{1}{2}\beta \int \mathrm{d}^2\mathbf{x}(\nabla\,\theta)^2 \qquad (133)$$

The classical field equation is $\Delta\theta = 0$. We require that θ be constant at infinity. Due to the fundamental property of harmonic functions, either θ is constant or has singularities. Assuming only isolated singularities $z_i^{\pm}$ around which θ varies by $\pm 2\pi$ as one turns anticlockwise, one finds

$$e^{i\theta} = e^{i\theta_\infty} \prod_{i=1}^{n} \frac{(z - z_i^+)/|z - z_i^+|}{(z - z_i^-)/|z - z_i^-|} \tag{134}$$

Vortices of intensity ± 1 are centered at the points $z_i^{\pm}$. Intensities of higher magnitude are obtained when several z_i^+ (or z_i^-) coincide. The angle θ is a multivalued function, but $e^{i\theta}$ (or the spin) is well-defined everywhere, except at the singular points.

The classical action (133) is infinite for singular configurations such as (134). However, within the framework of the approximations that were made, it is consistent to regularize the action by excluding from the integration domain a small region – e.g. a disk of radius r_0 – centered on each singular point, and small with respect to the distances between vortices. Let D be the remaining integration domain. In order to simplify the notation, we relabel the points $z_i^{\pm}$ as z_j, denoting their intensities as q_j, and write

$$F = \rho e^{i\theta} = e^{i\theta_\infty} \prod_{i=1}^{n} \frac{z - z_i^+}{z - z_i^-} = e^{i\theta_\infty} \prod_{j=1}^{2n}(z - z_j)^{q_j} \tag{135}$$

In the domain D, θ and $\ln\rho$ are harmonic functions and, as a consequence of the Cauchy–Riemann equations, $(\nabla\ \theta)^2 = (\nabla \ln\rho)^2 = \nabla (\ln\rho \, \nabla \ln\rho)$. The classical action can now be evaluated for sufficiently small r_0. If C_j is the circle surrounding z_j,

$$S(\{q_j, z_j\}) = \tfrac{1}{2}\beta \int_D d^2x (\nabla\ \theta)^2 = -\tfrac{1}{2}\beta \sum_j \oint_{C_j} ds \ln\rho \, \mathbf{n} \cdot \nabla \ln\rho$$

$$= -2\pi\beta \sum_{(ij)} q_i q_j \ln|z_i - z_j| + \sum_i q_i^2 \pi\beta \ln 1/r_0 \tag{136}$$

In this formula, distances are expressed in terms of the lattice spacing. However, the complete expression is independent from this choice of unit, due to the relation $\sum q_j = 0$ which agrees with the scale invariance of the action.

Consider the case of a vortex pair, at distance r_{12}. Its statistical weight is $\exp(-2\pi\beta \ln r_{12}/r_0)$ (according to (136)). In a box of linear dimension L, the global contribution of one vortex pair to the total free energy is, for dimensional reasons, of the form

$$\exp F_{\text{pair}} \sim \int_{|\mathbf{x}_1 - \mathbf{x}_2| > r_0} d^2\mathbf{x}_1 d^2\mathbf{x}_2 \exp\left(-2\pi\beta \ln \frac{|\mathbf{x}_1 - \mathbf{x}_2|}{r_0}\right)$$

$$\sim L^4 \exp\left(-2\pi\beta \ln L/r_0\right) \tag{137}$$

$$F_{\text{pair}} \sim (4 - 2\pi\beta) \ln L$$

If $2\pi\beta > 4$, this contribution is negligible in the limit $L \to \infty$. On the other hand, if $2\pi\beta < 4$, an instability appears. The creation of well-separated vortices is favoured and disorder increases. Hence a first estimate of the critical temperature is

$$2\pi\beta_{\text{c}} = 4 \tag{138}$$

More generally, choosing a length scale a, the action is rewritten near a stationary configuration as

$$S = S(q_i, \mathbf{x}_i) + \tfrac{1}{2}\beta \int d^2\mathbf{x}(\nabla\, \theta_{\text{fluc}})^2 \qquad \text{with} \quad \theta = \theta_{\text{conf}} + \theta_{\text{fluc}}$$

Summing over all vortex sectors yields an approximate partition function

$$Z = Z_{\text{sw}} \sum_n \frac{1}{n!^2} \exp\left(-2\pi\beta n \ln \frac{a}{r_0}\right)$$

$$\times \int \prod_j d^2\mathbf{x}_j \exp\left\{2\pi\beta \sum_{(ij)} q_i q_j \ln \frac{|\mathbf{x}_i - \mathbf{x}_j|}{a}\right\} \tag{139}$$

The (regularized) term $\pi\beta \ln a/r_0$ plays the role of the chemical potential for the $2n$ vortices of intensity $q_j = \pm 1$ located at $\mathbf{x}_j$. The factor $n!$ for each species takes into account their indistinguishability, and the contribution of Gaussian fluctuations is factorized. Up to this term Z_{sw}, the partition function describes a gas of classical charged particles, with a Coulomb interaction (in two dimensions) and globally neutral. In this interpretation, the kinetic term has been integrated and absorbed into a redefinition of the chemical potential. The corresponding charges in this picture are $\sqrt{2\pi}q_j$. The preceding heuristic considerations can be translated in this interpretation. In a low temperature phase, charges are bound and one has a dielectric medium. Above the critical temperature, one has a plasma of free charges.

4.2.3 The Villain action

It is also possible to obtain a Coulomb model, starting from a slightly modified version of the action due to Villain. In the expansion (119)–(120) of the exponentiated action, the coefficients are replaced by their asymptotic form for large β, i.e. $\exp(-n^2/2\beta)$. The Boltzmann weight is replaced by the periodic Jacobi function, solution of the heat equation on a circle,

$$z(\theta) = \sum_{n=-\infty}^{+\infty} \exp\left(-\frac{n^2}{2\beta} + in\theta\right) = \sqrt{2\pi\beta} \sum_{p=-\infty}^{+\infty} \exp\left(-\frac{1}{2}\beta(\theta - 2\pi p)^2\right)$$

$$(140)$$

Prove that the two forms given in (140) are equivalent, using the Poisson summation formula.

$$f(x) = \int_{-\infty}^{+\infty} \frac{dk}{2\pi} e^{ikx} \tilde{f}(k) \qquad \rightarrow \qquad \sum_{-\infty}^{+\infty} f(n) = \sum_{-\infty}^{+\infty} \tilde{f}(2\pi p)$$

$$(141)$$

The second form given in (140) is useful for large β; it appears as a sum of Gaussians of width $1/\sqrt{\beta}$ centered on the values $2\pi p$ (this ensures the periodicity in θ). The partition function reads

$$Z_V = \int \prod_i \frac{d\theta_i}{2\pi} \prod_{(ij)} z(\theta_i - \theta_j)$$

$$(142)$$

This form is suitable for a duality transformation. Integrating over the angles θ, one finds

$$Z_V = \sum_{\substack{\{n_{ij}\} \\ (\partial n)_i = 0}} \exp\left(-\frac{1}{2\beta} \sum_{(ij)} n_{ij}^2\right)$$

$$(143)$$

In this expression, a relative integer $n_{ij} \equiv -n_{ji}$ has been assigned to each link, and the condition $(\partial n)_i = 0$ means $\sum_j n_{ij} = 0$, as previously. On a two-dimensional square lattice, these conditions can be easily solved. To each plaquette (elementary square), or, equivalently, to its center which is a node of the dual lattice (isomorphic to the initial one), one assigns an integral variable m_a. Each link ij carries the difference between those variables

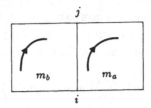

Fig. 11   Dual variables for the two-dimensional Villain model.

relative to the right and left adjacent plaquettes, $n_{ij} = m_a - m_b$ (figure 11).

In these variables, the Villain model is transformed by duality into a Gaussian discrete model

$$Z_V = \sum_{\{m_a\}} \exp\left(-\frac{1}{2\beta}\sum_{(ab)}(m_a - m_b)^2\right) \qquad (144)$$

This solution of the equations $(\partial n)_i = 0$ in the form $n_{ij} = m_a - m_b$, leaves an arbitrariness $m_a \to m_a + m$; m is any integer. This ambiguity leads to an infinity which should be avoided, e.g. by omitting the summation over one of the variables m_a. The duality has here three effects. The global symmetry group $O(2)$ is replaced by its dual, the additive group of integers. The lattice is replaced by its dual, which is here isomorphic to the original one. Finally, β is changed to $1/\beta$.

Configurations can be visualized as surfaces in three dimensions joining plaquettes at height m_a above the lattice basis. The Gaussian discrete model describes thus a fluctuating surface, as does a so-called *SOS* (solid on solid) model in which the term $(m_a - m_b)^2$ is replaced by $|m_a - m_b|$, with an action which measures the deviation with respect to a planar surface. For small β, configurations with equal m_a are favoured. The cost of a unit step is $e^{-2/\beta}$. As β increases and approaches β_c, the surface, which was a smooth one up to microscopic fluctuations, changes structure in such a way that two distant points have uncorrelated heights. The surface is said to become *"rough"* (*roughening transition*). This model is used in particular to describe the shape of crystals in equilibrium with a solution.

For sufficiently large β, the discrete Gaussian model behaves nearly as its continuous counterpart, i.e. as a spin wave system. This can be exhibited by substituting to the sum over discrete variables an integration over continuous ones φ, according to

$$\sum_{m=-\infty}^{+\infty} f(m) = \int d\varphi f(\varphi) \sum_{m=-\infty}^{+\infty} \delta(\varphi - m) = \int d\varphi f(\varphi) \sum_{q=-\infty}^{+\infty} e^{2i\pi q\varphi}$$

(145)

$$Z_V = \sum_{\{q_a\}} \int \prod_a d\varphi_a \exp\left\{ -\frac{1}{2\beta} \sum_{(ab)} (\varphi_a - \varphi_b)^2 + 2i\pi \sum_a q_a \varphi_a \right\}$$

(146)

In this expression, the variables $\{\varphi_a\}$ describe continuous Gaussian fluctuations and the discrete ones $\{q_a\}$ the intensity of vortices. The factorization described in the preceding subsection is explicitly realized, and the formulae become identical after a change of notations $\varphi = \beta\theta_{\text{fluc}}$. Global translational invariance of the fields φ is insured by the neutrality condition $\sum_a q_a = 0$. Other configurations have a vanishing weight in the thermodynamic limit.

This expression also allows one to understand the relation between the XY-model with yet another version, the so-called *sine-Gordon* model. The name (of somewhat dubious taste) arises from a nonlinear modification of the Klein–Gordon equation. For large β, vortices have an almost negligible effect, and the effect of the terms involving the variables q_a essentially introduce a periodic modulation in the integral over the fields φ_a. The action takes the sine-Gordon form

$$\frac{1}{2\beta} \sum_{(ab)} (\varphi_a - \varphi_b)^2 - \sum_a 2\cos 2\pi\varphi_a + \cdots$$

which is also suitable for a study of the XY-model.

Alternatively, integrating in equation (146) over the variables φ and taking into account neutrality, one obtains, without any approximation, a Coulomb gas of vortices

$$Z_V = Z_{\text{sw}} \sum_{\{q_a\}} \exp\left\{ 2\pi\beta \sum_{(ab)} q_a \Gamma(\mathbf{x}_a - \mathbf{x}_b) q_b \right\}$$ (147)

This is equivalent, to a very good approximation, to

$$Z_V = Z_{\text{sw}} \sum_{\{q_a\}} \exp \left\{ 2\pi\beta \sum_{(ab)} q_a \ln \frac{|\mathbf{x}_a - \mathbf{x}_b|}{a} q_b - \pi\beta \sum_a q_a^2 \ln \frac{a}{r_0} \right\}$$

(148)

The Villain model thus justifies the semiclassical derivation of the previous subsection.

As the effective chemical potential $\pi\beta \ln a/r_0$ is large (or the activity $y = \exp -\pi\beta \ln a/r_0$ is small), the first nontrivial configuration is a vortex pair of intensity ± 1 located respectively at $\mathbf{x}_1$ and $\mathbf{x}_2$. The corresponding correlation function is negative, since the intensities are opposite, and behaves as

$$\langle q(\mathbf{x}_1) q(\mathbf{x}_2) \rangle = -y^2 \exp -2\pi\beta \ln(|\mathbf{x}_1 - \mathbf{x}_2|/a)$$
$$\simeq -y^2 \left(a/|\mathbf{x}_1 - \mathbf{x}_2| \right)^{2\pi\beta}$$

(149)

when $|\mathbf{x}_1 - \mathbf{x}_2| \gg a$ and $y \to 0$. According to the neutrality condition requiring that

$$\sum_{\mathbf{x}_1} \langle q(\mathbf{x}_1) q(\mathbf{x}_2) \rangle = 0$$

one also has

$$\langle q^2(\mathbf{0}) \rangle \simeq y^2 \sum_{\mathbf{x} \neq 0} \left(a/|\mathbf{x}| \right)^{2\pi\beta}$$

(150)

As $\beta \to \infty$, few vortices are present and their correlations decrease rapidly with the relative distance. This describes a dielectric medium of neutral bound states. It is natural to perform an expansion in powers of y, considered as a parameter independent of β.

Let us now examine the effects of vortices on the spin correlations, already computed in the spin wave approximation (equations (128)–(132)).

4.2.4   Correlations

In the framework of the Villain model, let us compute a correlation

$$\left\langle e^{ip(\theta_1 - \theta_2)} \right\rangle = Z_V' / Z_V$$

(151)

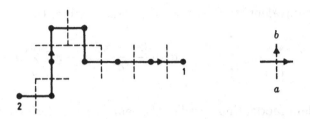

Fig. 12   The line C joining $\mathbf{x}_2$ to $\mathbf{x}_1$ in the computation of correlations. On the dual links, $\eta_{ab} = \pm 1$, according to the indicated orientation prescription.

In this ratio, Z'_V has a form similar to the sum (143) obtained after integration over the angles θ, i.e.

$$Z'_V = \sum_{\substack{\{n_{ij}\} \\ (\partial n)_i = p(\delta_{i,1} - \delta_{i,2})}} \exp\left(-\frac{1}{2\beta}n_{ij}^2\right) \tag{152}$$

In order to define the dual expression, one has to solve the condition $(\partial n)_i = p(\delta_{i,1} - \delta_{i,2})$. Let us draw on the lattice an arbitrary line C joining the points 1 and 2, and oriented, e.g. from 2 to 1. On the links of the dual lattice, we define an antisymmetric function η_{ab}, vanishing if the link ab does not cross C and equal to ± 1 according to the relative orientations of C and ab if they cross (figure 12). Accordingly, it is possible to define integral variables m_a on the dual lattice sites such that

$$n_{ij} = m_a - m_b + p\eta_{ab} \tag{153}$$

As previously, summing over all these variables but one,

$$Z'_V = \sum_{\{m_a\}} \exp -\frac{1}{2\beta} \sum_{(ab)} (m_a - m_b + p\eta_{ab})^2$$

$$= \sum_{\{q_a\}} \int \prod_a \mathrm{d}\varphi_a \exp\left\{ -\frac{1}{2\beta} \sum_{(ab)} (\varphi_a - \varphi_b + p\eta_{ab})^2 + 2\mathrm{i}\pi \sum_a q_a \varphi_a \right\} \tag{154}$$

Integration over the φ variables is readily performed. Defining

$$\eta_a = \sum_b \eta_{ab}$$

$$q'_a = -\frac{p}{2\mathrm{i}\pi\beta}\eta_a \tag{155}$$

and using the fact that $\sum_a q'_a = 0$ (due to the antisymmetry of η_{ab}), one gets

$$Z'_V = Z_{sw} \exp\left\{-\frac{p^2}{2\beta}\sum_{(ab)}\eta^2_{ab}\right\}$$

$$\times \sum_{\{q_a\}} \exp\left\{2\pi\beta\sum_{(ab)}(q_a + q'_a)\Gamma(\mathbf{x}_a - \mathbf{x}_b)(q_b + q'_b)\right\}$$

$$(156)$$

Taking into account the definitions (155) and the neutrality condition, the correlation factorizes as

$$\left\langle e^{ip(\theta_1-\theta_2)}\right\rangle = G_{\text{wave}}(\mathbf{x}_1 - \mathbf{x}_2)G_{\text{Coulomb}}(\mathbf{x}_1 - \mathbf{x}_2) \qquad (157)$$

with

$$\begin{cases} G_{\text{wave}}(\mathbf{x}_1 - \mathbf{x}_2) = \exp\left(-\frac{p^2}{2\beta}\sum_{(ab)}\eta^2_{ab} - \frac{p^2}{2\pi\beta}\sum_{ab}\eta_a\Gamma(\mathbf{x}_a - \mathbf{x}_b)\eta_b\right) \\[2ex] G_{\text{Coulomb}}(\mathbf{x}_1 - \mathbf{x}_2) = \left\langle \exp ip \sum_a q_a \sum_b \Gamma(\mathbf{x}_a - \mathbf{x}_b)\eta_b\right\rangle_{\text{Coulomb}} \end{cases}$$

$$(158)$$

The last mean value involves the weights corresponding to (148). The label of the first factor G_{wave} is justified as follows. The exponent on the r.h.s. is written as $-(p^2/2\beta)\gamma(\mathbf{x}_a - \mathbf{x}_b)$ such that

$$\gamma(\mathbf{x}_a - \mathbf{x}_b) = \sum_{(ab)}\eta^2_{ab} - \sum_{a,b}\eta_a G(\mathbf{x}_a - \mathbf{x}_b)\eta_b \qquad (159)$$

with

$$G(\mathbf{x}) = \int_{-\pi}^{\pi} \frac{d^2\mathbf{q}}{(2\pi)^2}\frac{e^{i\mathbf{q}\cdot\mathbf{x}}}{4 - 2\sum_{\mu=1}^{2}\cos q_\mu}$$

The infrared divergence is automatically subtracted by virtue of the neutrality condition. The function η_a is a discretized version of a dipole potential on the curve C, as indicated in a particular case on figure 13. We perform the computations in this particular case with $\mathbf{x}_1$ and $\mathbf{x}_2$ along a principal axis of the lattice. With $r = |\mathbf{x}_1 - \mathbf{x}_2|$, one has

$$\gamma(r) = r - \sum_{x,x'=1}^{r-1} \{2G(x - x', 0) - G(x - x', 1) - G(x - x', -1)\}$$

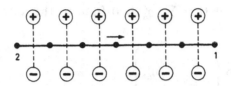

Fig. 13   The dipole potential η_a.

Since
$$4G(x,y) - [G(x+1,y) + G(x-1,y) + G(x,y+1) + G(x,y-1)] = \delta_{x,0}\delta_{y,0}$$

this is also
$$\gamma(r) = r - \sum_{x,x'=0}^{r-1} \left\{ \delta_{x,x'} - 2G(x-x',0) + G(x-x'+1,0) + G(x-x'-1,0) \right\}$$
$$= \sum_{x=0}^{r-1} \left\{ G(r-1-x,0) - G(r-x,0) + G(-x,0) - G(-x-1,0) \right\}$$
$$= 2\left\{ G(0) - G(r) \right\} = \Gamma(r)/\pi$$

Returning to G_{wave}, one obtains as expected
$$G_{\text{wave}}(\mathbf{x}_1 - \mathbf{x}_2) = \exp\left[-\frac{p^2}{2\pi\beta}\Gamma(|\mathbf{x}_1 - \mathbf{x}_2|) \right] \qquad (160)$$

In the evaluation of G_{Coulomb}, we use a cumulant expansion. To first order, $\langle q_a \rangle = 0$, and therefore

$G_{\text{Coulomb}}(\mathbf{x}_1 - \mathbf{x}_2) \simeq$

$$\exp\left(-\tfrac{1}{2}p^2 \sum_{a,b} \langle q(\mathbf{x}_a)q(\mathbf{x}_b)\rangle \sum_{a'} \Gamma(\mathbf{x}_a - \mathbf{x}_{a'})\eta_{a'} \sum_{b'} \Gamma(\mathbf{x}_b - \mathbf{x}_{b'})\eta_{b'} \right) \quad (161)$$

The correlation $\langle q(\mathbf{x}_a)q(\mathbf{x}_b)\rangle$ has already been computed (equations (149)–(150)). Sums over a' or b' are estimated in the limit where $\mathbf{x}_a$ is far from the curve C, compared to the lattice spacing. In this limit, we use the formalism of analytic functions. If C is along the x axis,

$$\sigma_a = \sum_{a'} \Gamma(\mathbf{x}_a - \mathbf{x}_{a'})\eta_{a'} \simeq \int_{x_2}^{x_1} dx \frac{\partial}{\partial y} \ln|z_a - z|_{z=x}$$
$$= -\int_{x_2}^{x_1} dx \frac{\partial}{\partial x} \text{Im} \ln(z_a - x) = -\text{Im } \ln\left(\frac{z_a - z_1}{z_a - z_2} \right) \qquad (162)$$

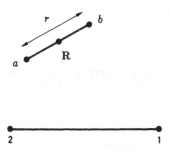

Fig. 14   Notations used in the computation of the effect of a vortex pair on the spin correlation function.

where $z_i = x_i$. Let z_R be the center of the segment ab. Pairs are strongly bound since the correlation $\langle q_a q_b \rangle$ decreases rapidly with the relative distance of the points ab. It is thus possible to expand $\sigma_{a,b}$ as

$$\sigma(\mathbf{R}) \pm \frac{1}{2}(\mathbf{x}_a - \mathbf{x}_b).\, \boldsymbol{\nabla}\, \sigma(\mathbf{R})$$

Hence

$$G_{\text{Coulomb}}(\mathbf{x}_1 - \mathbf{x}_2) \simeq \exp -\tfrac{1}{2}p^2 \sum_{a,b} \langle q(\mathbf{x}_a)q(\mathbf{x}_b)\rangle\, \sigma(\mathbf{x}_a)\sigma(\mathbf{x}_b)$$

$$\simeq \exp\left(\tfrac{1}{8}p^2 \sum_{\mathbf{r}} \langle q(\mathbf{0})q(\mathbf{r})\rangle \sum_{\mathbf{R}} [\mathbf{r}.\,\boldsymbol{\nabla}\,\sigma(\mathbf{R})]^2\right)$$

$$\tag{163}$$

If (X, Y) stand for the coordinates of $\mathbf{R}$,

$$\frac{\partial\sigma}{\partial X} = \frac{Y - y_1}{|\mathbf{R} - \mathbf{x}_1|^2} - \frac{Y - y_2}{|\mathbf{R} - \mathbf{x}_2|^2} \qquad \frac{\partial\sigma}{\partial Y} = -\frac{X - x_1}{|\mathbf{R} - \mathbf{x}_1|^2} + \frac{X - x_2}{|\mathbf{R} - \mathbf{x}_2|^2}$$

$$\tag{164}$$

Furthermore, $\langle q(\mathbf{0})q(\mathbf{r})\rangle$ is essentially a function of $r = |\mathbf{r}|$ only, so that one can perform an average over the orientation of $\mathbf{r}$,

$$\sum_{\mathbf{R}} [\mathbf{r}.\,\boldsymbol{\nabla}\,\sigma(\mathbf{R})]^2 \quad \rightarrow \quad \tfrac{1}{2}r^2 \sum_{\mathbf{R}} [\boldsymbol{\nabla}\,\sigma(\mathbf{R})]^2$$

According to equation (164), we have

$$[\nabla\sigma(R)]^2 = [\nabla\Gamma(R - x_1) - \nabla\Gamma(R - x_2)]^2$$

As $\Delta\Gamma(R) = 2\pi\delta_{R,0}$,

$$\tfrac{1}{2}r^2 \sum_R [\nabla\, \sigma(\mathbf{R})]^2$$

$$= -\tfrac{1}{2}r^2 \sum_R [\Gamma(\mathbf{R} - \mathbf{x}_1) - \Gamma(\mathbf{R} - \mathbf{x}_2)]\Delta[\Gamma(\mathbf{R} - \mathbf{x}_1) - \Gamma(\mathbf{R} - \mathbf{x}_2)]$$

$$= 2\pi r^2 \Gamma(\mathbf{x}_1 - \mathbf{x}_2)$$

Hence the final expression is

$$G_{\text{Coulomb}}(\mathbf{x}_1 - \mathbf{x}_2) \simeq \exp\left(\tfrac{1}{4}\pi p^2 \Gamma(\mathbf{x}_1 - \mathbf{x}_2) \sum_{|\mathbf{r}|>a} \mathbf{r}^2 \langle q(\mathbf{0})q(\mathbf{r})\rangle\right)$$

$$(165)$$

This result is very similar to the one given by the spin wave approximation (160), and differs by the coefficient of the exponent. Multiplying (160) and (165), one gets

$$\left\langle e^{ip(\theta_1 - \theta_2)}\right\rangle \simeq \exp\left\langle -\frac{p^2}{2\pi\beta_{\text{eff}}}\Gamma(\mathbf{x}_1 - \mathbf{x}_2)\right\rangle$$

$$\frac{1}{\beta_{\text{eff}}} = \frac{1}{\beta} - \frac{1}{2}\pi^2 \sum_{|\mathbf{r}|>a} \mathbf{r}^2 \langle q(\mathbf{0})q(\mathbf{r})\rangle$$

$$(166)$$

The effect of taking into account the vortices is to increase the effective temperature β^{-1}, since equations (149)–(150) lead to

$$\frac{1}{\beta_{\text{eff}}} = \frac{1}{\beta} + \pi^3 y^2 \int_a^\infty \frac{dr}{a} \left(\frac{a}{r}\right)^{2\pi\beta-3} \qquad y = \exp\left(-\pi\beta\ln\frac{a}{r_0}\right)$$

$$(167)$$

in which we reinstated the cutoff factor a. As a consequence, the correlation, which behaves as $r^{-p^2/2\pi\beta_{\text{eff}}}$, decrease more rapidly than in the spin wave approximation. The integral (167) has an infrared divergence for $2\pi\beta = 4$, a signal of the failure of the perturbative calculation when vortices become widely separated. This yields the same estimate as before for the critical temperature. We can now establish heuristically the renormalization group equations governing the critical behaviour.

Derive similar equations for higher order correlation functions.

4.2.5   *Renormalization flow*

We have just computed the effects of vortices on the correlation function to lowest order in the activity y, and found that the corresponding effective temperature increases. The parameters y and β are dimensionless, and are defined at the microscopic length scale a, the lattice spacing. The question is now to obtain the values y_λ and β_λ at a larger scale λa, in order that the correlation functions pertaining to this new scale retain their original values. We have to assume that y_λ remains sufficiently small, in order that the above calculation retains its validity. In formula (167), the integral from a to infinity is split into two parts, one from a to λa, and the second from λa to infinity. With $x = r/a$, this yields

$$\int_1^\infty dx\; x^{3-2\pi\beta} = \int_1^\lambda dx\; x^{3-2\pi\beta} + \int_\lambda^\infty dx\, x^{3-2\pi\beta}$$

$$= \frac{\lambda^{4-2\pi\beta} - 1}{4 - 2\pi\beta} + \lambda^{4-2\pi\beta} \int_1^\infty dx\; x^{3-2\pi\beta}$$

$$\underset{\lambda \to 1}{\sim} \ln\lambda + [1 + (4 - 2\pi\beta)\ln\lambda] \int_1^\infty dx\, x^{3-2\pi\beta}$$

Using this expression in the neighborhood of $\lambda = 1$, one finds

$$\frac{1}{\beta_{\text{eff}}} = \frac{1}{\beta} + \pi^3 y^2 \ln\lambda + \pi^3 y^2 [1 + (4 - 2\pi\beta)\ln\lambda] \int_1^\infty dx\; x^{3-2\pi\beta}$$

As $\lambda \to 1, \ln\lambda \sim \lambda - 1$ is small and the formula can be recast into a similar form

$$\frac{1}{\beta_{\text{eff}}} = \frac{1}{\beta_\lambda} + \pi^3 y_\lambda^2 \int_1^\infty dx\; x^{3-2\pi\beta_\lambda} \qquad\qquad (168a)$$

with

$$\frac{1}{\beta_\lambda} = \frac{1}{\beta} + \pi^3 y^2 \ln\lambda$$

$$y_\lambda = y + (2 - \pi\beta)y \ln\lambda \qquad\qquad (\lambda \to 1) \qquad\qquad (168b)$$

These relations are rewritten in differential form

$$\begin{cases} \lambda \dfrac{d}{d\lambda}\beta_\lambda^{-1} = \pi^3 y_\lambda^2 \\[2mm] \lambda \dfrac{d}{d\lambda}y_\lambda = (2 - \pi\beta_\lambda)y_\lambda \end{cases} \qquad\qquad (169)$$

Although the preceding reasoning might look somehow dubious, since we obtain the variation of two parameters using a single

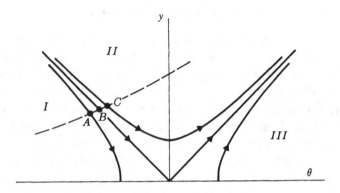

Fig. 15   The Kosterlitz-Thouless renormalization flow.

condition, the result agrees with a more elaborate analysis. The system (169) describes the renormalization flow of the bare coupling constants as the cut-off scale varies, and has been analyzed by Kosterlitz. Initial conditions are, for $\lambda = 1, \beta_1 = \beta$ and $y_1 = y = \exp(-\pi\beta \ln a/r_0)$. The system is only valid for small y_λ. Additional terms involving higher powers in y should become relevant outside this domain. The equations simplify when one introduces the reduced variables

$$\Theta_\lambda = 2 - \pi\beta_\lambda$$
$$Y_\lambda = 2\pi y_\lambda \tag{170}$$

in which case

$$\begin{cases} \lambda\dfrac{d}{d\lambda}\Theta_\lambda = Y_\lambda^2(1 - \dfrac{1}{2}\Theta_\lambda)^2 \simeq Y_\lambda^2 \\[2mm] \lambda\dfrac{d}{d\lambda}Y_\lambda^2 = 2\Theta_\lambda Y_\lambda^2 \end{cases} \tag{171a}$$

Near the fixed point $\Theta = 0, Y = 0$, the integral curves are hyperbolae

$$\Theta_\lambda^2 - Y_\lambda^2 = Cst \tag{171b}$$

In view of the second equation (171a), as λ grows, the axis $Y = 0$ is attractive for $\Theta < 0$ and repulsive for $\Theta > 0$. The renormalization flow is displayed in figure 15. In the domain I, renormalization drives the representative point to a critical one characterized by $Y_\infty = 0, \Theta_\infty \simeq -\sqrt{\Theta^2 - Y^2} < 0$. This is a scale invariant low temperature fixed point, and the correlations decrease according to a power law $|\mathbf{x}|^{-1/2\pi\beta_\infty}$. In regions II and III, the renormalization flow escapes rapidly from the domain of validity for the approximations. The effective activity for vortices Y_λ increases almost linearly with Θ, and one approaches a high temperature fixed point with free vortices (the plasma phase of the Coulomb gas, or the smooth phase in the language of surfaces), characterized by an exponential decay of correlations. The separating line is $Y = -\Theta$, along which the representative point flows towards the limiting point $Y = 0, \Theta = 0$. The figure also displays the initial conditions for $\lambda = 1, y = \exp(-\pi\beta \ln a/r_0)$, as a dashed curve

$$Y = 2\pi e^{(\Theta-2)\ln a/r_0} \simeq 2\pi e^{-2\ln a/r_0}(1 + \Theta \ln \frac{a}{r_0} + \cdots) \qquad (172)$$

in which $\ln a/r_0 = \ln 2\sqrt{2}e^\gamma \simeq 1.6$. This can be improved in a more detailed treatment. In any case, this is a nonuniversal curve, depending on the details of the model. For the Villain model, the intersection of the curve (172) with $Y = -\Theta$ leads to an estimate of the critical temperature

$$Y_c = -\Theta_c = \pi\beta_c - 2 \simeq \left(\ln \frac{a}{r_0} + \frac{e^{2\ln a/r_0}}{2\pi}\right)^{-1} \qquad (173)$$

The parametrization in λ of the renormalization flow trajectories near the origin can also be found. In region I, we have $\Theta_\lambda^2 = Y_\lambda^2 + \Theta_\infty^2$. From the initial values $\Theta_1 = \Theta, Y_1 = Y$, one gets $\Theta_\infty^2 = \Theta^2 - Y^2$. Solving (171), one obtains ($\Theta_\infty < 0$)

$$(I) \quad \begin{cases} \Theta_\lambda = \Theta_\infty \dfrac{1 + (\lambda/\lambda_i)^{2\Theta_\infty}}{1 - (\lambda/\lambda_i)^{2\Theta_\infty}} \\[3mm] Y_\lambda = \sqrt{\Theta_\lambda^2 - \Theta_\infty^2} \\[3mm] \lambda_i^{-2\Theta_\infty} = \dfrac{\Theta - \Theta_\infty}{\Theta + \Theta_\infty} > 0 \end{cases} \qquad (174)$$

As $\Theta_\infty < 0$, the terms depending on initial conditions decay as power laws in λ, which is characteristic of scale invariant theories.

In region II, let $Y_0 > 0$ be the height of the hyperbola above the point $\Theta = 0, Y_\lambda^2 - \Theta_\lambda^2 = Y_0^2 = Y^2 - \Theta^2$. One introduces a variable

ψ varying from ψ_i to $\frac{1}{2}\pi$, with

$$(II) \quad \begin{cases} \Theta_\lambda = Y_0 \tan\psi & \lambda = \exp\dfrac{\psi - \psi_i}{Y_0} \\[2mm] Y_\lambda = \dfrac{Y_0}{\cos\psi} & \tan\psi_i = \dfrac{\Theta}{\sqrt{Y^2 - \Theta^2}} \end{cases} \quad (175)$$

The effective parameter Θ_λ varies monotonically from 0 to infinity, while Y_λ remains positive, increases indefinitely as $\lambda \to \infty$ and has a minimum if $\psi_i < 0$. The large positive values correspond to a domain where the approximation is not valid. However, if one starts from an initial value very close to the critical point, with $\Theta < 0, \psi_i < 0$, the renormalization flow can be followed in a region where Y_λ remains small, and the approximation can still be trusted. This is justified until one reaches a point close to the minimum of the curve Y_λ, i.e. with Y_0 corresponding to $\psi = 0$, up to a value of λ of the order of

$$\lambda \simeq e^{-\psi_i/Y_0} \qquad (176)$$

This value of λ may be identified, up to a factor, with ξ/a, where ξ is the correlation length in the disordered phase near the end point of the critical line, since it defines the scale up to which the system remains close to the critical regime. When one approaches this point, Y_0 vanishes as

$$Y_0^2 \sim (\Theta - \Theta_c) \simeq (\beta_c - \beta) \qquad (177)$$

while $\sin\psi_i = \Theta/Y$ is very close to -1, i.e. $\psi_i \sim -\frac{1}{2}\pi$. Hence,

$$\xi/a \quad \simeq \quad A \exp\frac{B}{\sqrt{\beta_c - \beta}} \qquad (178)$$

where A and B are positive constants. Thus the correlation length increases faster than any power when one approaches the critical point. In this sense, the exponent ν is infinite, and the singular part of the free energy is expected to behave for $\beta_c > \beta$ as

$$F_{\text{sing}} \simeq \xi^{-2} \simeq Cst \exp\frac{-2B}{\sqrt{\beta_c - \beta}} \qquad \beta \to \beta_c - 0 \qquad (179)$$

In particular, all derivatives are continuous at β_c. This means that the XY transition is of infinite order.

The critical point is represented as point B in figure 15. On the one hand, the temperature is smaller than the estimate

$2\pi\beta_c = 4$. On the other hand, the effect of vortices is to damp the correlations. The two effects go in opposite directions and finally the critical correlation function behaves as $r^{-1/2\pi\beta_{\rm eff}} = r^{-1/4}$ up to logarithms. Following Kosterlitz, we may analyze the effect of changing the cutoff by an infinitesimal amount, $a \to a + \varepsilon a$, in the vortex contribution. According to (168b), we have for large $r = |\mathbf{x}|$

$$G_{\rm Coulomb}(\mathbf{x}, a) \sim G_{\rm Coulomb}(\mathbf{x}, a + \varepsilon a) \ \exp -\frac{\pi^2}{2}y^2\varepsilon \ln \frac{r}{a} \qquad (180)$$

Here, the dependence on a arises also from y and β. Equation (180) means that we trust the perturbative calculation only insofar as it provides us a means to estimate the effect in the change of scale, and we conceal the rest into the correlation computed with a cutoff $a + \varepsilon a$. This can in turn be iterated, provided we use the scale dependent quantities β_λ, y_λ until $\lambda a \sim r$, at which point the vortex contribution becomes of order unity, with a vortex separation of order of the renormalized lattice spacing. Using $Y_\lambda = 2\pi y_\lambda$ as above, this means that, provided the activity remains small, the total correlation function reads

$$G(\mathbf{x}) \simeq Cst \left(\frac{a}{|\mathbf{x}|}\right)^{1/2\pi\beta} \exp -\tfrac{1}{8}\int_1^{r/a} \frac{d\lambda}{\lambda} Y_\lambda^2 \ln \frac{r}{\lambda a} \qquad (181)$$

the first factor being the spin wave contribution. Recall from equations (171) that, for small Y_λ and $\Theta_\lambda, Y_\lambda^2 d\lambda/\lambda = d\Theta_\lambda$. Let us look at the critical theory, characterized by the fact that initial conditions satisfy $\Theta_1 + Y_1 = 0$, a relation which remains valid for any scale λ. We have therefore

$$\Theta^{-1} - \Theta_\lambda^{-1} = \ln \lambda \qquad (182)$$

both Θ and Θ_λ being negative. In this case, the exponent in (181) takes the form

$$-\frac{1}{8}\int_1^{r/a} d\Theta_\lambda \ln \frac{|\mathbf{x}|}{\lambda a} \underset{r/a\to\infty}{\sim} \frac{\Theta}{8}\ln\frac{|\mathbf{x}|}{a} + \frac{1}{8}\ln\ln\frac{|\mathbf{x}|}{a} + Cst \qquad (183)$$

It is good to note that changing the upper limit by a finite factor would not affect the leading behaviour. Working to leading order in Θ (as we did in the renormalization flow equations), we can (and should) approximate $(a/|\mathbf{x}|)^{1/2\pi\beta} = (a/|\mathbf{x}|)^{1/4(1-\Theta/2)} \simeq (a/|\mathbf{x}|)^{1/4+\Theta/8+\cdots}$, so that, combining with (183), we find at criticality

$$G(\mathbf{x}) \sim \mathrm{Cst} \left(\frac{a}{|\mathbf{x}|}\right)^{1/4} (\ln |\mathbf{x}|)^{1/8} \tag{184}$$

Since $d = 2$, this means that the exponent η is given by

$$\eta = \tfrac{1}{4} \tag{185}$$

a result surprisingly identical to the one of Ising model, except for the logarithmic correction. The latter is seen to arise from the fact that the initial nonuniversal model is described by a curve in the (Y, Θ) plane, which does not go through the point $(0,0)$. The logarithmic term reflects the presence of a marginal operator conjugate to the vortex activity.

If one uses ξ^{-1} instead of $\beta - \beta_c$ as a parameter to characterize the approach to the critical point (ν is infinite), one expects (cf. equation (42)) that the susceptibility in the disordered phase behaves as

$$\chi \sim \xi^{2-\eta} \sim \exp \frac{b}{\sqrt{\beta_c - \beta}} \qquad \beta \to \beta_c - 0 \tag{186}$$

and diverges more rapidly than any power of $\beta_c - \beta$. Finally, the exponent δ describing the response to an external field is also given by scaling relations (which can be verified through a direct calculation)

$$\delta = \frac{d + 2 - \eta}{d - 2 + \eta} = 15 \tag{187}$$

again as in the Ising model.

The renormalization group ideas have therefore allowed Kosterlitz and Thouless to give a very precise description of the transition of the XY-model. There exists an exactly soluble model (in the family of six vertex models solved by Lieb) belonging to the same universality class which confirms the preceding analysis, as shown by van Beijeren.

Appendix 4.A   Two-dimensional systems with continuous symmetry

4.A.1   Magnetization inequality

We present the proof given by Hohenberg, Mermin and Wagner concerning the vanishing of the order parameter for a system with a continuous symmetry, in dimension less than or equal to two. We only discuss a n-vector model with nearest neighbour ferromagnetic interactions. However, the following proof can be easily extended to a larger class of models (short-range ferromagnetic interactions, constraint $\mathbf{S}^2 = 1$ released,...). The goal is to establish a strong enough inequality on the magnetization in the presence of an external uniform field. One uses Schwarz's inequality and the rotational invariance of the spin integration measure.

We consider a large, but finite lattice with N sites. Fourier analysis uses discrete momenta. The limit $N \to \infty$ will be implicitly taken at the end. The n-component spins $\mathbf{S_x}$ interact through the action

$$S = \beta \left(\sum_{(\mathbf{x},\mathbf{y})} \mathbf{S_x}.\mathbf{S_y} + \sum_{\mathbf{x}} \mathbf{H}.\mathbf{S_x} \right) \qquad (A.1)$$

and the statistical weight is $Z^{-1} \prod_{\mathbf{x}} d^n \mathbf{S_x} \delta(\mathbf{S_x^2} - 1) e^S$. The external field $\mathbf{H}$ is chosen along the nth coordinate axis in internal isotopic space, with $H = |\mathbf{H}|$. The magnetization

$$M(H) = \langle S_\mathbf{x}^n \rangle = \frac{1}{N} \left\langle \sum_{\mathbf{x}} S_\mathbf{x}^n \right\rangle \qquad (A.2)$$

is positive, as can be seen easily (using e.g. a high temperature expansion, which is convergent on a finite lattice). With the positive statistical weight, any pair of finite sets of observables A_i and B_i (functions of the spin variables) satisfies the Schwarz inequality

$$\left| \left\langle \sum_i A_i^* B_i \right\rangle \right|^2 \le \left\langle \sum_i A_i^* A_i \right\rangle \left\langle \sum_i B_i^* B_i \right\rangle \qquad (A.3)$$

In order to use the rotational invariance of the measure, we consider the infinitesimal generators $L_{\alpha\beta} = S^\alpha \partial/\partial S^\beta - S^\beta \partial/\partial S^\alpha$.

They obey the following identities

$$\int d^n S \delta(S^2 - 1) L_{\alpha\beta} f(S) = 0$$

$$\int d^n S \delta(S^2 - 1) \left[g(S) L_{\alpha\beta} f(S) + f(S) L_{\alpha\beta} g(S) \right] = 0 \qquad (A.4)$$

The second one follows from the first if one remarks that L is a first order differential operator. The first one is a consequence of the rotational invariance of the measure, expressed by the relation

$$0 = \frac{d}{d\theta} \int d^n S \delta(S^2 - 1) f(S^1, ..., S^\alpha \cos\theta + S^\beta \sin\theta, ..., S^\beta \cos\theta - S^\alpha \sin\theta, ..., S^n)$$

We also use the notation $L_{x,i} = S_x^n \partial/\partial S_x^i - S_x^i \partial/\partial S_x^n$ for the rotations in the plane (n, i), and $\mathbf{L_x} \equiv \{ L_{x,i} \}$, i varying here from 1 to $n - 1$.

Inequality $(A.3)$ is applied to

$$\begin{cases} \mathbf{A(k)} = \sum_{\mathbf{x}} e^{i\mathbf{k}.\mathbf{x}} \mathbf{S}_{\mathbf{x}}^{\perp} \\ \\ \mathbf{B(k)} = -\sum_{\mathbf{x}} e^{i\mathbf{k}.\mathbf{x}} \mathbf{L}_{\mathbf{x}}(S) \end{cases} \qquad (A.5)$$

In $\mathbf{B(k)}$, the operator $\mathbf{L_x}$ acts on the action S. From equations $(A.4)$, one deduces, whatever the functional C is,

$$\langle C\mathbf{B(k)} \rangle = \left\langle \sum_{\mathbf{x}} e^{i\mathbf{k}.\mathbf{x}} \mathbf{L_x}(C) \right\rangle \qquad (A.6)$$

This relation allows one to recast the various terms of $(A.3)$ in the form

$$\langle \mathbf{A(k)}^* \mathbf{B(k)} \rangle = \sum_{\mathbf{x}} \langle \mathbf{L_x}.\mathbf{S}_{\mathbf{x}}^{\perp} \rangle = (n - 1) \sum_{\mathbf{x}} \langle S_{\mathbf{x}}^n \rangle = (n - 1) N M(H)$$

$$\langle \mathbf{A(k)}^* \mathbf{A(k)} \rangle = \sum_{\mathbf{x},\mathbf{y}} e^{i\mathbf{k}.(\mathbf{x}-\mathbf{y})} \langle \mathbf{S}_{\mathbf{x}}^{\perp}.\mathbf{S}_{\mathbf{y}}^{\perp} \rangle$$

$$\langle \mathbf{B(k)}^* \mathbf{B(k)} \rangle = -\sum_{\mathbf{x},\mathbf{y}} e^{i\mathbf{k}.(\mathbf{x}-\mathbf{y})} \langle \mathbf{L_x}.\mathbf{L_y}(S) \rangle$$

$$= \beta \left\{ N(n-1) H M(H) + \sum_{(\mathbf{x},\mathbf{y})} 2(1 - \cos \mathbf{k}.(\mathbf{x}-\mathbf{y})) \right.$$

$$\left. \langle (n-1) S_{\mathbf{x}}^n S_{\mathbf{y}}^n + \mathbf{S}_{\mathbf{x}}^{\perp}.\mathbf{S}_{\mathbf{y}}^{\perp} \rangle \right\}$$

Hence the inequality

$$(n-1)^2 M^2(H) \leq \beta \frac{1}{N} \sum_{\mathbf{x},\mathbf{y}} e^{i\mathbf{k}.(\mathbf{x}-\mathbf{y})} \left\langle \mathbf{S}_{\mathbf{x}}^{\perp}.\mathbf{S}_{\mathbf{y}}^{\perp} \right\rangle \left((n-1)HM(H) \right.$$

$$\left. + \frac{1}{N} \sum_{(\mathbf{x},\mathbf{y})} 2(1 - \cos \mathbf{k}.(\mathbf{x}-\mathbf{y})) \left\langle (n-1)S_{\mathbf{x}}^n S_{\mathbf{y}}^n + \mathbf{S}_{\mathbf{x}}^{\perp}.\mathbf{S}_{\mathbf{y}}^{\perp} \right\rangle \right)$$

Each factor in the r.h.s. is positive. In the second brace, the mean value (which is multiplied by a positive factor) can be bounded by

$$\left\langle (n-2)S_{\mathbf{x}}^n S_{\mathbf{y}}^n + \mathbf{S}_{\mathbf{x}}.\mathbf{S}_{\mathbf{y}} \right\rangle \leq (n-2) + 1 = n - 1$$

and one has

$$\frac{1}{N} \sum_{(\mathbf{x},\mathbf{y})} 2(1 - \cos \mathbf{k}.(\mathbf{x}-\mathbf{y})) = 2 \sum_{\mu=1}^{d} (1 - \cos k_\mu)$$

Therefore, the inequality reads

$$\frac{\beta^{-1}(n-1)M^2(H)}{2 \sum_{\mu=1}^{d}(1 - \cos k_\mu) + HM(H)} \leq \frac{1}{N} \sum_{\mathbf{x},\mathbf{y}} e^{i\mathbf{k}.(\mathbf{x}-\mathbf{y})} \left\langle \mathbf{S}_{\mathbf{x}}^{\perp}.\mathbf{S}_{\mathbf{y}}^{\perp} \right\rangle$$

We now take the thermodynamic limit and integrate over $\mathbf{k}$ in the Brillouin zone

$$\beta^{-1}(n-1)M^2(H) \int \frac{d^d\mathbf{k}}{(2\pi)^d} \frac{1}{2 \sum_{\mu=1}^{d}(1 - \cos k_\mu) + HM(H)}$$

$$\leq \left\langle \left(\mathbf{S}_{\mathbf{x}}^{\perp} \right)^2 \right\rangle \leq 1 \quad (A.7)$$

The result is valid for any integral dimension. Of course, this upper bound is quite loose, but is sufficient to exclude a spontaneous magnetization $M = \lim_{H \to 0} M(H)$ in dimensions 1 (as expected) and 2 (which is the required theorem). Indeed, the integral of the free propagator with a square mass $HM(H) \leq H$ remains finite as $H \to 0$ only when the dimension is strictly greater than 2. In dimension 1 or 2, it diverges, so that the magnetization should vanish in order to satisfy the inequality. More precisely, if $d = 1$, the integral is $\frac{1}{2}[HM(H) + H^2 M^2(H)/4]^{-\frac{1}{2}}$. Hence

$$0 \leq M(H) \leq \left(\frac{2\beta}{n-1} \right)^{2/3} H^{1/3} \left(1 + \frac{H^2}{4} \right)^{1/3} \qquad (d=1)$$

$$(A.8)$$

and M vanishes with H, whatever the value of β. This result is not sufficiently strong to reproduce the known linear behaviour of M with respect to H, with a coefficient (susceptibility) diverging as $\beta \to \infty$. However, it is sufficient to eliminate the possibility of a phase with a nonvanishing order parameter. The interesting case leading to a new result is $d = 2$. The integral is logarithmically divergent as H vanishes, and, for sufficiently small H,

$$0 \le M(H) \le \sqrt{\frac{4\pi\beta}{n-1}} \frac{1}{\sqrt{\text{Cst} - \ln H}} \qquad (d = 2) \qquad (A.9)$$

Again, a spontaneous magnetization is excluded at any temperature. We note that this inequality is uniformly valid both in H and β.

4.A.2   Correlation inequality

The ideas developed in the context of the two-dimensional XY-model have been used by McBryan and Spencer to deduce a polynomial decrease of the correlations in the absence of external field. The proof can be generalized to fields with more than two components; however, the corresponding bound is very weak, since in this case the correlations decrease exponentially, but is of course sufficient to establish the absence of a spontaneous magnetization.

Consider the two-point function, in the two-component case

$$0 \le \left\langle \mathbf{S_x}.\mathbf{S_y} \right\rangle = Z^{-1}\text{Re} \int \prod_y d\theta_y$$

$$\times \exp\left\{ \beta \sum_{(\mathbf{x'},\mathbf{x''})} \cos(\theta_{\mathbf{x'}} - \theta_{\mathbf{x''}}) + \text{i}(\theta_{\mathbf{x}} - \theta_{\mathbf{y}}) \right\} \qquad (A.10)$$

As above, we assume that the volume is large, but finite, and use periodic boundary conditions to be specific. For brevity, we denote by $G(\mathbf{x} - \mathbf{y})$ the lattice solution of $-\Delta G(\mathbf{x} - \mathbf{y}) = \delta_{\mathbf{x},\mathbf{y}}$, with an implicit subtraction which will drop out in the subsequent estimates. One wants to make use of the fact that, for finite N, the integrand in $(A.10)$ is not only periodic in each variable θ, of period 2π, but also analytic. One can therefore displace the contour of integration by setting $\theta_{\mathbf{x'}} \to \theta_{\mathbf{x'}} + \text{i}\psi_{\mathbf{x'}}$, where $\psi_{\mathbf{x'}}$ is chosen as

$$\psi_{\mathbf{x}'} = \beta^{-1}\left[G(\mathbf{x}' - \mathbf{x}) - G(\mathbf{x}' - \mathbf{y})\right] \qquad (A.11)$$

The imaginary part $\psi_{\mathbf{x}'}$ is well-defined, and bounded in $\mathbf{x}'$ by Cst/β. After this substitution, we can majorize the phases by unity, and replace $(A.10)$ by the weaker inequality

$$0 \le \left\langle \mathbf{S_x \cdot S_y} \right\rangle \le Z^{-1}$$
$$\times \, e^{-(\psi_{\mathbf{x}} - \psi_{\mathbf{y}})} \int \prod_{\mathbf{y}} d\theta_{\mathbf{y}} e^{\beta \sum_{(\mathbf{x}', \mathbf{x}'')} \cos(\theta_{\mathbf{x}'} - \theta_{\mathbf{x}''}) \cosh(\psi_{\mathbf{x}'} - \psi_{\mathbf{x}''})}$$

and hence

$$0 \le \left\langle \mathbf{S_x \cdot S_y} \right\rangle \le e^{-(\psi_{\mathbf{x}} - \psi_{\mathbf{y}})} \left\langle e^{\beta \sum_{(\mathbf{x}', \mathbf{x}'')} \cos(\theta_{\mathbf{x}'} - \theta_{\mathbf{x}''})(\cosh(\psi_{\mathbf{x}'} - \psi_{\mathbf{x}''}) - 1)} \right\rangle$$

If β is sufficiently large, the uniform bound on ψ implies that

$$\beta \sum_{(\mathbf{x}', \mathbf{x}'')} (\cosh(\psi_{\mathbf{x}'} - \psi_{\mathbf{x}''}) - 1) \le \tfrac{1}{2}\beta[1 + \mathcal{O}(\beta^{-2})] \sum_{(\mathbf{x}', \mathbf{x}'')} (\psi_{\mathbf{x}'} - \psi_{\mathbf{x}''})^2$$

Moreover,

$$\sum_{(\mathbf{x}', \mathbf{x}'')} (\psi_{\mathbf{x}'} - \psi_{\mathbf{x}''})^2 = -\sum_{\mathbf{x}'} \psi_{\mathbf{x}'} \sum_{\mathbf{x}''(\mathbf{x}')} (\psi_{\mathbf{x}''} - \psi_{\mathbf{x}'}) = \beta^{-1}(\psi_{\mathbf{x}} - \psi_{\mathbf{y}})$$

The cosine in the exponential is bounded by 1, and hence, for $\beta > \beta(\varepsilon)$,

$$0 \le \left\langle \mathbf{S_x \cdot S_y} \right\rangle \le \exp\left(-\tfrac{1}{2}(1 - \varepsilon)(\psi_{\mathbf{x}} - \psi_{\mathbf{y}})\right)$$

or

$$0 \le \left\langle \mathbf{S_x \cdot S_y} \right\rangle \le \exp\left(-\tfrac{1}{2}(1 - \varepsilon)[G(\mathbf{0}) - G(\mathbf{x} - \mathbf{y})]/\beta\right) \qquad (A.12)$$

The quantity $G(\mathbf{0}) - G(\mathbf{x}) = \Gamma(\mathbf{x})/2\pi$ behaves as $\ln(|\mathbf{x}|/r_0)/2\pi$ for large $|\mathbf{x}|$, and is in any case positive. For large separations, one obtains the realistic bound for the XY-model

$$0 \le \left\langle \mathbf{S_x \cdot S_y} \right\rangle \le \left(\frac{r_0}{|\mathbf{x} - \mathbf{y}|}\right)^{(1-\varepsilon)/2\pi\beta} \qquad \beta > \beta(\varepsilon) \qquad (A.13)$$

This inequality again excludes the possibility of a spontaneous magnetization, since $\lim_{|\mathbf{x}-\mathbf{y}|\to\infty} \left\langle \mathbf{S_x \cdot S_y} \right\rangle = \left\langle \mathbf{S_x} \right\rangle^2 = 0$, at least at low temperature. The spontaneous magnetization is a nondecreasing function of β and thus always vanishes. The reasoning can be generalized to $n \geq 3$ components and yields as an upper bound a polynomial decrease.

To summarize, these proofs are related to the impossibility of defining a genuine free massless field in two dimensions, due to the infrared divergence of fluctuations. This idea has been formalized by S. Coleman. This does not prevent one from using such a field, in any finite volume, and even some of its derived quantities such as derivatives or exponentials, in an infinite volume, as will be discussed in chapter 9 (in volume 2).

Appendix 4.B   Phenomenological renormalization

In low enough dimension, in practice two or three, it is possible to develop a renormalization method in real space based on a hypothesis of quasiscale invariance for large, but finite systems. We shall return in much greater detail to this subject in chapter 9 (volume 2) devoted to conformal invariance in two-dimensional systems. Let us consider for instance the eigenvalues of the transfer matrix of a system with transverse size L^{d-1}. The logarithm of the largest eigenvalue is the free energy per volume unit, times L^{d-1} (up to a factor). Moreover, the ratio of the two largest eigenvalues (assumed nondegenerate) defines a correlation length, according to

$$\exp(-1/\xi_L) = \lambda_L^{(1)}/\lambda_L^{(2)} \qquad (B.1)$$

As $L \to \infty, \xi_L$ tends towards the correlation length of the infinite system ξ_∞, at least for $\beta < \beta_c$. When the lengths L, ξ_L, ξ_∞ are large with respect to the lattice spacing, one expects a relation

$$\xi_L = \xi_\infty f(L/\xi_\infty) \qquad (B.2)$$

where the function $f(x)$ approaches unity when $x \to \infty$ and ignores the lattice spacing. Conversely, if ξ_∞ increases indefinitely near a critical point, ξ_L remains finite, so that $f(x) \sim Ax$ when x vanishes. In the particular example of the Ising model where ξ_L and ξ_∞ depend on two essential parameters $\theta \sim (T - T_c)/T_c$ and H, one has

$$\frac{\xi_{L'}(\theta', H')}{\xi_L(\theta, H)} = \frac{\xi_\infty(\theta', H')}{\xi_\infty(\theta, H)} \frac{f(L'/\xi_\infty(\theta', H'))}{f(L/\xi_\infty(\theta, H))} \qquad (B.3)$$

Let us choose θ' and H', other parameters being kept fixed, in such a way that

$$\frac{L'}{\xi_\infty(\theta', H')} = \frac{L}{\xi_\infty(\theta, H)} \tag{B.4}$$

In the relation $(B.3)$, the unknown function f can be eliminated and one finds

$$\frac{\xi_{L'}(\theta', H')}{\xi_L(\theta, H)} = \frac{\xi_\infty(\theta', H')}{\xi_\infty(\theta, H)} = \frac{L'}{L} \tag{B.5}$$

As L and L' are large (but finite), this formula establishes a relation between θ', H' and θ, H, depending on the finite lengths L and L'. If the function $\xi_L(\theta, H)$ is known, one gets a renormalization relation. Since ξ_∞ only diverges as $H \to 0, \theta \to 0$, the relation between (θ', L') and (θ, L) defined by

$$\frac{\xi_{L'}(\theta', 0)}{\xi_L(\theta, 0)} = \frac{L'}{L} = \frac{\xi_\infty(\theta', 0)}{\xi_\infty(\theta, 0)} \tag{B.6}$$

has a fixed point located at 0, with an asymptotic behaviour

$$\theta' = \left(\frac{L}{L'}\right)^{1/\nu} \theta \tag{B.7}$$

when L and L' tends simultaneously to infinity, with a fixed ratio. The effective correlation length ξ_L increases linearly with L at the fixed point $\theta = 0$. Similarly, we can retrieve the magnetic exponent by studying the behaviour at $\theta = 0$, where $\xi_\infty(0, H)$ diverges as $|H|^{-1/y_H}$ with $y_H = \frac{1}{2}(d + \frac{\gamma}{\nu})$. The transformation $(B.5)$ has $H = 0$ as a fixed point and generates a behaviour

$$H' = \left(\frac{L}{L'}\right)^{y_H} H \tag{B.8}$$

This method has both advantages and defects. On the positive side, it does not require the introduction of any new interaction by renormalization. On the other hand, one has just one equation to determine several variables (here θ' and H'). In the present case, one fortunately knows that the critical point lies at $H = 0$; it is hence possible to determine first the critical temperature, then the magnetic behaviour at this temperature.

The method requires the determination of $\xi_L(\theta, H)$. Hence, up to now, only two- (or perhaps three-) dimensional systems can be studied in practice.

Examine the convergence of the method for the two-dimensional Ising model, using two-dimensional strips of width L with periodic boundary conditions. Using the expressions of chapter 2, show that the correlation length is given by

$$\xi_L^{-1}(\beta) = \tfrac{1}{2}[(\gamma_1 + \gamma_3 + \cdots + \gamma_{2L-1}) - (\gamma_0 + \gamma_2 + \cdots + \gamma_{2L-2})]$$

$$\cosh \gamma_r = \frac{\cosh^2 2\beta}{\sinh 2\beta} - \cos \frac{r\pi}{L}$$

$$(B.9)$$

Defining $\beta_c(L, L')$ and $\nu(L, L')$ through

$$\frac{\xi_L(\beta_c)}{\xi_{L'}(\beta_c)} = \frac{L'}{L} \equiv u$$

$$\nu^{-1} = \frac{\ln \left(\dfrac{d\xi_L(\beta_c)}{d\beta} \bigg/ \dfrac{d\xi_{L'}(\beta_c)}{d\beta} \right)}{\ln(L/L')} - 1$$

$$(B.10)$$

one finds

$$\beta_c(L, L') = \tfrac{1}{2}\ln(1 + \sqrt{2}) - \frac{\pi^3}{192} \frac{u(u+1)}{L^3} + \cdots \qquad (B.11)$$

$$\nu(L, L') = 1 - \frac{\pi^2 \ln 2}{24L^2} \frac{(u^2 - 1)}{\ln u} + \cdots \qquad (B.12)$$

This shows that even with strips of small width, one can get sensible results. The optimal choice of u lies near 1, that is $L = L' \pm 1$.

The technique can be applied to various two-dimensional problems, such as percolation, localization, XY-model,... Equation $(B.2)$ and its generalization to other observables needs a justification which can be given in the framework of the renormalization group, at least in dimension less than or equal to four. Indeed this relation is not valid any more in higher dimensions (in general above the upper critical dimension). This is to be expected, since the mean field approximation predicts a transition ignoring fluctuations, even for a finite size system.

We illustrate the technique on the spherical model studied in chapter 3, i.e. for the $n \to \infty$ limit of the n-vector model. We saw that the correlation length of an infinite system satisfies

$$\beta = \int \frac{d^d q}{(2\pi)^d} \frac{1}{\xi_\infty^{-1} + 2\sum_{\mu=1}^d (1 - \cos q_\mu)} \qquad (B.13)$$

In a strip of width L, transverse components of the momentum take discrete values which are multiples of $2\pi/L$. Using Poisson's relation, a formula analog to $(B.13)$ can be written for the correlation length of this system

$$\beta = \int \frac{d^d q}{(2\pi)^d} \frac{1}{\xi_L^{-2} + 2\sum_{\mu=1}^d (1 - \cos q_\mu)} \sum_{\mathbf{m}_\perp} (2\pi)^{d-1} \delta^{(d-1)} (\mathbf{q}_\perp - \frac{2\pi}{L} \mathbf{m}_\perp)$$

$$= \int \frac{d^d q}{(2\pi)^d} \frac{1}{\xi_L^{-2} + 2\sum_{\mu=1}^d (1 - \cos q_\mu)} \sum_{\mathbf{n}_\perp} e^{i\mathbf{q}_\perp \cdot \mathbf{n}_\perp L}$$

$$(B.14)$$

where the $(d-1)$-dimensional vectors $\mathbf{m}_\perp$ and $\mathbf{n}_\perp$ have integer coordinates. Isolating in $(B.14)$ the term $\mathbf{n}_\perp = 0$ and subtracting $(B.13)$, one obtains

$$(\xi_\infty^{-2} - \xi_L^{-2}) \int \frac{d^d q}{(2\pi)^d} \left(\sum_{\mathbf{n}_\perp}' \frac{e^{i\mathbf{q}_\perp \cdot \mathbf{n}_\perp L}}{\xi_L^{-2} + 2\sum_{\mu=1}^d (1 - \cos q_\mu)} + \right.$$

$$\left. \frac{1}{\left[\xi_\infty^{-2} + 2\sum_{\mu=1}^d (1 - \cos q_\mu) \right] \left[\xi_L^{-2} + 2\sum_{\mu=1}^d (1 - \cos q_\mu) \right]} \right) = 0$$

$$(B.15)$$

In the limit $\beta \to \beta_c$ and $L \to \infty$, ξ_∞ increases indefinitely and ξ_L is large. Under these conditions, one can replace $2\sum_1^d (1 - \cos q_\mu)$ by $\mathbf{q}^2$, when the integrals are dominated by small momenta $\mathbf{q}$. This is indeed the case for $2 < d < 4$. This approximation gives, for the second term,

$$\int \frac{d^d q}{(2\pi)^d} \sum_{\mathbf{n}_\perp}' \frac{e^{i\mathbf{q}_\perp \cdot \mathbf{n}_\perp L}}{\xi_L^{-2} + \mathbf{q}^2} = L^{2-d} \int_0^\infty dt\, e^{-t(L/\xi_L)^2} u(t) \qquad (B.16)$$

We use the notation

$$u(t) = \sum_{\mathbf{n}_\perp}' \frac{e^{-\mathbf{n}_\perp^2/4t}}{(4\pi t)^{d/2}} \quad \sim \quad \begin{cases} \dfrac{2(d-1)}{(4\pi t)^{d/2}} e^{-1/4t} & t \to 0 \\[2ex] \dfrac{1}{(4\pi t)^{1/2}} & t \to \infty \end{cases} \qquad (B.17)$$

The small and large t behaviour explicitly shows the convergence of the integral $(B.16)$. With a similar approximation for the first integral in $(B.15)$, one finds, in the critical region and for large L,

$$\left[\left(\frac{L}{\xi_\infty}\right)^2 - \left(\frac{L}{\xi_L}\right)^2\right]\left(\frac{\xi_L}{L}\right)^{4-d}\frac{\Gamma(2-d/2)}{(4\pi)^{d/2}}\times$$

$$\int_0^1 dt\left[t\left(\frac{\xi_L}{\xi_\infty}\right)^2 + (1-t)\right]^{d/2-2} + \int_0^\infty dt\, e^{-t(L/\xi_L)^2}u(t) = 0$$

$$(B.18)$$

This relation, which is of the form $\varphi(L/\xi_L, L/\xi_\infty) = 0$, can be solved as $\xi_L/\xi_\infty = f(L/\xi_\infty)$. The resulting function f has the required properties. In particular, at the critical point where ξ_∞ diverges, $\xi_L(\beta_c)$ grows linearly with L, according to

$$\xi_L(\beta_c) = sL \qquad (B.19)$$

and s is given by the implicit equation

$$s^{2-d}\left\{\frac{-\Gamma(1-d/2)}{(4\pi)^{d/2}}\right\} = \int_0^\infty dt e^{-t/s^2}u(t) \qquad (B.20)$$

In particular, one finds in three dimensions

$$\sum_{n_1,n_2}'\frac{s}{\sqrt{n_1^2+n_2^2}}\exp\left(-\frac{\sqrt{n_1^2+n_2^2}}{s}\right) = 1 \qquad s\simeq 0.539 \quad (d=3)$$

$$(B.21)$$

When the dimension d approaches 4, the l.h.s. of $(B.20)$ develops a pole and s increases. As the integral diverges in the large t domain, one can replace $u(t)$ by its large t behaviour. Hence, with $d = 4 - \varepsilon$,

$$\xi_L(\beta_c) = \left(\frac{1}{4\pi^2\varepsilon}\right)^{1/3}L \qquad (d = 4-\varepsilon) \qquad (B.22)$$

Above four dimensions, relation $(B.2)$ is no more valid. Approximate scale invariance is broken and the cutoff factor, here the lattice spacing, plays a role. In equation $(B.15)$, it is not possible anymore to replace the denominators $2\sum_1^d(1-\cos q_\mu)$ by $\mathbf{q}^2$, since this operation introduces ultraviolet divergences. For example, when $d = 4$, one finds

$$\left(\frac{L}{\xi_L}\right)^2\ln\xi_L^2 - \left(\frac{L}{\xi_\infty}\right)^2\ln\xi_\infty^2 = (4\pi)^2\int_0^\infty dt\, e^{-t(L/\xi_L)^2}u(t) \quad (d=4)$$

$$(B.23)$$

At the critical temperature, the linear growth of $\xi_L(\beta_c)$ is modified by a logarithmic factor

$$\xi_L(\beta_c) = L \left(\frac{\ln L}{4\pi^2} \right)^{1/3} \qquad (d = 4) \qquad (B.24)$$

Finally, as $d > 4$, there is no infrared divergence in the first integral $(B.15)$ as $\xi_\infty^{-1} \to 0$ and $\xi_L^{-1} \to 0$. One can then replace these quantities by 0 in the integrand. Introducing the value of the Green function at the origin

$$g = \int \frac{d^d q}{(2\pi)^d} \frac{1}{\left[2 \left(1 - \sum_{\mu=1}^d \cos q_\mu \right) \right]^2} \qquad (d > 4) \qquad (B.25)$$

one obtains the following relation

$$\left(\frac{L}{\xi_\infty} \right)^2 = \left(\frac{L}{\xi_L} \right)^2 - \frac{1}{2g} \left(\frac{\xi_L}{L} \right)^{4-d} \qquad (d > 4) \qquad (B.26)$$

This is of the form

$$\frac{\xi_L}{\xi_\infty} = f \left(\frac{L}{\xi_\infty} L^{(d-4)/3} \right) \qquad (d > 4) \qquad (B.27)$$

In particular, for $\beta = \beta_c, \xi_L(\beta_c)$ grows with L faster than linearly

$$\xi_L(\beta_c) = L(2gL^{d-4})^{1/3} \qquad (d > 4) \qquad (B.28)$$

Finally, as $d \to \infty, \xi_L(\beta_c)$ diverges, as is suggested by the mean field approximation.

(1) Show that, for the spherical model, in the broken symmetry phase, the behaviour is given by

$$\xi_L(\beta) = 2(\beta - \beta_c)L^{d-1} \qquad \beta > \beta_c \qquad (B.29)$$

(2) Conversely, for a system with a discrete symmetry, it is expected that ξ_L grows exponentially with L for $\beta > \beta_c$. Why?

(3) Is the ratio $s = \xi_L(\beta_c)/L$ universal? Estimate this ratio for two-dimensional models.

As we shall see later, if a two-dimensional critical theory is conformally invariant, it is possible to find the form of the correlation function for various geometries, as for instance strips with periodic boundary conditions. One can use a conformal mapping of the complex plane $z = x + iy$ on the strip $w = w_1 + iw_2, -\frac{1}{2}L < w_1 < \frac{1}{2}L$ of the form

$$z = \exp -2i\pi w/L$$

$$z_a - z_b = -2\mathrm{i}\sin\left(\pi\frac{w_a - w_b}{L}\right)\exp\left(-\mathrm{i}\pi\frac{w_a + w_b}{L}\right)$$

This mapping is unique, up to a global transformation of the plane $z \to \alpha z + \beta$ which only modifies the correlation function by a multiplicative factor

$$\langle\varphi(\mathbf{x}_a)\varphi(\mathbf{x}_b)\rangle = \frac{\mathrm{Cst}}{|\mathbf{x}_a - \mathbf{x}_b|^\eta} = \frac{\mathrm{Cst}}{|z_a - z_b|^\eta}$$

If we choose in the strip $w_a = \mathrm{i}t$ (with real positive t), $w_b = 0$, one finds

$$\langle\varphi(t)\varphi(0)\rangle_{\mathrm{strip}} \sim \frac{\mathrm{Cst}}{|\exp(2\pi t/L) - 1|^\eta} \underset{t\to\infty}{\sim} \mathrm{e}^{-2\pi\nu t/L} \qquad (B.30)$$

Therefore, in two dimensions, $\xi_L(\beta_c) = L/2\pi\eta$ indeed grows linearly with L at the critical point, with a universal coefficient $s = 1/2\pi\eta$.

Notes

Ideas of scale invariance were developed by B. Widom, *J. Chem. Phys.* **43**, 3892, 3898 (1965); A.Z. Patashinskii and V.L. Pokrovskii, *Zh. Eksper. Teor. Fiz.* **50**, 439 (1966); L.P. Kadanoff, *Physics* **2**, 263 (1966), *Rev. Mod. Phys.* **39**, 395 (1967); M.E. Fisher, *Rept. Prog. Phys.* **30**, 615 (1967); K.G. Wilson, *Phys. Rev.* **D2**, 1473, 1478 (1970), *Phys. Rev.* **B4**, 3174, 3184 (1971); F.J. Wegner, *Phys. Rev.* **B5**, 4529 (1972) and *Lecture Notes in Physics* **37**, 171 (1973) Springer Verlag, Berlin.

A general view on critical phenomena prior to the explosion of the renormalization group is found in H.E. Stanley, *Introduction to Phase Transitions and Critical Phenomena* Clarendon Press, Oxford (1971). Further important references will be quoted after the next chapter.

The q-state model is due to R.B. Potts, *Proc. Camb. Phil. Soc.* **48**, 106 (1952). The corresponding algebraic aspects were developed by H.N.V. Temperley and E.H. Lieb, *Proc. Roy. Soc. London* **A322**, 251 (1971).

For real space renormalization methods, see L. Kadanoff and A. Houghton, *Phys. Rev.* **B11**, 377 (1975); Th. Niemejer and J.M.J. Van Leeuwen in Domb and Green vol. VI (1976); A.A. Migdal, *Z. Eksper. Teor. Fiz.* **69**, 810, 1457 (1975); L.P. Kadanoff, *Ann. Phys.* **100**, 359 (1976). Combinations of real space renormalization and Monte-Carlo methods are treated in the volume *Real-space Renormalization*, T.W. Burkhardt and J.M.J. Van Leeuwen eds, Topics in Current Physics **30**, Springer Verlag, Berlin (1982). Julia sets of zeroes for partition functions on hierarchical lattices are taken from a joint work with B. Derrida, L. de Sèze and J.-M. Luck and are reviewed in *Critical Phenomena* 1983 Brasov school, V. Ceausescu *et al* eds, Birkhaüser, Boston (1985). Phenomenological renormalization discussed in the appendix originates in the work of M.E. Fisher and M.N. Barber, *Phys. Rev. Lett.* **28**, 1516 (1972) on finite size scaling and was applied by many authors following M.P. Nightingale, *Physica* **83A**, 561 (1976). For a field theoretical treatment, see E. Brézin, *J. Physique* **43**, 15 (1982).

The absence of spontaneous magnetization for two-dimensional systems with continuous symmetries is proved by P.C. Hohenberg, *Phys. Rev.* **158**, 383 (1967); N.D. Mermin and H. Wagner, *Phys. Rev. Lett.* **17**, 1133 (1966); N.D. Mermin, *Journ. Math. Phys.* **8**, 1061 (1967). The pathologies of massless two dimensional scalar fields as well as their positive aspects are studied by S. Coleman, *Comm. Math. Phys.* **31**, 259 (1973) and *Phys. Rev.* **D11**, 2088 (1975). Part of appendix A is based on the work of O. McBryan and T. Spencer, *Comm. Math. Phys.* **53**, 299 (1977).

The role of vortices in the XY-system was discussed by V.L. Berezinskii, *Soviet Physics JETP*, **32**, 493 (1971) and the nature of the transition by J.M. Kosterlitz, *J. Phys.* **C7**, 1046 (1974). Our discussion is patterned after the work of J.V. José, L.P. Kadanoff, S. Kirkpatrick and D.R. Nelson, *Phys. Rev.* **B16**, 1217 (1977) except for minor modifications. The spin wave plus Coulomb gas version is due to J. Villain, *J. Physique* **36**, 581 (1975). A systematic field theoretic analysis in the framework of the sine-Gordon model is provided by D.J. Amit, Y.Y. Goldschmidt and G. Grinstein, *J. Phys.* **A13**, 585 (1980), with references to previous works.

Interestingly, the same XY-model can be interpreted in various ways which allow experimental checks. It is in the same universal class as the solid on solid model of W.K. Burton, N. Cabrera and F.C. Franck, *Phil. Trans. Roy. Soc.* **A243**, 299 (1951) which describes the transition from a flat to a rough interface. As discussed in the text, it is also related to the dielectric plasma transition of the classical Coulomb gas; see S.T. Chui and

J.D. Weeks, *Phys. Rev.* **B14**, 4978 (1976); J.D. Weeks in *Ordering in Strongly Fluctuating Condensed Matter Systems*, T. Riste ed., Plenum Press, New York (1980). One can also mention the theory of superconducting films discussed for instance by B.I. Halperin and D.R. Nelson, *J. Low Temp. Phys.* **36**, 599 (1979) or the universal jump in the He^4 superfluid density, D.R. Nelson and J.M. Kosterlitz, *Phys. Rev. Lett.* **39**, 1201 (1977). The list is not limitative.

5

CONTINUOUS FIELD
THEORY AND THE
RENORMALIZATION GROUP

Continuous field models were originally systematically elaborated (and still are) in the context of particle physics. Most relevant to the study of critical phenomena is the derivation of a renormalization group flow, characterized by a set of Callan–Symanzik equations for Green functions, with regular coefficients derived from perturbation theory. This is supplemented by the idea due to Fisher and Wilson of an expansion in powers of the deviation from the strict renormalizability dimension, i.e. four in the case of the φ^4 model, but can also be presumed to work directly in the physical dimension three (and even possibly two). We devote this chapter to a general presentation and a survey of some applications. An appendix gives a short introduction to multicritical phenomena.

5.1   The Lagrangian and dimensional analysis

5.1.1   Introduction

We want to investigate universal critical properties. A discrete lattice model appears as a regularizing intermediate stage, which allows a precise meaning to be given to the functional integrals of the field theory. Following the scheme suggested by the mean field approximation, one is tempted to start directly from a continuous model, in which the lattice is replaced by a d-dimensional continuous Euclidean space, and also, even for models with a discrete symmetry, the dynamical variables are replaced by continuous fields. From the original formulation, we just retain the idea of a cutoff factor Λ large with respect to all momenta, which ensures that some possibly divergent expressions remain finite.

The (positive) statistical weight of a configuration is the exponential of an action which is the integral over space of a *local*

Lagrangian. The finiteness of the interaction range is replaced by the notion of *locality*, which means in practice the introduction of as limited a number of derivative terms as is possible. The Lagrangian may be considered as the relevant leading term in an expansion allowing the calculation of fluctuation effects around a mean field solution. Alternatively, it can be postulated *a priori* to arise from averages over domains large with respect to the microscopic scale, but small compared to the correlation lengths when they increase indefinitely.

These considerations establish a connection with continuous field theory. Conversely, using this analogy with statistical mechanics in the Euclidean domain, one can have access to global properties. In a first stage, we limit ourselves to a scalar model with Z_2 symmetry ($n = 1$), similar to the Ising model. Denoting by φ the fluctuating field, the Lagrangian, invariant under the substitution $\varphi \to -\varphi$, is expanded as

$$\mathcal{L}(\varphi) = \tfrac{1}{2}(\partial\varphi)^2 + \tfrac{1}{2}m_0^2\varphi^2 + \frac{1}{4!}g_0\varphi^4 + \cdots \tag{1}$$

A multiplicative factor has been absorbed in the definition of the field φ, so that the coefficient of the derivative term (*kinetic* term) is normalized to $\tfrac{1}{2}$. Note the Euclidean invariance of the various terms, when φ is assumed to behave as a scalar field. Omitted terms contain higher powers or more derivatives of φ. We will have to justify the fact that they are indeed negligible in the critical domain.

In this construction, the coefficients m_0^2 and g_0 are regular functions of the basic parameters, in particular the temperature. By tuning these parameters, it is possible to increase indefinitely the correlation length. The dependence of the latter on the variable m_0^2 thus yields a linear temperature scale near the critical point and allows us to obtain the critical exponent ν.

Given the Lagrangian (1), the partition function Z is expressed as a functional integral

$$Z = \int \mathcal{D}\varphi \exp\left(-\int d^d\mathbf{x}\mathcal{L}(\varphi)\right) \tag{2}$$

analogous to the corresponding quantity in the discrete case. This symbolic expression acquires a precise meaning, using a mathematical construction, for an integral dimension less than or equal to three. As we shall see below, the essential reason is that a finite number of subtractions is sufficient to compensate the short distance singularities and to make the theory finite. From the present point of view, however, we assume the existence of a microscopic structure, characterized for instance by the lattice spacing, and the continuous theory is used to describe phenomena at large distances. Hence, possible divergences can be avoided from a physical point of view by introducing in the integrations over the momenta a cutoff factor $\Lambda \sim a^{-1}$. The renormalization theory will organize the potentially divergent expressions into a finite number of factors relating the macroscopic theory to the microscopic parameters. The exploitation of these relations leads to the renormalization group and finally allows the calculation of the critical behaviour. One should therefore consider that, besides the parameters m_0^2 and g_0 explicitly written in the Lagrangian, the model depends on the regularization procedure and on its cutoff parameter Λ.

When one neglects fluctuation effects, the Lagrangian (1) describes a second order transition in terms of the parameter m_0^2. In the high temperature symmetric phase ($m_0^2 > 0$), the field fluctuates around a zero value. When $m_0^2 < 0$ (with of course $g_0 > 0$), there are fluctuations around a nonzero value of φ (spontaneous magnetization), corresponding to the minimization of $\frac{1}{2}m_0^2\varphi^2 + g_0\varphi^4/4!$, i.e. $M^2 \equiv \langle\varphi\rangle^2 = -6m_0^2/g_0$. It is easily seen that this corresponds to the mean field description studied in previous chapters.

Fluctuations induce a modification of critical properties when the dimension is less or equal to four. Wilson and Fisher have suggested an expansion in a parameter $\varepsilon = 4 - d$, assumed small and positive. Continuity (and even more differentiability) properties of the critical exponents as functions of ε should be presupposed. This requires a precise definition of the dimensional continuation, an example of which has been given for the spherical model in chapter 3.

In the sequel, we shall use extensively a formal perturbative method in the coupling constant g_0 in order to justify the renormalizability properties, even though we cannot assume at first that it is a small parameter. As we shall see later, this is not true in the four-dimensional case. The only fully soluble model uses Gaussian fluctuations, which provide a first approximation to the deviations to mean field. For instance, if, in the high temperature phase, one neglects g_0 and takes m_0^2 proportional to the deviation θ with respect to the critical temperature, the Fourier transform of the two point correlation function is

$$G_0(\mathbf{p}, \theta) = \int \mathrm{d}^d \mathbf{x}\, e^{i\mathbf{p}\cdot\mathbf{x}} \langle \varphi(\mathbf{x})\varphi(\mathbf{0}) \rangle = \frac{1}{\mathbf{p}^2 + \theta} = \frac{1}{\theta}\left(\frac{1}{1 + \mathbf{p}^2/\theta}\right) \tag{3}$$

This can be rewritten

$$G_0(\mathbf{p}, \theta) = \theta^{-\gamma} g(p\theta^{-\nu}) \tag{4}$$

with

$$\gamma = 1 \qquad \nu = \tfrac{1}{2} \tag{5}$$

being the critical exponents given by the mean field approximation. The susceptibility is indeed proportional to the correlation function at zero momentum. The function g, regular at the origin, has an argument of the form $p\xi$, ξ being the correlation length proportional to $\theta^{-\nu}$. Hence the study of the 2-point correlation function allows one to obtain two critical exponents, and the scaling properties – which will be justified in the framework of the continuous model – then allow the calculation of all other exponents. The effects of anharmonic terms for $g_0 > 0$ have now to be studied.

5.1.2   *Generating functionals and dimensional analysis*

The relative dimensionless probabilities are written $\exp(-S)$, where the action S, which is also dimensionless, is the energy divided by kT in the thermodynamical language. Dimensions are conventionally expressed in momentum units (to be more precise, in wavenumber units; the practice of using the language of quantum mechanics with $\hbar = 1$ is, however, in general use). Correspondingly, lengths have dimension -1, so that the dimension of the Lagrangian is

$$[\mathcal{L}] = d \tag{6}$$

The coefficient of the kinetic term being a pure number, one is led to assign to the field a dimension

$$[\varphi] = \tfrac{1}{2}(d - 2) \tag{7}$$

In particular, φ is dimensionless in space dimension 2 (as are the angles in the XY-model) and of dimension 1 for $d = 4$.

As a consequence,

$$[m_0] = 1 \qquad\qquad [g_0] = 4 - d \qquad\qquad (8)$$

More generally, the coupling constant g_{n_1,n_2} of an additional term $\partial^{n_2}\varphi^{n_1}$ has a dimension

$$\left[g_{n_1,n_2}\right] = -\delta_{n_1,n_2} = d - \tfrac{1}{2}n_1(d-2) - n_2 \qquad\qquad (9)$$

The number δ_{n_1,n_2} is called the *canonical degree* of the corresponding *operator*. It should of course not be confused with the usual notation for the Kronecker symbol. It is common usage, by reference to quantum field theory, to speak of operator for a local quantity built out of fields and their derivatives, even though in the statistical context they might more naturally be referred to as local observables. The degree given by equation (9) is 2 in dimension two when n_2 vanishes, whatever n_1 is. In space dimension four, the terms explicitly written in the Lagrangian (1) are the only ones with a nonpositive degree. In dimension three, there is only one additional term φ^6 satisfying both Euclidean invariance and $\varphi \to -\varphi$ symmetry. However, this term is dominated by the φ^4 term for small fluctuations of φ around 0, except for the case where g_0 vanishes. The inclusion of such a term allows one to study a change from a second to a first order transition. This *tricritical* phenomenon is discussed in appendix A.

The microscopic scale parameter is the cutoff factor $\Lambda \sim 1/a$. Operators with a positive degree behave as $\Lambda^{-\delta}$ and seem therefore negligible as $\Lambda \to \infty$. This naive view will nevertheless have to be slightly corrected. For example, terms with a positive degree obviously play a role in the determination of the critical temperature, which is not universal. But the universal dominant critical behaviour is not sensitive to these operators. Hence, at least near $d = 4$, the large distance behaviour is governed by the φ^4 Lagrangian (1), with at most a redefinition of the *bare* parameters m_0^2 and g_0. We will show indeed that renormalization acts as a filter, suppressing the irrelevant terms (with a positive degree) and keeping only a minimal number of relevant parameters.

This discussion which justifies the use of renormalizable local Lagrangians for the description of critical phenomena is quite natural in particle physics. Indeed, we have as yet no knowledge of very high energy processes (i.e. at very small distances). The ability to construct efficient models describing phenomena at a given scale is based on a quasi-decoupling between this scale and those of much higher energies. The latter only yields the values of some parameters which cannot be determined in an effective low energy theory as, for instance, the fine structure constant of electrodynamics.

Correlation functions (or Green functions, or also Schwinger functions in this context) are obtained by coupling φ to an external source $h(\mathbf{x})$, adding to the Lagrangian a term

$$\mathcal{L}_{\text{source}} = -h(\mathbf{x})\varphi(\mathbf{x}) \tag{10}$$

The dimension of the external field $h(\mathbf{x})$ is

$$[h] = \tfrac{1}{2}(d+2) \tag{11}$$

With the additional source term (10), the partition function becomes a generating functional $Z(h)$ for the correlation functions

$$Z(h) = \int \mathcal{D}\varphi \exp\left(-\int d^d\mathbf{x}\mathcal{L}(\varphi) + \int d^d\mathbf{x}h\varphi\right)$$
$$= Z(0)\left\langle \exp\left(\int d^d\mathbf{x}\, h\varphi\right)\right\rangle \tag{12}$$

The n-field (n-point) correlation functions are defined as

$$\langle\varphi(\mathbf{x}_1)\dots\varphi(\mathbf{x}_n)\rangle_h = Z(h)^{-1}\frac{\delta^n}{\delta h(\mathbf{x}_1)\dots\delta h(\mathbf{x}_n)}Z(h) \tag{13}$$

When $h \to 0$ and in the symmetric phase, only correlations involving an even number of points do not vanish. They are translationally invariant, and we will denote them by

$$\langle\varphi(\mathbf{x}_1)\dots\varphi(\mathbf{x}_n)\rangle = Z(h)^{-1}\frac{\delta^n}{\delta h(\mathbf{x}_1)\dots\delta h(\mathbf{x}_n)}Z(h)\Big|_{h=0}$$
$$= \int\prod_1^n\left(\frac{d^d\mathbf{p}_k}{(2\pi)^d}e^{-i\mathbf{p}_k\cdot\mathbf{x}_k}\right)(2\pi)^d\delta(\textstyle\sum\mathbf{p}_k)G_n(\mathbf{p}_1,\dots,\mathbf{p}_n) \tag{14}$$

Conversely,

$$Z(h)/Z(0) = \sum_{n=0}^{\infty}\frac{1}{n!}\int d^d\mathbf{x}_1\dots d^d\mathbf{x}_n h(\mathbf{x}_1)\dots h(\mathbf{x}_n)\langle\varphi(\mathbf{x}_1)\dots\varphi(\mathbf{x}_n)\rangle \tag{15}$$

The momentum conservation distribution $\delta(\sum \mathbf{p})$ has been extracted, so that in momentum space G_n is defined on the linear manifold $\sum \mathbf{p} = 0$. Its *canonical* dimension is

$$[G_n] = d - n\,[h] = d - \tfrac{1}{2}n(d+2) \tag{16}$$

In particular, $[G_2] = -2$ in any dimension, as in the free field example (3).

The *free energy* $\ln Z(h)$ (for convenience, our definition differs by a factor $-1/\beta$ from usual thermodynamics) is the generating functional for the *connected* correlation functions. In the zero field limit,

$$\langle \varphi(\mathbf{x}_1)...\varphi(\mathbf{x}_n)\rangle_c = \left.\frac{\delta^n}{\delta h(\mathbf{x}_1)...\delta h(\mathbf{x}_n)}\ln Z(h)\right|_{h=0} \tag{17}$$

These connected functions have of course the same dimension as the general correlation functions. The connection between the two types is given by *cumulants* which allow the exponentiation of the expansion (15). The terminology comes from the fact that only connected diagrams appear in the perturbative expansion. These results will be recalled in chapter 7 (volume 2), and we assume in this chapter that they are familiar to the reader. The free energy is an extensive quantity (proportional to the volume) when fields are defined on a (large) finite domain.

From the free energy $\ln Z(h)$, one obtains the generating functional of *one particle irreducible* Green functions (or *vertex functions*) $\Gamma(\varphi)$ by applying a Legendre transformation. The corresponding variable (spatially varying magnetization) is commonly denoted by the same symbol φ as the integration field variable in equation (2). In the perturbative theory, the contributing diagrams cannot be disconnected by cutting an internal line, and propagators corresponding to the external lines are omitted (*amputation*). This is the origin of the qualification *one-particle irreducible*. The quantity $\Gamma(\varphi)$ is defined through

$$\Gamma(\varphi) = \sup_h \left\{ \int d^d\mathbf{x}\, h(\mathbf{x})\varphi(\mathbf{x}) - \ln Z(h) \right\} \tag{18}$$

In practice, one only checks the extremality condition, so that the transformation

$$\Gamma(\varphi) + \ln Z(h) = \int d^d x\, h(\mathbf{x})\varphi(\mathbf{x})$$

$$\frac{\delta \ln Z(h)}{\delta h(\mathbf{x})} = \varphi(\mathbf{x}) = \langle\varphi(\mathbf{x})\rangle_h \tag{19}$$

is reciprocal.

(1) Obtain the Legendre transformation for a Gaussian integral.
(2) Prove that the Legendre transformation (19) is reciprocal

$$\frac{\delta\Gamma(\varphi)}{\delta\varphi(\mathbf{x})} = h(\mathbf{x})$$

A nonzero limit of $\delta \ln Z(h)/\delta h$ as h becomes uniformly small with a fixed sign signals a spontaneous breaking of the Z_2 symmetry. The corresponding value of φ is the (uniform) spontaneous magnetization which occurs in the low temperature phase.

When h tends to be uniform, the free energy appears as the integral of a constant density. The same is true for $\Gamma(\varphi)$ when φ is uniform, and we write in this case

$$\Gamma(\varphi) = \int d^d x\, V(\varphi) \tag{20}$$

The effective potential $V(\varphi)$, which takes into account the fluctuations, is a generalization of the corresponding term $\frac{1}{2}m_0^2\varphi^2 + g_0\varphi^4/4!$ in the Lagrangian (1).

Returning to the general case, the vertex functions are obtained by expanding $\Gamma(\varphi)$ in increasing powers of φ, as one does with $\ln Z(h)$ to get the connected functions

$$\Gamma(\varphi) = \sum \frac{1}{n!}\int d^d x_1...d^d x_n\, \varphi(\mathbf{x}_1)...\varphi(\mathbf{x}_n)\Gamma_n(\mathbf{x}_1,\dots,\mathbf{x}_n)$$

$$\Gamma_n(\mathbf{x}_1,\dots,\mathbf{x}_n) = \left.\frac{\delta^n\Gamma(\varphi)}{\delta\varphi(\mathbf{x}_1)\cdots\delta\varphi(\mathbf{x}_n)}\right|_{\varphi=0}$$

$$= \int \prod_1^n \left(\frac{d^d p_k}{(2\pi)^d}e^{-i\mathbf{p}_k\cdot\mathbf{x}_k}\right)(2\pi)^d\delta(\sum \mathbf{p}_k)\Gamma_n(\mathbf{p}_1,\dots,\mathbf{p}_n)$$

$$\tag{21}$$

Functions and their Fourier transforms share abusively the same notation. In the expansion (21) around $\varphi = 0$, only terms with n even do not vanish. This is no longer the case if $\Gamma(\varphi)$ is expanded around another value (for example the spontaneous magnetization of the ordered phase).

The dimension of the argument φ of Γ is the same as the one of the field φ, i.e. $\frac{1}{2}(d-2)$, while $\Gamma(\varphi)$ is of course dimensionless, and hence

$$
\begin{aligned}
[\Gamma_n(\mathbf{x}_1, ..., \mathbf{x}_n)] &= \tfrac{1}{2}n(d+2) \\
[\Gamma_n(\mathbf{p}_1, ..., \mathbf{p}_n)] &= d - \tfrac{1}{2}n(d-2) = [G_n(\mathbf{p}_1, ..., \mathbf{p}_n)] + 2n
\end{aligned}
\tag{22}
$$

The relation between the dimensions of the vertex functions and of the Green functions corresponds to the amputation of the external line factors, as can be seen by making the relation (19) explicit. In particular, $[\Gamma_2(\mathbf{p})] = 2$, $[\Gamma_4(\mathbf{p})] = 4 - d$. More generally, by comparing relations (9) and (22), one sees that vertex functions generalize the coupling constants of the Lagrangian.

It is also possible to introduce other source terms analogous to (10), coupled to more complex combinations than the field φ. In particular, as the control parameter is m_0^2, one is naturally led to consider a source term coupled to $\frac{1}{2}\varphi^2$. The corresponding Green functions *with insertion of* φ^2 reads

$$
\langle \varphi(\mathbf{x}_1) \cdots \varphi(\mathbf{x}_n) \tfrac{1}{2}\varphi^2(\mathbf{y}_1) \cdots \tfrac{1}{2}\varphi^2(\mathbf{y}_r) \rangle = \int (2\pi)^d \delta \left(\sum \mathbf{p}_k + \sum \mathbf{q}_l \right)
$$

$$
G_{n,r}(\mathbf{p}_k, \mathbf{q}_l) \prod_1^n \left(\frac{d^d \mathbf{p}_k}{(2\pi)^d} e^{-i\mathbf{p}_k \cdot \mathbf{x}_k} \right) \prod_1^r \left(\frac{d^d \mathbf{q}_l}{(2\pi)^d} e^{-i\mathbf{q}_l \cdot \mathbf{y}_l} \right)
$$

$$
\left[G_{n,r}(\mathbf{p}, \mathbf{q}) \right] = [G_n] - 2r = d - \tfrac{1}{2}n(d+2) - 2r
$$

$$
\tag{23}
$$

The same dimensional considerations also apply to the connected functions. In particular, the specific heat is related to the connected function $G_{0,2}$ at zero momentum, the dimension of which is $d - 4$. Vertex functions with insertions of φ^2 are defined by a Legendre transformation on the source of the φ field only. Their Fourier transforms $\Gamma_{n,r}(\mathbf{p}, \mathbf{q})$ have a dimension

$$
\left[\Gamma_{n,r} \right] = [\Gamma_n] - 2r = d - \tfrac{1}{2}n(d-2) - 2r
\tag{24}
$$

In all these expressions, the dimensions are called *canonical*, in order to distinguish them from the *anomalous* or *dynamical* dimensions arising from nontrivial critical properties. The original cutoff factor Λ introduced by regularization will be absorbed by the renormalization theory into suitable large scale quantities.

We have only presented here the simplest Lagrangian model corresponding to the scalar φ^4 theory. The generalization to fields taking their values in a vector or matrix space, or a Grassmann algebra, does not present conceptual difficulties. Technical subtleties may however arise in a proper definition of the path integral measure, and this is the source of possible anomalies in the implementation of various symmetries.

5.2   The perturbative method

5.2.1   Diagrammatic series

The free field Gaussian model gives an accurate description of the critical behaviour when $d > 4$. Near $d = 4$, it is therefore natural, despite some difficulties which will indeed arise, to perform a perturbative expansion with respect to the anharmonic terms of the Lagrangian, i.e. in powers of the bare coupling constant g_0. The analysis of this expansion will provide some clues about how to derive general properties. Let us split the Lagrangian and the action according to

$$\mathcal{L}_0 = \tfrac{1}{2}\{(\partial\varphi)^2 + m_0^2\varphi^2\} \qquad S_0 = \int d^d\mathbf{x}\,\mathcal{L}_0(\varphi)$$
$$\mathcal{L}_I = \frac{1}{4!}g_0\varphi^4 \qquad\qquad S_I = \int d^d\mathbf{x}\,\mathcal{L}_I(\varphi) \tag{25}$$

and expand the generating functional in a power series in the bare coupling constant g_0 as

$$Z(h) = \int \mathcal{D}\varphi \exp - \left\{S_0 + S_I - \int d^d\mathbf{x}\,h\varphi\right\}$$
$$= \sum_{n=0}^{\infty} \frac{(-1)^n}{n!} \int \mathcal{D}\varphi \left(\int d^d\mathbf{x}\mathcal{L}_I(\varphi)\right)^n \exp - \left\{S_0 - \int d^d\mathbf{x}\,h\varphi\right\}$$
$$= \sum_{n=0}^{\infty} \frac{(-1)^n}{n!} \left(\int d^d\mathbf{x}\mathcal{L}_I(\frac{\delta}{\delta h})\right)^n \int \mathcal{D}\varphi \exp - \left\{S_0 - \int d^d\mathbf{x}\,h\varphi\right\}$$

The last expression involves a Gaussian functional integral which can be expressed in terms of the free propagator G_0 written in (3) (with θ replaced by m_0^2). Indeed we recall that

$$\int \mathcal{D}\varphi \exp - \left\{ S_0 - \int d^d\mathbf{x} h\varphi \right\}$$

$$= Z_0 \exp \tfrac{1}{2} \int d^d\mathbf{x} d^d\mathbf{y} h(\mathbf{x}) G_0(\mathbf{x} - \mathbf{y}) h(\mathbf{y})$$

$$Z_0 = \det{}^{1/2} G_0(\mathbf{x} - \mathbf{y}) \qquad (26)$$

To define precisely the normalization factor Z_0, one requires both a short-range regularization and an infrared cutoff (since $\ln Z_0$ is proportional to the volume). Of course, Z_0 is not $Z(h = 0)$. Hence we get

$$Z(h) = Z_0 \exp \left(- \int d^d\mathbf{x} \, \mathcal{L}_I \left(\frac{\delta}{\delta h} \right) \right)$$

$$\exp \tfrac{1}{2} \int d^d\mathbf{x} d^d\mathbf{y} h(\mathbf{x}) G_0(\mathbf{x} - \mathbf{y}) h(\mathbf{y}) \quad (27)$$

The unrenormalized perturbative series is obtained by expanding the first exponential and by performing, term by term, the differentiation operations. These implicitly contain Wick's theorem in the form

$$\langle \varphi(\mathbf{x}_1)...\varphi(\mathbf{x}_{2n}) \rangle_0 = \frac{\delta^{2n}}{\delta h(\mathbf{x}_1)...\delta h(\mathbf{x}_{2n})} \frac{1}{2^n n!}$$

$$\left. \left(\int d^d\mathbf{x} d^d\mathbf{y} h(\mathbf{x}) G_0(\mathbf{x} - \mathbf{y}) h(\mathbf{y}) \right)^n \right|_{h=0}$$

$$= \sum_{\substack{\text{distinct} \\ \text{terms}}} G_0(\mathbf{x}_{a_1} - \mathbf{x}_{a_2})...G_0(\mathbf{x}_{a_{2n-1}} - \mathbf{x}_{a_{2n}})$$

$$(28)$$

where the last sum contains $(2n)!/2^n n!$ distinct terms (taking into account the invariance of G_0 under parity, $G_0(\mathbf{x}) = G_0(-\mathbf{x})$).

It is convenient to express the perturbative rules directly in momentum space for the vertex functions. We recall that these functions are amputated from the external line factors corresponding to the arguments φ of $\Gamma(\varphi)$. To each term of (27) is associated a representative diagram combining vertices (from each monomial of $\mathcal{L}_I$) and lines (from each factor G_0 in (28)). Only the irreducible (and hence connected) diagrams are considered in the calculation of Γ_n.

For the φ^4 interaction, each vertex is associated with a factor $-g_0$ and a momentum conservation δ-function $(2\pi)^d \delta(\Sigma_p)$ for the sum Σ_p of internal and external momenta entering the vertex. A line joining a vertex to itself (*tadpole*) does not contribute to Σ_p. Each internal line is associated with a d-dimensional momentum $\mathbf{p}$ with a propagation factor $G_0(\mathbf{p}) = (\mathbf{p}^2 + m_0^2)^{-1}$. One has to integrate over internal momenta with the measure $d^d\mathbf{p}/(2\pi)^d$.

These Feynman rules are complemented by a combinatorial factor resulting, in some cases, in the incomplete cancellations of the factorials introduced in the definition of the interaction $(1/4!)$ and in the expansion of the exponential. This *symmetry factor* divides the above contribution and is the order of the symmetry group of the diagram. This group corresponds to the permutations of *internal* lines and vertices leaving the diagram invariant (external lines are tied to sources and should not be permuted). Note that tadpoles yield a factor $\frac{1}{2}$ due to these symmetries. If there is some hesitation while applying these rules, one is well advised to return to relation (27). Examples of symmetry factors can be found in the following exercise. From the coefficient of $\ln Z$ in (19), vertex functions have an additional overall factor -1.

In a zero-dimensional theory, the partition function is the simple integral

$$z(g) = \int \frac{d\varphi}{(2\pi)^{1/2}} \exp - \left\{ \frac{\varphi^2}{2} + g\frac{\varphi^4}{4!} \right\} \tag{29}$$

It allows a verification of symmetry factors for vacuum diagrams, i.e. without external lines. Since all other factors in the Feynman rules are equal to unity, the coefficient of $(-g)^n$ is the sum of symmetry factors over all diagrams of order n

$$z(g) = \sum_{n=0}^{\infty} (-g)^n z_n \tag{30}$$

$$z_n = \frac{1}{4!^n} \int \frac{d\varphi}{(2\pi)^{1/2}} \frac{\varphi^{4n}}{n!} e^{-\frac{1}{2}\varphi^2} = \frac{4^n \Gamma(2n + \frac{1}{2})}{4!^n n! \Gamma(\frac{1}{2})} = \frac{(4n-1)!!}{4!^n n!}$$

This result can be checked using the symmetry factors of the diagrams displayed in figure 1. Indeed,

$$n = 1 \quad \frac{1}{8} = \frac{1}{8}$$

$$n = 2 \quad \frac{5 \cdot 7}{2^7 3} = \frac{1}{8^2 2!} + \frac{1}{2^4} + \frac{1}{2 \cdot 4!}$$

$$n = 3 \quad \frac{5 \cdot 7 \cdot 11}{2^{10} 3} = \frac{1}{8^3 3!} + \frac{1}{2^4 8} + \frac{1}{8 \cdot 2 \cdot 4!} + \frac{1}{2^5} + \frac{1}{2^4 3} + \frac{1}{2^2 3!} + \frac{1}{2^4 3}$$

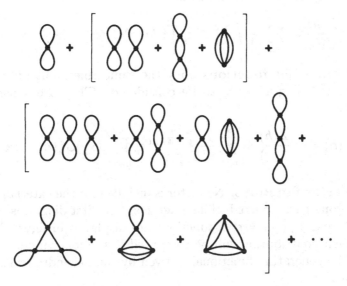

Fig. 1   Vacuum diagrams contributing to $Z(h = 0)$ up to and including order g^3.

and so on. In these expressions, the successive terms correspond respectively to the diagrams of figure 1.

The coefficient z_n increases as n^n, and, as a result, the series (30) is divergent, in fact asymptotic as $g \to 0$. The point $g = 0$ is an essential singularity of $z(g)$ with a cut on the negative real axis, showing an instability for $g < 0$. These properties, which are clear for the simple integral (29), may be generalized to functional integrals. For the same reasons, one expects that the perturbative expansion leads to divergent, asymptotic series. However, signs alternate and there exist resummation methods (such as the one using a Borel transformation).

Study the perturbative expansion of the integral (29) with an external source added.

Let us now summarize the sets of diagrams for the computation of the various Green functions.

i) **General correlation functions.**   The contributions of external lines should be kept. Diagrams may be disconnected, but connected parts without external lines should be omitted (vacuum diagrams are eliminated by division by $Z(h = 0)$).

The generating functional is

$$\frac{Z(h)}{Z(0)} = \sum_{n=0}^{\infty} \frac{1}{n!} \int \prod_{k=1}^{n} d^d \mathbf{x}_k \prod_{k=1}^{n} h(\mathbf{x}_k) G_n(\mathbf{x}_1, ..., \mathbf{x}_n) \qquad (31a)$$

ii) **Connected functions** obey the same rules, but only connected diagrams are to be considered. They are generated by

$$F(h) = \ln \frac{Z(h)}{Z(0)} = \sum_{n=0}^{\infty} \frac{1}{n!} \int \prod_{k=1}^{n} d^d \mathbf{x}_k \prod_{k=1}^{n} h(\mathbf{x}_k) G_n^c(\mathbf{x}_1, ..., \mathbf{x}_n)$$
$$(31b)$$

iii) **Vertex functions.** No factor is included for the external lines of one-particle irreducible diagrams. The first diagrams of the 2- and 4-point vertex functions are displayed in figure 2 and their contribution will be computed in the following sections. The generating functional is given by the Legendre transform of $F(h)$

$$\Gamma(\varphi) = \operatorname*{Sup}_{h} \left\{ \int d^d \mathbf{x} h \varphi - F(h) \right\}$$
$$= \sum_{n=0}^{\infty} \frac{1}{n!} \int \prod_{k=1}^{n} d^d \mathbf{x}_k \prod_{k=1}^{n} \varphi(\mathbf{x}_k) \Gamma_n(\mathbf{x}_1, \dots, \mathbf{x}_n)$$
$$(31c)$$

Besides the enumeration of diagrams and the determination of the combinatorial factors (a relatively easy task at low orders), one has to perform integrations. The computation of the corresponding Feynman integrals is in general difficult and involves many technical tricks. In most cases, these integrals cannot be expressed in terms of usual special functions.

5.2.2   *Loop expansion*

An n-point Green function depends, in its momentum representation, on n d-dimensional vectors, the sum of which vanishes due to translational invariance. If rotational invariance is preserved by the regularization procedure – which we assume to be the case – the function can be expressed in terms of the $n(n-1)/2$ invariants $\mathbf{p}_i \cdot \mathbf{p}_j$, $i \le j \le n-1$. For an integral dimension such that $n - 1 \ge d + 1$, there are necessarily additional linear relations

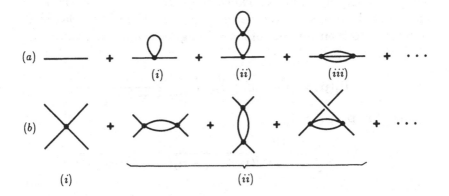

(a) ... (i) ... (ii) ... (iii) ...

(b) ... (i) ... (ii) ...

Fig. 2   First diagrams for the vertex functions. (a) 2-point function up to two-loop contributions, (b) 4-point function up to one-loop contributions.

between the vectors $\mathbf{p}_i$ which can be obtained by writing that the matrix $\left\{\mathbf{p}_i \cdot \mathbf{p}_j\right\}$, with $1 \leq i, j \leq n-1$, is at most of rank d.

Let us consider a connected n-point diagram of order v (with v vertices). In the perturbative theory around the fundamental state $\langle \varphi \rangle = 0$, the number n of external lines is necessarily even. Let l be the number of internal lines. Note that some of them are contracted on the same vertex (tadpole), with an associated factor

$$\int \frac{\mathrm{d}^d \mathbf{p}}{(2\pi)^d} \frac{1}{\mathbf{p}^2 + m_0^2}$$

which is independent of the remaining part of the diagram and which converges only due to the presence of the cutoff factor. There exists a prescription (*Wick's ordering*) which eliminates these terms, but we will not use it here. With a φ^4 interaction, there are $4v$ lines incident on the vertices. Taking into account the fact that an internal line is incident on two vertices, we get

$$4v = n + 2l \tag{32a}$$

There are l vectorial integration variables, and v conservation rules at the vertices, one of which expresses the total external momenta conservation. Hence, the number B of independent (vectorial) integration variables, called the *number of loops*, is

$$B = l - v + 1 = v - \tfrac{1}{2}n + 1 \tag{32b}$$

Diagrams without loops are called *trees*. As an example, the lowest order diagrams for the vertex functions have been classified according to their number of loops in figure 2, and their contributions are

$$\Gamma_2(\mathbf{p}) = \mathbf{p}^2 + m_0^2 + \tfrac{1}{2}g_0 \int \frac{d^d\mathbf{q}}{(2\pi)^d} \frac{1}{\mathbf{q}^2 + m_0^2} + \cdots$$

$$\Gamma_4(\mathbf{p}_1, \mathbf{p}_2, \mathbf{p}_3, \mathbf{p}_4) = g_0$$

$$-\tfrac{1}{2}g_0^2 \sum_{i=2}^{4} \int \frac{d^d\mathbf{q}}{(2\pi)^d} \frac{1}{[(\mathbf{p}_1 + \mathbf{p}_i - \mathbf{q})^2 + m_0^2][\mathbf{q}^2 + m_0^2]} + \cdots$$

$$(33)$$

The number of loops B may be interpreted as the power of $\hbar$ in a semiclassical expansion. We have indeed $(\hbar)^{B-1} = \hbar^I \hbar^{-v}$, hence each vertex is assigned a factor $1/\hbar$ and each propagator a factor $\hbar$. This is equivalent to dividing the action by $\hbar$. The perturbative expansion of $Z(h)$ appears as the semiclassical expansion of the functional integral around the saddle point $\varphi = 0$.

Let us compute the generating functional $\Gamma(\varphi)$ up to and including one-loop corrections. The partition function $Z(h)$ is obtained by determining the saddle point of $\exp\{-S + \int d^d\mathbf{x}\, h\varphi\}$, i.e.

$$(-\Delta + m_0^2)\varphi_c + \frac{1}{3!}g_0\varphi_c^3 = h \tag{34}$$

One considers the solution which vanishes with h

$$\varphi_c[h] = \frac{1}{-\Delta + m_0^2}h - \frac{1}{3!}g_0\left(\frac{1}{-\Delta + m_0^2}h\right)^3 + \cdots \tag{35}$$

The next order is obtained by expanding the action around φ_c, and performing the corresponding Gaussian integral

$$\frac{Z(h)}{Z(0)} = \exp\left\{-S(\varphi_c[h]) + \int d^d\mathbf{x}\, h\varphi_c[h]\right\}$$

$$\times \det^{-1/2}\left[\frac{\delta^2 S(\varphi_c[h])}{\delta\varphi(\mathbf{x})\delta\varphi(\mathbf{y})}\right] \det^{1/2}\left[\frac{\delta^2 S_0}{\delta\varphi(\mathbf{x})\delta\varphi(\mathbf{y})}\right] \tag{36}$$

The Legendre transformation gives

$$\Gamma(\varphi) = S(\varphi) - \tfrac{1}{2}\operatorname{Tr}\ln(G_\varphi G_0^{-1}) + \cdots \tag{37}$$

where $G_\varphi(\mathbf{x}, \mathbf{y})$ is the inverse kernel of $\delta^2 S / \delta\varphi(\mathbf{x})\delta\varphi(\mathbf{y})$. This reads

$$G_\varphi = [-\Delta + m_0^2 + \tfrac{1}{2}g_0\varphi^2]^{-1}$$
$$G_\varphi G_0^{-1} = 1 - G_0\tfrac{1}{2}g_0\varphi^2 + G_0\tfrac{1}{2}g_0\varphi^2 G_0\tfrac{1}{2}g_0\varphi^2 + \cdots \qquad (38)$$
$$= [1 + G_0\tfrac{1}{2}g_0\varphi^2]^{-1}$$

Hence formula (37) is rewritten as

$$\Gamma(\varphi) = S(\varphi) + \tfrac{1}{2}\operatorname{Tr}\ln\left(1 + \tfrac{1}{2}G_0 g_0\varphi^2\right)$$

$$= S(\varphi) + \sum_{p=1}^\infty \frac{(-1)^{p+1} g_0^p}{2p} \operatorname{Tr}\left(\tfrac{1}{2}G_0\varphi^2\right)^p$$

$$= S(\varphi) + \tfrac{1}{2}g_0\frac{1}{2!}\int \frac{d^d\mathbf{p}}{(2\pi)^d}\varphi(\mathbf{p})\varphi(-\mathbf{p})\int \frac{d^d\mathbf{q}}{(2\pi)^d}\frac{1}{\mathbf{q}^2 + m_0^2}$$

$$- \tfrac{1}{2}g_0^2\frac{1}{4!}\int \prod_{k=1}^{3}\frac{d^d\mathbf{p}_k}{(2\pi)^d}\varphi(\mathbf{p}_1)\varphi(\mathbf{p}_2)\varphi(\mathbf{p}_3)\varphi(\mathbf{p}_4 = -\mathbf{p}_1 - \mathbf{p}_2 - \mathbf{p}_3)$$

$$\times \sum_{i=2}^{4}\int \frac{d^d\mathbf{q}}{(2\pi)^d}\frac{1}{[(\mathbf{p}_1 + \mathbf{p}_i - \mathbf{q})^2 + m_0^2][\mathbf{q}^2 + m_0^2]} + \cdots$$
$$(39)$$

As $\varphi(\mathbf{x})$ tends towards a constant φ, $\Gamma(\varphi)$ becomes the integral of the effective potential $V(\varphi)$

$$\Gamma(\varphi) = \int d^d\mathbf{x}\, V(\varphi) \qquad (40)$$

With the one-loop correction, $V(\varphi)$ is given by

$$V(\varphi) = \tfrac{1}{2}m_0^2\varphi^2 + \frac{1}{4!}g_0\varphi^4 + \tfrac{1}{2}\int \frac{d^d\mathbf{p}}{(2\pi)^d}\ln\left[1 + \frac{1}{\mathbf{p}^2 + m_0^2}\frac{1}{2}g_0\varphi^2\right] + \cdots$$
$$(41)$$

The steepest descent method, including the corrections due to the fluctuations around the saddle point $\varphi_c[h]$, is thus identical to the perturbative theory. Terms with more loops are obtained from irreducible diagrams using the propagator G_φ given in (38). Apart from technical complications, this method can be generalized in various ways, e.g. to degenerate saddle points or to fields with internal degrees of freedom.

Besides the Green functions considered so far, one is also interested in mixed correlations including insertions of local operators, as already mentioned above. For instance, the insertion of $\frac{1}{2}\varphi^2$,

$$\left\langle \varphi(\mathbf{x}_1)...\varphi(\mathbf{x}_n)\frac{1}{2}\varphi^2(\mathbf{y}) \right\rangle = \int \prod_{k=1}^{n} \frac{d^d \mathbf{p}_k}{(2\pi)^d} \frac{d^d \mathbf{q}}{(2\pi)^d} (2\pi)^d$$

$$\delta(\sum \mathbf{p}_k + \mathbf{q}) \, e^{-i(\sum \mathbf{p}_k \cdot \mathbf{x}_k + \mathbf{q} \cdot \mathbf{y})} G_{n,1}(\mathbf{p}_1, ..., \mathbf{p}_n; \mathbf{q})$$

(42)

arises formally from a limiting procedure

$$\left\langle \varphi(\mathbf{x}_1)...\varphi(\mathbf{x}_n)\frac{1}{2}\varphi^2(\mathbf{y}) \right\rangle = \lim_{\mathbf{z} \to 0} \frac{1}{2} \langle \varphi(\mathbf{x}_1)...\varphi(\mathbf{x}_n)\varphi(\mathbf{y}+\mathbf{z})\varphi(\mathbf{y}-\mathbf{z}) \rangle$$

(43)

that is, in momentum representation,

$$G_{n,1}(\mathbf{p}_1, ..., \mathbf{p}_n; \mathbf{q}) = \frac{1}{2} \int \frac{d^d \mathbf{k}}{(2\pi)^d} G_{n+2}(\mathbf{p}_1, ..., \mathbf{p}_n, \tfrac{1}{2}\mathbf{q}+\mathbf{k}, \tfrac{1}{2}\mathbf{q}-\mathbf{k})$$

(44)

The insertion of $\frac{1}{2}\varphi^2$ at zero momentum is equivalent to a derivative with respect to m_0^2. The insertion of more terms $\frac{1}{2}\varphi^2$ or other combinations of local operators leads to analogous expressions. The corresponding perturbative expansions are easily derived.

5.2.3   *Evaluation of integrals and dimensional continuation*

Several techniques have been developed for the evaluation of Feynman integrals and for the study of their analytical properties. We give here only some illustrative examples and focus on the dimensional continuation.

The introduction of vertex functions is a first step in reducing the complexity, since other Green functions can be deduced by algebraic operations. Note that there still remain articulated diagrams – i.e. which can be disconnected by omission of one vertex, see chapter 7 (in volume 2) – which lead to factorizable contributions.

As we are interested by dimensional continuation which, of course, makes sense only for scalars, it is important to use only scalar invariant combinations in the integrands. The identity

$$\frac{1}{(q^2 + m_0^2)^n} = \frac{1}{\Gamma(n)} \int_0^\infty d\alpha \, \alpha^{n-1} e^{-\alpha(q^2 + m_0^2)}$$

(45)

allows one to substitute exponentials for polynomials in the denominators. These exponentials are Gaussian with respect to the momenta and the corresponding integrations can be carried out in any dimension. This method introduces conjugate variables α with inverse-square mass dimension. In a relativistic context, they are interpreted as proper times.

Let us take as an example the first (constant) correction to the 2-point function, which yields a shift of the critical temperature due to fluctuations. From equation (33), (diagram i of figure $2a$)

$$\Gamma_2^a = \tfrac{1}{2}g_0 \int \frac{d^d\mathbf{q}}{(2\pi)^d} \frac{1}{\mathbf{q}^2 + m_0^2} = \tfrac{1}{2}g_0 \int \frac{d^d\mathbf{q}}{(2\pi)^d} \int_0^\infty d\alpha \, e^{-\alpha(\mathbf{q}^2 + m_0^2)}$$

$$= \frac{g_0}{2(4\pi)^{d/2}} \int_0^\infty d\alpha \, \alpha^{-d/2} e^{-\alpha m_0^2}$$

$$(46)$$

The Gaussian representation allows one to integrate over the internal momentum for an arbitrary dimension d. Rotational invariance has provided the area of the unit sphere

$$S_d = \frac{2\pi^{d/2}}{\Gamma(d/2)} \qquad (47)$$

and the remaining radial integral yields in particular the power $\alpha^{-d/2}$. The dimension of $d\alpha\,\alpha^{-d/2}$ is the same as the original one of $d^d\mathbf{q}(\mathbf{q}^2 + m_0^2)^{-2}$. In the integral over the parameter α, two regions may generate divergences. The infrared region $\alpha \to \infty$ will give divergences if the masses vanish for $d \leq 2$. The ultraviolet region $\alpha \to 0$ is not sensitive to masses and gives singularities when $d \geq 2$. The integral is never convergent if initially $m_0^2 = 0$. However, with $m_0^2 > 0$, the integral is well-defined for $d < 2$ and can be analytically continued as a meromorphic function of d with simple poles for $d = 2, 4, \ldots$

$$\Gamma_2^a = \frac{g_0 m_0^{d-2}}{2(4\pi)^{d/2}} \Gamma(1 - d/2) \qquad (48)$$

As the dimension of g_0 is $4 - d$, $g_0 m_0^{d-2} = (g_0 m_0^{d-4})m_0^2$ has the correct dimension 2 of the irreducible 2-point function. For an even integral dimension, the expression (48) is meaningless due to ultraviolet singularities. Therefore, an ultraviolet regularization – a last remnant from the lattice – should be used. For this purpose, it is sufficient to remove from the integral over α a domain $(0, \Lambda^{-2})$. More generally, one introduces in the integral a function $\chi_\Lambda(\alpha)$ equal to unity as $\alpha \geq \Lambda^{-2}$ and vanishing with a sufficient number of its derivatives at $\alpha = 0$. The arbitrariness on this function reflects the arbitrariness on the regularization procedure which should not modify the universal critical properties at distances large with respect to $a \sim \Lambda^{-1}$. With the above choice for the function $\chi_\Lambda(\alpha)$, i.e. Heaviside's step function $\theta(\alpha - \Lambda^{-2})$, we find

$$\Gamma_2^a \to \frac{g_0}{2(4\pi)^{d/2}} \int_{\Lambda^{-2}}^\infty d\alpha\, \alpha^{-d/2} e^{-m_0^2 \alpha} = \frac{g_0 m_0^{d-2}}{2(4\pi)^{d/2}} \times$$

$$\left\{ \frac{2}{d-2} \left(\frac{\Lambda}{m_0}\right)^{d-2} e^{-m_0^2/\Lambda^2} - \frac{4}{(d-2)(d-4)} \left(\frac{\Lambda}{m_0}\right)^{d-4} e^{-m_0^2/\Lambda^2} \right.$$

$$\left. + \frac{4}{(d-2)(d-4)} \int_{m_0^2/\Lambda^2}^\infty dx\, x^{2-d/2} e^{-x} \right\}$$

$$(49)$$

This expression can now be evaluated near $d = 4$ in the limit $\Lambda/m_0 \gg 1$. After expanding $\exp(-m_0^2/\Lambda^2)$, we get, with γ standing for Euler's constant (see below equation (59)),

$$\Gamma_2^a \underset{d\to 4}{\sim} \frac{g_0 m_0^{d-2}}{(4\pi)^{d/2}} \frac{1}{(d-2)}$$

$$\times \left\{ (\Lambda/m_0)^{d-2} - (\Lambda/m_0)^{d-4} + 2\frac{1 - (\Lambda/m_0)^{d-4}}{d-4} + \gamma + \cdots \right\} \quad (50)$$

For $d = 4$, this is

$$\Gamma_2^a = \frac{g_0 m_0^2}{2(4\pi)^2} \left\{ \frac{\Lambda^2}{m_0^2} + \gamma - 1 - \ln \frac{\Lambda^2}{m_0^2} \right\} \qquad (d = 4) \qquad (51)$$

Thanks to the cutoff factor Λ, the pole at $d = 2$ translates into a quadratically ultraviolet divergent term, and the one at $d = 4$, into a logarithmically divergent term.

Of course, the above integral is particularly simple. However, the various phenomena just encountered generalize to more complex cases. Let us look for instance at the two-loop diagrams for the 2-point function. The two irreducible diagrams *ii* and *iii* of figure 2a yield the contributions

$$\Gamma_2^b = -\tfrac{1}{4} g_0^2 \int \frac{d^d q}{(2\pi)^d} \frac{1}{q^2 + m_0^2} \int \frac{d^d k}{(2\pi)^d} \frac{1}{(k^2 + m_0^2)^2}$$

$$= -\tfrac{1}{4} g_0^2 \frac{m_0^{2d-6}}{(4\pi)^d} \Gamma(1 - d/2) \Gamma(2 - d/2)$$

$$(52)$$

$$\Gamma_2^c(\mathbf{p}) = -\tfrac{1}{6}g_0^2 \int \frac{d^d q d^d k}{(2\pi)^{2d}} \frac{1}{[q^2 + m_0^2][k^2 + m_0^2][(\mathbf{q} - \mathbf{p} + \mathbf{k})^2 + m_0^2]}$$

$$= -\tfrac{1}{6}g_0^2 \int_0^\infty d\alpha_1 d\alpha_2 d\alpha_3 \, e^{-m_0^2(\alpha_1 + \alpha_2 + \alpha_3)}$$

$$\int \frac{d^d q d^d k}{(2\pi)^{2d}} e^{-(\alpha_1 q^2 + \alpha_2 k^2 + \alpha_3 (q+k-p)^2)} \tag{53}$$

$$= -\frac{g_0^2}{6(4\pi)^d} \int_0^\infty \frac{d\alpha_1 d\alpha_2 d\alpha_3}{(\alpha_2 \alpha_3 + \alpha_3 \alpha_1 + \alpha_1 \alpha_2)^{\frac{d}{2}}}$$

$$\exp - \left\{ m_0^2 (\alpha_1 + \alpha_2 + \alpha_3) + \frac{\alpha_1 \alpha_2 \alpha_3 \mathbf{p}^2}{\alpha_1 \alpha_2 + \alpha_2 \alpha_3 + \alpha_3 \alpha_1} \right\}$$

The homogeneity properties with respect to the variable α_i can be exploited by introducing a factor

$$1 = \int_0^\infty d\lambda \delta(\alpha_1 + \alpha_2 + \alpha_3 - \lambda)$$

Then α_i is changed to $\lambda \alpha_i$, and the integral over λ, which is convergent for $d < 3$, is performed. Hence

$$\Gamma_2^c(\mathbf{p}) = -\frac{g_0^2}{6(4\pi)^d} \int_0^1 \frac{d\alpha_1 d\alpha_2 d\alpha_3 \delta(\alpha_1 + \alpha_2 + \alpha_3 - 1)}{(\alpha_2 \alpha_3 + \alpha_3 \alpha_1 + \alpha_1 \alpha_2)^{d/2}}$$

$$\times \int_0^\infty d\lambda \lambda^{2-d} \exp -\lambda \left\{ m_0^2 + \frac{\alpha_1 \alpha_2 \alpha_3 \mathbf{p}^2}{\alpha_1 \alpha_2 + \alpha_2 \alpha_3 + \alpha_3 \alpha_1} \right\}$$

$$= -\frac{g_0^2}{6(4\pi)^d} \Gamma(3 - d) \int_0^1 \frac{d\alpha_1 d\alpha_2 d\alpha_3 \delta(\alpha_1 + \alpha_2 + \alpha_3 - 1)}{(\alpha_2 \alpha_3 + \alpha_3 \alpha_1 + \alpha_1 \alpha_2)^{d/2}}$$

$$\times \left\{ m_0^2 + \frac{\alpha_1 \alpha_2 \alpha_3 \mathbf{p}^2}{\alpha_1 \alpha_2 + \alpha_2 \alpha_3 + \alpha_3 \alpha_1} \right\}^{d-3} \tag{54}$$

Analyse the ultraviolet convergence of the diagram using dimensional arguments.

For $d \geq 3$, keeping $m_0^2 > 0$ prevents infrared divergences, but ultraviolet ones appear. At $d = 3$, the divergence is logarithmic and $\Gamma_2^c(\mathbf{p}) - \Gamma_2^c(0)$ remains finite. In dimension four, the value at zero momentum is quadratically divergent (as is Γ_2^a) and the first derivative $d\Gamma_2^c(0)/d\mathbf{p}^2$ is still logarithmically divergent. This leads to the so-called wavefunction renormalization which appears in φ^4 theory at the two-loop order.

Introducing the cutoff factor Λ in the α integrals, let us compute these divergent terms at $d = 4$

$$\Gamma_2^c(\mathbf{p}) = -\frac{1}{6(4\pi)^4} g_0^2 \left\{ A + B\mathbf{p}^2 + C(\mathbf{p}^2) \right\} \qquad (55)$$

where the finite function $C(\mathbf{p}^2)$ vanishes with its first derivative at the origin. Using the notation

$$\eta = \frac{m_0^2}{\Lambda^2} \qquad (56)$$

we get from (53)

$$\frac{A}{m_0^2} = \eta^{-1} \int_1^\infty d\alpha_1 \int_1^\infty d\alpha_2 \int_1^\infty d\alpha_3 \frac{e^{-\eta(\alpha_1 + \alpha_2 + \alpha_3)}}{(\alpha_1 \alpha_2 + \alpha_2 \alpha_3 + \alpha_3 \alpha_1)^2} \qquad (57)$$

The symmetric integral is computed in the sector $\alpha_3 > \alpha_2 > \alpha_1 > 1$. Introducing the variables $\alpha_3/\alpha_2 = z$, $\alpha_2/\alpha_1 = y$, $\alpha_1 = x$,

$$\frac{A}{m_0^2} = 6\eta^{-1} \int_1^\infty \frac{dx}{x^2} \int_1^\infty \frac{dy}{y} \int_1^\infty dz \frac{e^{-\eta x(1+y+yz)}}{(1+z+yz)^2} = 6 \int_\eta^\infty \frac{du}{u^2} e^{-u} \psi(u)$$

where $u = \eta x$ and

$$\psi(u) = \int_1^\infty \frac{dy}{y} \int_1^\infty dz \frac{e^{-uy(1+z)}}{(1+z+yz)^2}$$

The singular behaviour of A as $\eta \to 0$ arises from the small u integration domain where

$$\psi(u) = \psi_0 + \psi_1 u \ln u + \psi_2 u + \mathcal{O}(u^2 \ln^2 u)$$

A useful table of integrals is

$$\int_\eta^\infty \frac{du}{u^2} e^{-u} = \frac{1}{\eta} + \ln \eta + \textit{finite part}$$

$$\int_\eta^\infty \frac{du}{u} \ln u \, e^{-u} = -\tfrac{1}{2} \ln^2 \eta + \textit{finite part} \qquad (58)$$

$$\int_\eta^\infty \frac{du}{u} e^{-u} = -\ln \eta + \textit{finite part}$$

The final result is

$$\frac{A}{m_0^2} = 6\eta^{-1} \psi_0 - 3\psi_1 \ln^2 \eta + 6(\psi_0 - \psi_2) \ln \eta + \textit{finite part}$$

with

$$\psi_0 = \int_1^\infty \frac{dy}{y} \int_1^\infty dz \frac{1}{(1+z+yz)^2} = \int_1^\infty dy \frac{1}{y(1+y)(2+y)} = \tfrac{1}{2} \ln \tfrac{4}{3}$$

To evaluate ψ_1 and ψ_2, let us study

$$\frac{d\psi(u)}{du} = \psi_1 \ln u + (\psi_1 + \psi_2) + \mathcal{O}(u \ln^2 u)$$

$$= -\int_1^\infty dy \int_1^\infty dz \frac{1+z}{[1+z(1+y)]^2} e^{-uy(1+z)}$$

The last integral is split into two terms φ_1 and φ_2, corresponding to the decomposition

$$\frac{1+z}{[1+z+yz]^2} = \frac{y}{(1+y)[1+z+yz]^2} + \frac{1}{(1+y)[1+z+yz]}$$

$$\varphi_1(u) = -\int_1^\infty dy \frac{y}{1+y} \int_1^\infty dz \frac{e^{-uy(1+z)}}{(1+z+zy)^2}$$

$$\xrightarrow[u \to 0]{} -\int_1^\infty dy \frac{y}{1+y} \int_1^\infty \frac{dz}{(1+z+zy)^2} = 2\ln \tfrac{2}{3} + \tfrac{1}{2}$$

$$\varphi_2(u) = -\int_1^\infty \frac{dy}{1+y} \int_1^\infty \frac{dz}{1+z+zy} e^{-uy(1+z)}$$

$$= -\int_1^\infty \frac{dy}{(1+y)^2} e^{-uy + \frac{uy}{1+y}} \int_{uy(2+y)/(1+y)}^\infty \frac{e^{-v} dv}{v}$$

Using Euler's constant

$$\gamma = -\Gamma'(1) = -\int_0^\infty dt \, \ln t \, e^{-t} \tag{59}$$

one finds

$$\varphi_2(u) = \int_1^\infty \frac{dy}{(1+y)^2} \exp\left(-uy + \frac{uy}{1+y}\right)$$

$$\left\{ \exp\left(-\frac{uy(2+y)}{1+y}\right) \ln \frac{uy(2+y)}{1+y} + \gamma \right\} + \mathcal{O}(u \ln^2 u)$$

$$= (\ln u + \gamma) \int_1^\infty \frac{dy}{(1+y)^2} + \int_1^\infty \frac{dy}{(1+y)^2} \ln \frac{y(2+y)}{1+y}$$

$$+ \mathcal{O}(u \ln^2 u)$$

Therefore one gets

$$\varphi_2(u) = \tfrac{1}{2}(\ln u + \gamma - 1 + 3\ln 3 - \ln 2) + \mathcal{O}(u \ln^2 u)$$

Gathering all these results,

$$\frac{d\psi}{du} = \psi_1 \ln u + \psi_1 + \psi_2 + \mathcal{O}(u \ln^2 u)$$

$$= \varphi_1(u) + \varphi_2(u) = \tfrac{1}{2} \ln u + \tfrac{1}{2}(\gamma - \ln 3 + 3\ln 2) + \mathcal{O}(u \ln^2 u)$$

Hence

$$\psi_1 = \tfrac{1}{2} \qquad\qquad \psi_2 = \tfrac{1}{2}(\gamma - 1 - \ln 3 + 3\ln 2)$$

The first divergent coefficient finally reads

$$\frac{A}{m_0^2} = (3\ln \tfrac{4}{3})\frac{\Lambda^2}{m_0^2} - \tfrac{3}{2}\ln^2 \frac{\Lambda^2}{m_0^2} + 3[\ln 2 - 1 + \gamma]\ln \frac{\Lambda^2}{m_0^2} + finite\ part \tag{60}$$

Now we turn to the coefficient B in equation (55). One has

$$\begin{aligned}
B &= -\int_{\Lambda^{-2}}^{\infty} \frac{d\alpha_1 d\alpha_2 d\alpha_3 \alpha_1 \alpha_2 \alpha_3}{(\alpha_1\alpha_2 + \alpha_2\alpha_3 + \alpha_3\alpha_1)^3} e^{-m_0^2(\alpha_1+\alpha_2+\alpha_3)} \\
&= -6\int_{\eta}^{\infty}\frac{dx}{x}e^{-x}\int_1^{\infty} dy e^{-xy}\int_1^{\infty} dz\frac{ze^{-xyz}}{(1+z+yz)^3} \tag{61}\\
&= -6\int_{\eta}^{\infty}\frac{dx}{x}e^{-x}\tilde{\psi}(x) = 6\tilde{\psi}(0)\ln \eta + finite\ part
\end{aligned}$$

where $\tilde{\psi}(0)$ is finite

$$\begin{aligned}
\tilde{\psi}(0) &= \int_1^{\infty} dy \int_1^{\infty} dz\frac{z}{(1+z+yz)^3} = \int_1^{\infty} dz \int_1^{\infty} \frac{zdy}{(1+z+zy)^3} \\
&= \tfrac{1}{2}\int_1^{\infty}\frac{dz}{(1+2z)^2} = \tfrac{1}{12}
\end{aligned}$$

Consequently,

$$B = -\tfrac{1}{2}\ln \frac{\Lambda^2}{m_0^2} + finite\ part \tag{62}$$

This completes the computation of singular terms of $\Gamma_2^c(\mathbf{p}^2)$ in four dimensions.

(1) Show that, for $d = 3$, one can write

$$\Gamma_2^c(\mathbf{p}^2) = -\frac{g_0^2}{6(4\pi)^3}\left\{\bar{A} + \bar{B}\mathbf{p}^2 + \bar{C}(\mathbf{p}^2)\right\} \tag{63}$$

$$\bar{A} = 2\pi \ln \frac{\Lambda^2}{m_0^2} + finite\ part \qquad \bar{B} = -\frac{2\pi}{27}m_0^{-2} \qquad \bar{C} = \mathcal{O}(\mathbf{p}^4)$$

In this case, only $\bar{A}$ is logarithmically divergent, and the terms within the braces are dimensionless since g_0 has the dimension of a mass.

(2) If $m_0^2 = 0$, the integral defining $\Gamma_2^c(\mathbf{p})$ converges for $2 < d < 3$. Show that its analytic continuation reads

$$\Gamma_2^c(\mathbf{p}^2, m_0^2 = 0) = -\frac{1}{6(4\pi)^d} g_0^2 (\mathbf{p}^2)^{d-3} \frac{\Gamma(3-d)\Gamma^3(d/2-1)}{\Gamma(3(d/2-1))} \quad (64)$$

which behaves as

$$\Gamma_2^c(\mathbf{p}^2, m_0^2 = 0) \underset{d \to 4}{\sim} -\frac{1}{6(4\pi)^d} g_0^2 \mathbf{p}^2 \frac{(\mathbf{p}^2)^{d-4}}{2(d-4)} \quad (65)$$

Let us call $G(\mathbf{p}^2)$ the coefficient of $-g_0^2 \mathbf{p}^2 / 6(4\pi)^d$. The quantity $G(\mathbf{p}^2) - G(\mathbf{P}^2)$ has a finite limit as $d \to 4$

$$\lim_{d \to 4} G(\mathbf{p}^2) - G(\mathbf{P}^2) = \tfrac{1}{2} \ln \frac{\mathbf{p}^2}{\mathbf{P}^2}$$

This result may be compared to the expression (62) for the coefficient B. In the four-dimensional regularized theory, we have indeed

$$\Gamma_2^c(\mathbf{p}^2, m_0^2, \Lambda^2) = -\frac{1}{6(4\pi)^4} \left\{ A(0, m_0^2, \Lambda^2) + \mathbf{p}^2 G(\mathbf{p}^2, m_0^2, \Lambda^2) \right\}$$

with

$$G(\mathbf{p}^2, m_0^2, \Lambda^2) = \tfrac{1}{2} \ln \frac{m_0^2}{\Lambda^2} + g(\frac{\mathbf{p}^2}{m_0^2})$$

for large Λ^2, with g finite and dimensionless just as G. In the region $m_0^2 \ll \mathbf{p}^2$, $\mathbf{P}^2 \ll \Lambda^2$, we introduce $G(\mathbf{p}^2, m_0^2, \Lambda^2) - G(\mathbf{P}^2, m_0^2, \Lambda^2) = g(\mathbf{p}^2/m_0^2) - g(\mathbf{P}^2/m_0^2)$. The preceding result shows that the limit $m_0^2 \to 0$ can be taken, and this implies that, for $\mathbf{p}^2 \gg m_0^2$

$$g(\mathbf{p}^2/m_0^2) \to \tfrac{1}{2} \ln(\mathbf{p}^2/m_0^2)$$

In this domain $m_0^2 \ll \mathbf{p}^2 \ll \Lambda^2$,

$$G(\mathbf{p}^2, m_0^2, \Lambda^2) \sim \tfrac{1}{2} \ln \frac{m_0^2}{\Lambda^2} + \tfrac{1}{2} \ln \frac{\mathbf{p}^2}{m_0^2} = \tfrac{1}{2} \ln \frac{\mathbf{p}^2}{\Lambda^2}$$

Let us also compute the 4-point function at the one-loop order. The contributions of the corresponding diagrams are successively (figure 2b)

$$\Gamma_4^a = g_0 \quad (66)$$

$$\Gamma_4^b(\mathbf{p}_1, \dots, \mathbf{p}_4) = -\tfrac{1}{2} g_0^2 \sum_{i=2}^{4} \int \frac{d^d \mathbf{q}}{(2\pi)^d} \frac{1}{[\mathbf{q}^2 + m_0^2][(\mathbf{p}_1 + \mathbf{p}_i - \mathbf{q})^2 + m_0^2]}$$

$$(67)$$

Let us denote the integral $F(\mathbf{p}^2 = (\mathbf{p}_1 + \mathbf{p}_i)^2)$. It reads

$$
\begin{aligned}
F(\mathbf{p}^2) &= \frac{1}{(4\pi)^{d/2}} \int_0^\infty \frac{d\alpha_1 d\alpha_2}{(\alpha_1 + \alpha_2)^{d/2}} \exp - \left\{ m_0^2(\alpha_1 + \alpha_2) + \frac{\alpha_1\alpha_2\mathbf{p}^2}{\alpha_1 + \alpha_2} \right\} \\
&= \frac{\Gamma(2 - d/2)}{(4\pi)^{d/2}} \int_0^1 \frac{d\alpha}{[m_0^2 + \mathbf{p}^2\alpha(1 - \alpha)]^{2-d/2}}
\end{aligned}
$$

$$(68)$$

In the case $m_0^2 = 0$, one gets

$$
F(\mathbf{p}^2, m_0^2 = 0) = (\mathbf{p}^2)^{d/2-2} \frac{\Gamma(2 - d/2)\Gamma^2(d/2 - 1)}{(4\pi)^{d/2}\Gamma(d - 2)}
$$

$$(69)$$

and one notices an infrared divergence at $d = 2$. Near $d = 4$, the quantity $F(\mathbf{p}^2, m_0^2 = 0) - F(\mu^2, m_0^2 = 0)$ has a limit

$$
\lim_{d \to 4} F(\mathbf{p}^2, m_0^2 = 0) - F(\mu^2, m_0^2 = 0) = -\frac{1}{(4\pi)^2} \ln \frac{\mathbf{p}^2}{\mu^2}
$$

$$(70)$$

The same logarithmic divergence can be computed for $d = 4$ in the regularized theory where

$$
\begin{aligned}
&F(\mathbf{p}^2, m_0^2, \Lambda^2) \\
&= \frac{1}{(4\pi)^2} \int_{\Lambda^{-2}}^\infty \frac{d\alpha_1 d\alpha_2}{(\alpha_1 + \alpha_2)^2} \exp - \left\{ m_0^2(\alpha_1 + \alpha_2) + \frac{\alpha_1\alpha_2\mathbf{p}^2}{\alpha_1 + \alpha_2} \right\}
\end{aligned}
$$

$$(71)$$

As $\Lambda^2 \to \infty$,

$$
\begin{aligned}
&F(\mathbf{p}^2 = 0, m_0^2, \Lambda^2) \\
&= \frac{1}{(4\pi)^2} \left\{ \ln \frac{\Lambda^2}{m_0^2} - \ln 2 - 1 - \gamma + \mathcal{O}(\frac{m_0^2}{\Lambda^2} \ln \frac{\Lambda^2}{m_0^2}) \right\}
\end{aligned}
$$

$$(72)$$

The dimensionless function F can be cast into the form

$$
F(\mathbf{p}^2, m_0^2, \Lambda^2) = \frac{1}{(4\pi)^2} \ln \frac{\Lambda^2}{\mathbf{p}^2} + f(\frac{m_0^2}{\mathbf{p}^2})
$$

The combination $F(\mathbf{p}^2, m_0^2, \Lambda^2) - F(\mu^2, m_0^2, \Lambda^2)$ has a finite limit for $m_0^2 \to 0$

$$
\lim_{m_0^2 \to 0} F(\mathbf{p}^2, m_0^2, \Lambda^2) - F(\mu^2, m_0^2, \Lambda^2) = -\frac{1}{(4\pi)^2} \ln \frac{\mathbf{p}^2}{\mu^2}
$$

and this expression agrees with (70). This example shows explicitly the relations between ultraviolet behaviour and massless theory.

Rather than giving more examples, let us summarize the method. One first uses a parametric representation of the propagators. Then the Gaussian integration over the momenta is performed. Finally, the ultraviolet behaviour as $\alpha \to 0$ is analyzed in each sector. The large α behaviour is controlled by internal masses such as m_0^2.

5.2.4   Group theoretical factors.

Let us look at the n-vector model with an $O(n)$ internal symmetry, described by dynamical variables $\boldsymbol{\Phi} = \{\Phi_1, \ldots, \Phi_n\}$ with an interaction $(\boldsymbol{\Phi}^2)^2$. The correlation functions now depend not only on external momenta, but also on indices $i_1, \ldots, i_n$ attached to the external fields. They are symmetric under simultaneous permutations of these indices and of the momenta. The contributions of diagrams contain extra polynomials in n of degree at most equal to the number of loops. They arise from contractions and summations on the internal indices. For instance, in the symmetric phase, the two-point function is diagonal in this internal space (see diagrams of figure 2a)

$$
\begin{aligned}
\Gamma_{2,ij}(\mathbf{p}^2) = \delta_{ij} \Big\{ &\mathbf{p}^2 + m_0^2 + \\
&+ g_0 \frac{n+2}{6} \int \frac{d^d q}{(2\pi)^d} \frac{1}{\mathbf{q}^2 + m_0^2} \\
&- g_0^2 \frac{(n+2)^2}{36} \int \frac{d^d q}{(2\pi)^d} \frac{1}{\mathbf{q}^2 + m_0^2} \int \frac{d^d k}{(2\pi)^d} \frac{1}{(\mathbf{k}^2 + m_0^2)^2} \\
&- g_0^2 \frac{n+2}{18} \int \frac{d^d q d^d k}{(2\pi)^{2d}} \frac{1}{[\mathbf{q}^2 + m_0^2][\mathbf{k}^2 + m_0^2][(\mathbf{q} + \mathbf{k} - \mathbf{p})^2 + m_0^2]} \\
&+ \cdots \Big\}
\end{aligned}
\tag{73}
$$

The coefficients have been obtained by contracting indices of internal and external lines, following Feynman's rules. The zero-dimensional model yields sum rules on these coefficients. For instance, for the 2-point function,

$$
\langle \Phi_i \Phi_j \rangle = \frac{\delta_{ij}}{\Gamma} = \frac{\int d^n \boldsymbol{\Phi} \; \Phi_i \Phi_j \exp - \{\frac{1}{2}\boldsymbol{\Phi}^2 + \frac{1}{4!}g_0(\boldsymbol{\Phi}^2)^2\}}{\int d^n \boldsymbol{\Phi} \exp - \{\frac{1}{2}\boldsymbol{\Phi}^2 + \frac{1}{4!}g_0(\boldsymbol{\Phi}^2)^2\}}
\tag{74}
$$

hence

$$\Gamma = \frac{1}{2}n \frac{\sum_{p=0}^{\infty} \frac{1}{p!} \left(-\frac{g_0}{6}\right)^p \Gamma(n/2 + 2p)}{\sum_{p=0}^{\infty} \frac{1}{p!} \left(-\frac{g_0}{6}\right)^p \Gamma(n/2 + 2p + 1)}$$

$$= 1 + g_0 \frac{n+2}{6} - g_0^2 \frac{(n+2)(n+4)}{36} + \cdots \qquad (75)$$

which coincides with (73) when integrals are replaced by unity.

The representation

$$\Gamma = 1 + \frac{g_0}{3} \frac{\int_0^{\infty} dx\, x^{n/2+1} \exp - \left(x + \frac{g_0}{6} x^2\right)}{\int_0^{\infty} dx\, x^{n/2} \exp - \left(x + \frac{g_0}{6} x^2\right)}$$

shows that Γ can be expressed as a continuous fraction in the variable $u = g_0/6$

$$\Gamma = 1 + \cfrac{(n+2)u}{1 + \cfrac{(n+4)u}{1 + \cfrac{(n+6)u}{1 + \cdots}}} \qquad (76)$$

This gives

$$\Gamma = 1 + (n+2)u - (n+2)(n+4)u^2 + (n+2)(n+4)(2n+10)u^3$$
$$- (n+2)(n+4)[(2n+10)^2 + (n+6)(n+8)]u^4 + \cdots$$
$$(77)$$

In particular $\Gamma = 1$ for $n = -2$ and it is a rational fraction in g_0 for any $n = -2p$.

For the 4-point function, (figure 2*b*) we have

$$\Gamma_{4,i_1 i_2 i_3 i_4}(\mathbf{p}_1, \mathbf{p}_2, \mathbf{p}_3, \mathbf{p}_4) = \frac{1}{3} g_0 (\delta_{i_1 i_2} \delta_{i_3 i_4} + \delta_{i_1 i_3} \delta_{i_2 i_4} + \delta_{i_1 i_4} \delta_{i_2 i_3})$$
$$- \frac{1}{9} g_0^2 \sum_{\substack{3 \ terms}} \left((2 + \tfrac{1}{2}n)\delta_{i_1 i_2} \delta_{i_3 i_4} + \delta_{i_1 i_3} \delta_{i_2 i_4} + \delta_{i_1 i_4} \delta_{i_2 i_3} \right)$$
$$\times \int \frac{d^d q}{(2\pi)^d} \frac{1}{[q^2 + m_0^2][(\mathbf{p}_1 + \mathbf{p}_2 - \mathbf{q})^2 + m_0^2]} + \cdots$$
$$(78)$$

In particular, the four-point function is proportional to the combination $\sum \delta_{i_1 i_2} \delta_{i_3 i_4}$, for zero external momenta

$$\Gamma_{4,i_1 i_2 i_3 i_4}(\mathbf{0}, \mathbf{0}, \mathbf{0}, \mathbf{0}) = (\delta_{i_1 i_2} \delta_{i_3 i_4} + \delta_{i_1 i_3} \delta_{i_2 i_4} + \delta_{i_1 i_4} \delta_{i_2 i_3}) \times$$
$$\left\{ \frac{g_0}{3} - g_0^2 \frac{n+8}{18} \int \frac{d^d q}{(2\pi)^d} \frac{1}{(q^2 + m_0^2)^2} + \cdots \right\} \qquad (79)$$

Show that, in zero dimension, with $u = \frac{1}{6}g_0$

$$\Gamma_{4,i_1i_2i_3i_4} = (\delta_{i_1i_2}\delta_{i_3i_4} + \delta_{i_1i_3}\delta_{i_2i_4} + \delta_{i_1i_4}\delta_{i_2i_3})\left\{\Gamma^2 - \Gamma^3\frac{\Gamma-1}{(n+2)u}\right\}$$

(80)

where Γ is defined in (74). The quantity between braces can be developed as

$$\Gamma^2 - \Gamma^3\frac{\Gamma-1}{(n+2)u} = \frac{g_0}{3} - g_0^2\frac{n+8}{18} + g_0^3\frac{3n^2+46n+140}{108} + \cdots$$ (81)

There are of course numerous models with various internal symmetries. We will in particular encounter systems in which the dynamical variables are matrices.

5.2.5 *Power counting*

We want to generalize the observations made in the previous particular cases, which deal with the ultraviolet behaviour of the perturbative integrals. We assume $m_0^2 \neq 0$. In the Fourier representation, let us study the effect of a dilatation of all the momenta by the same factor. Consider a diagram with n external lines and v vertices. From (32) the number of independent integration variables is $B = l - v + 1 = 1 + v - \frac{1}{2}n$, if l denotes the number of internal lines.

In the case of a φ^4 interaction, a factor g_0 with a dimension $4 - d$ is associated to each vertex. The following discussion can be extended in an obvious way to the general case of local interactions. Similarly, the dimension two of the propagator associated to internal lines is to be modified accordingly if we study nonscalar fields. Let us also recall that the dimension of the vertex function Γ_n is given by

$$[\Gamma_n] = d_n = d - \tfrac{1}{2}n(d-2)$$

(82)

The integrand has l quadratic denominators and B d-dimensional infinitesimal volume elements. A necessary (but not sufficient) condition for ultraviolet convergence is that the degree

$$\omega = Bd - 2l = d - \tfrac{1}{2}n(d-2) - v(4-d) = d_n - v(4-d)$$ (83)

be negative. We assume of course $B \geq 1$ (i.e. $v \geq \frac{1}{2}n$) (there should be at least one integral!). The interpretation of equation

(83) is that the degree ω of the integral is such that its product with g_0^v has the degree of Γ_n. The diagram is *superficially convergent* if ω is negative. This is the case at sufficiently high orders for $d < 4$. Thus only a finite number of superficially divergent diagrams arises, corresponding to small values of n and v, for $d < 4$. The theory is then called *super-renormalizable.*

In four dimensions, the degree ω is independent of the number of vertices v and nonnegative for $n \leq 4$. The theory is called *renormalizable.* In higher dimension, the theory is not renormalizable and ultraviolet divergences become more and more severe with increasing perturbative order. We have already seen that the large-distance behaviour becomes trivial in this case.

The vertex functions with a nonnegative superficial degree of divergence to lowest nontrivial perturbative order $(d \geq n)$ are called *primarily divergent.* These are the 2- and 4-point functions for $d = 4$, and only the 2-point function for $d = 2$ or 3.

Classify the superficially divergent diagrams in 2 and 3 dimensions. Study more carefully the nature of the divergence.

	$\omega = 0$	$\omega = 1$
$d = 2$		
$d = 3$		

For a scalar model, an integral with a nonnegative superficial degree of divergence is divergent. However, the converse statement might in general be false. For instance, functions with more than 4 points may be divergent at $d = 4$, because some subintegrals may diverge. The main aspect of the perturbative discussion is the study of this point. The results are the following.

i) A divergence occurs *if and only if* there is at least one irreducible subdiagram with a nonnegative superficial degree. This corresponds to insertions of irreducible 2-point functions (*self-energy*) for $d < 4$. In dimension four, 4-point irreducible sub-diagrams should also be considered.

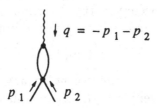

Fig. 3   First diagram contributing to $\Gamma_{2,1}$.

ii) If the degree of a diagram together with those of all its subdiagrams is negative, the corresponding integral converges as expected.

These ultraviolet divergences are at first controlled by introducing, at an intermediate stage, a regularization governed by a cutoff factor Λ, already present in lattice versions. Divergences yield positive powers of Λ and $\ln \Lambda$ in the computation of integrals. Then a consistent subtraction procedure is required, leading to finite results in the limit $\Lambda \to \infty$.

We have also to consider vertex functions with insertions of composite operators, such as $\frac{1}{2}\varphi^2$. Let us recall the dimension of the function $\Gamma_{n,r}$ with r such insertions

$$\left[\Gamma_{n,r}\right] = d_n - 2r \equiv d - \frac{1}{2}n(d-2) - 2r \tag{84}$$

The corresponding superficial degree of divergence to order v of a perturbative contribution is

$$\omega = d_n - 2r - v(4-d) \tag{85}$$

The superficial degree of divergence is improved, i.e. decreases, with increasing number of insertions. For instance, $\Gamma_{2,1}$ is logarithmically divergent to all orders in four dimensions, as exemplified in the expression of the one loop diagram (figure 3)

$$\Gamma_{2,1}^a(\mathbf{p}_1, \mathbf{p}_2, \mathbf{q}) = -\frac{1}{2}g_0 \int \frac{d^4k}{(2\pi)^4} \frac{1}{[k^2 + m_0^2][(q+k)^2 + m_0^2]} \tag{86}$$

Recall that $\Gamma_{2,1}(\mathbf{p}, -\mathbf{p}, 0)$ is the derivative of the 2-point function with respect to m_0^2, $\partial\Gamma_2^a/\partial m_0^2$.

The function $\Gamma_{0,2}$ occurs in the determination of the specific heat. Its superficial degree of divergence to order v is

$$\omega = -(1+v)(4-d) \qquad (87)$$

It is therefore ultraviolet convergent to any order for $d < 4$, logarithmically divergent to all orders (including 0) in four dimensions. Note that $\Gamma_{0,1}$ is a divergent constant related to the internal energy. However, it does not play any role in the contribution to subdiagrams since there is no external lines to be attached to a φ field.

This discussion also generalizes to insertions of more complex composite operators.

5.2.6   *Perturbative renormalization*

In dimension smaller than four, ultraviolet divergences only occur in the 2-point function, which also appears in subdiagrams pertaining to other correlation functions. In dimension four, there are also divergences in the 4-point function governing the coupling constant. We will proceed to renormalization according to the four-dimensional scheme. In lower dimension, some subtractions may appear unnecessary; however, they allow a smooth dimensional continuation and also play a role in the control of the infrared behaviour for $d < 4$.

In an abstract approach to renormalization, one defines subtractions in the integrands of the perturbative expressions in such a way as to obtain finite results without any auxiliary regularization. This technique has been developed by Bogoliubov, Parasiuk, Hepp and Zimmermann. It is, however, more instructive to use first a regularized theory. The perturbative expansion (considered as a series in the topological parameter counting the number of loops) is reexpressed in terms of *renormalized* (or *physical*) quantities. Here, these are the mass (the inverse correlation length) and the value of the 4-point vertex function for given values of the external momenta (the renormalized coupling constant). These renormalized quantities m and g are substituted for the *bare* quantities m_0 and g_0 initially present in the Lagrangian. Moreover, the field is rescaled by a factor $Z_1^{1/2}$: $\varphi = Z_1^{1/2}\varphi_R$. This process is called *wavefunction renormalization*. The corresponding correlation functions will remain finite when Λ increases indefinitely, and all divergences are then concentrated into the relations between m, g, Z_1 and m_0, g_0, Λ.

The *renormalized* Green functions (correlations of φ_R fields) depend on external momenta and on m and g. These two constants are defined using normalization conventions. In a massive theory, these conventions fix the value of some vertex functions and of some of their derivatives at zero momenta. That this is sufficient to define a finite theory is due to a point that we have not yet emphasized, i.e. that all potentially divergent terms are at most polynomials in the external momenta, as indicated by power counting. An important consequence is that the counterterms which appear in the Lagrangian when it is expressed in terms of m, g, and φ_R are local. The locality property is a crucial feature of field theory, in that it allows control of the renormalized quantities in terms of finitely many parameters.

If one wants to study directly the massless critical theory, the above conventions should be changed, owing to potential infrared divergences. One has to fix renormalization conditions at nonzero momenta.

In the massive theory, we impose for instance the three normalization conditions

$$\left\{ \begin{array}{c} \Gamma_2^R(\mathbf{0}) = m^2 \\[2mm] \dfrac{d\Gamma_2^R(\mathbf{0})}{d\mathbf{p}^2} = 1 \\[2mm] \Gamma_4^R(\mathbf{0}, \mathbf{0}, \mathbf{0}, \mathbf{0}) = m^{4-d}g \end{array} \right. \tag{88}$$

The normalization of Γ_4^R is such that the renormalized coupling constant g is dimensionless, in contradistinction with g_0. In four dimensions, these three conditions correspond precisely to the three primitive divergences, the quadratic and logarithmic ones of the 2-point function Γ_2 and the logarithmic one of the 4-point function Γ_4. The relation between *bare* regularized and *renormalized* Green functions reads for very large cutoff factor

$$\left\{ \begin{array}{l} G_n(\mathbf{p}, m_0, g_0, \Lambda) = Z_1^{n/2} G_n^R(\mathbf{p}, m, g) \\[2mm] \Gamma_n(\mathbf{p}, m_0, g_0, \Lambda) = Z_1^{-n/2} \Gamma_n^R(\mathbf{p}, m, g) \end{array} \right. \tag{89}$$

The first relation extends also to connected functions. The wavefunction renormalization factor Z_1 depends of course on Λ. Equations (88) and (89) allow the determination of the quantities

m_0^2, g_0 and Z_1 in terms of m^2, g and Λ as perturbative series in the number of loops

$$\begin{cases} m_0^2 = m_0^{2(0)} + m_0^{2(1)} + m_0^{2(2)} + \cdots \\ g_0 = g_0^{(0)} + g_0^{(1)} + g_0^{(2)} + \cdots \\ Z_1 = Z_1^{(0)} + Z_1^{(1)} + Z_1^{(2)} + \cdots \end{cases} \qquad (90)$$

The above normalization conventions have been chosen in such a way that the bare and renormalized constants are identical to lowest order $(B = 0)$, except for a trivial scaling factor in the coupling constant.

$$\begin{cases} m_0^{2(0)} = m^2 \\ g_0^{(0)} = m^{4-d}g \\ Z_1^{(0)} = 1 \end{cases} \qquad (91)$$

Hence, to this order, the functions Γ_2^R and Γ_4^R are

$$\begin{aligned} \Gamma_2^R &= \mathbf{p}^2 + m^2 \\ \Gamma_4^R &= m^{4-d}g \end{aligned} \qquad (92)$$

As a consequence, normalization conditions are homogeneous at higher orders. These operations can be summarized by expressing the Lagrangian in terms of the renormalized field, Z_1, m and g. It takes the form

$$\begin{aligned} \mathcal{L} &= \frac{1}{2}Z_1\left[(\partial\varphi_R)^2 + m_0^2\varphi_R^2\right] + \frac{1}{4!}g_0 Z_1^2 \varphi_R^4 \\ &= \frac{1}{2}\left[(\partial\varphi_R)^2 + m^2\varphi_R^2\right] + \frac{1}{4!}m^{4-d}g\varphi_R^4 + \\ &\quad \sum_{k=1}^{\infty}\left[Z_1^{(k)}\frac{1}{2}(\partial\varphi_R)^2 + (Z_1 m_0^2)^{(k)}\frac{1}{2}\varphi_R^2 + \frac{1}{4!}(g_0 Z_1^2)^{(k)}\varphi_R^4\right] \end{aligned}$$

$$\qquad (93)$$

This expression contains counterterms of order $B = 1, 2,...,$ the effect of which has to be included order by order. For instance, counterterms of order 1 are to be taken into account at the one-loop order, while the irreducible subdiagrams (here, vertex and propagators) are to be computed at a strictly lower order, here 0. For higher orders, this method of including counterterms explicitly solves a complex combinatorial problem, implying internal subtractions in the diagrams.

Let us compute as an example the renormalized quantities to order one.

$$\Gamma_2^{R(1)}(\mathbf{p}^2) = \tfrac{1}{2}m^{4-d}g \int_\Lambda \frac{d^d\mathbf{q}}{(2\pi)^d}\frac{1}{\mathbf{q}^2+m^2} + (Z_1 m_0^2)^{(1)} + Z_1^{(1)}\mathbf{p}^2$$

$$\Gamma_4^{R(1)}(\mathbf{p}_1,\mathbf{p}_2,\mathbf{p}_3,\mathbf{p}_4) = -\tfrac{1}{2}m^{8-2d}g^2 \tag{94}$$

$$\times \sum_{i=2}^4 \int_\Lambda \frac{d^d\mathbf{q}}{(2\pi)^d}\frac{1}{[\mathbf{q}^2+m^2][(\mathbf{q}-\mathbf{p}_1-\mathbf{p}_i)^2+m^2]} + (g_0 Z_1^2)^{(1)}$$

To first order, the renormalization conditions (88) require

$$\Gamma_2^{R(1)}(\mathbf{0}) = 0, \qquad \frac{d}{d\mathbf{p}^2}\Gamma_0^{R(1)}(\mathbf{0}) = 0, \qquad \Gamma_4^{R(1)}(\mathbf{0},\mathbf{0},\mathbf{0},\mathbf{0}) = 0$$

hence

$$Z_1^{(1)} = 0 \tag{95}$$

$$(Z_1 m_0^2)^{(1)} = m_0^{2(1)} = -\tfrac{1}{2}m^{4-d}g \int_\Lambda \frac{d^d\mathbf{q}}{(2\pi)^d}\frac{1}{\mathbf{q}^2+m^2}$$

and using our earlier results (71) and (72),

$$(g_0 Z_1^2)^{(1)} = g_0^{(1)} = \tfrac{3}{2}m^{8-2d}g^2 \int_\Lambda \frac{d^d\mathbf{q}}{(2\pi)^d}\frac{1}{(\mathbf{q}^2+m^2)^2}$$

$$= \begin{cases} \tfrac{3}{2}m^{4-d}g^2 \dfrac{1}{(4\pi)^{d/2}}\Gamma(2-d/2) & \text{for } d<4 \\[3mm] \tfrac{3}{2}g^2 \dfrac{1}{(4\pi)^2}(\ln\dfrac{\Lambda^2}{m^2} - \ln 2 - \gamma - 1) & \text{for } d=4 \end{cases} \tag{96}$$

Finally,

$$\Gamma_2^{R(1)}(\mathbf{p}) = 0$$

$$\Gamma_4^{R(1)}(\mathbf{p}_1,\mathbf{p}_2,\mathbf{p}_3,\mathbf{p}_4) = -\tfrac{1}{2}m^{8-2d}g^2 \sum_{i=2}^4 \int \frac{d^d\mathbf{q}}{(2\pi)^d}\frac{1}{\mathbf{q}^2+m^2} \tag{97}$$

$$\times \left\{ \frac{1}{(\mathbf{q}-\mathbf{p}_1-\mathbf{p}_i)^2+m^2} - \frac{1}{\mathbf{q}^2+m^2} \right\}$$

In dimension less than four, $g_0^{(1)}$ remains finite. The subtracted integrals define convergent renormalized functions and are expressed in terms of the physical quantities m and g. At the one-loop level, there is no wavefunction renormalization in the φ^4 theory; it appears only at the two-loop level, as we have already seen.

One can proceed similarly for higher orders. The counterterms defined at lower orders compensate the internal divergences of the diagrams and the normalization conditions define the new counterterms which compensate the global divergences. This is a recursive procedure which defines both the counterterms of the bare Lagrangian and the renormalized functions correctly subtracted. In a super-renormalizable theory, this process is finite and counterterms are polynomials in the coupling constant. It continues indefinitely in a renormalizable theory.

It is a rather delicate matter to analyze the procedure to all orders. The main result is that the successive terms of the expansion for the correlation functions are subtracted convergent integrals. Furthermore their structure is independent from the original regularization procedure. Indeed, one can develop an algorithm for the construction of their integrands using only the normalization conditions. The regularization has the advantage of yielding expressions for g_0, m_0 and Z_1 in terms of g, m and Λ, at least when it is consistently implemented. Of course, the cutoff factor disappears in the renormalized theory. As a consequence, the latter depends on one dimensionful parameter less than the bare regularized theory. The proof of these statements is at the heart of the renormalization theory and requires a sophisticated mathematical apparatus which will not be developed here.

It is useful to comment on the effects of the regularization and of the renormalization on the symmetries of a model. It is clear, for instance, that if a bare theory is seemingly invariant under dilatations, this symmetry is broken by regularization. We will study this effect in the renormalized theory, using the equations governing changes of scale. On the other hand, the global symmetry $\varphi \to -\varphi$ of the scalar case, and more generally the $O(n)$ symmetry of the n component vector model are broken neither by regularization nor by renormalization. There will exist a symmetric phase and spontaneous symmetry breaking will be a consequence of dynamics. The question of scale invariance is but one aspect of the quest for *anomalies* to apparent symmetries in the bare Lagrangian, or equivalently in the classical theory. Such anomalies may occur when functional integrals are more precisely defined.

The introduction of composite operators in the correlation functions leads to new divergences. Let us examine the insertions of $\frac{1}{2}\varphi^2$. In four dimensions, $\Gamma_{2,1}$ and $\Gamma_{0,2}$ are logarithmically divergent. The function $\Gamma_{2,1}$ is dimensionless and is normalized using its value at order 0

$$\Gamma_{2,1}^{R}(\mathbf{p} = 0) = 1 \tag{98}$$

where $\mathbf{p}$ represents symbolically all momenta. The relation between bare and renormalized functions introduces a new divergent constant Z_2, which we define through

$$\Gamma_{2,1}(\mathbf{p}, g_0, m_0, \Lambda) = Z_1^{-1} Z_2 \Gamma_{2,1}^{R}(\mathbf{p}, g, m) \tag{99}$$

Write the corresponding relation for correlation functions. The function $G_{2,1}$ is such that

$$G_{2,1} = G_2 \Gamma_{2,1} G_2 = Z_1 G_2^{R} Z_1^{-1} Z_2 \Gamma_{2,1}^{R} Z_1 G_2^{R} = Z_1 Z_2 G_{2,1}^{R}$$

When the momentum associated to $\frac{1}{2}\varphi^2$ vanishes, one has

$$G_{2,1}(\mathbf{p}_1, \mathbf{p}_2 = -\mathbf{p}_1, \mathbf{q} = 0) = -\frac{\partial}{\partial m_0^2} G_2(\mathbf{p}_1, \mathbf{p}_2 = -\mathbf{p}_1)$$

while

$$\Gamma_{2,1}(\mathbf{p}_1, \mathbf{p}_2 = -\mathbf{p}_1, \mathbf{q} = 0) = \frac{\partial}{\partial m_0^2} \Gamma_2(\mathbf{p}_1, \mathbf{p}_2 = -\mathbf{p}_1)$$

The relation (99) expresses the multiplicative renormalization of one insertion of $\frac{1}{2}\varphi^2$

$$\tfrac{1}{2}\varphi^2 = \tfrac{1}{2} Z_2 (\varphi^2)^{R} \tag{100}$$

Note that $(\varphi^2)^{R}(\mathbf{x})$ differs from $(\varphi^{R}(\mathbf{x}))^2$ due to additional divergences appearing when the arguments of two fields coincide. This results in a divergent ratio Z_1/Z_2. The same factor Z_2 renormalizes a single insertion of $\frac{1}{2}\varphi^2$ in *all* correlation functions.

As previously, Z_2 can be expanded according to the number of loops

$$Z_2 = 1 + Z_2^{(1)} + Z_2^{(2)} + \cdots \tag{101}$$

As $Z_1^{(1)} = 0$, equations (98) and (99) lead to the expression (figure 3)

$$1 + Z_2^{(1)} + \Gamma_{2,1}^{R(1)}(\mathbf{p}_1, \mathbf{p}_2, \mathbf{q})$$

$$= 1 - \tfrac{1}{2} m^{4-d} g \int_\Lambda \frac{d^d\mathbf{k}}{(2\pi)^d} \frac{1}{[\mathbf{k}^2 + m^2][(\mathbf{k} + \mathbf{q})^2 + m^2]} \tag{102}$$

$$Z_2^{(1)} = -\tfrac{1}{2}m^{4-d}g \int_\Lambda \frac{d^d k}{(2\pi)^d} \frac{1}{[k^2+m^2]^2}$$

$$= \begin{cases} -\dfrac{1}{2}g(4\pi)^{-d/2}\Gamma(2-d/2) & d < 4 \qquad (103) \\[2mm] -\dfrac{1}{2}g(4\pi)^{-2}\left(\ln \Lambda^2/m^2 - \ln 2 - 1 - \gamma\right) & d = 4 \end{cases}$$

$$\Gamma_{2,1}^{R(1)}(\mathbf{p}_1,\mathbf{p}_2,\mathbf{q}) = -\frac{1}{2}m^{4-d}g \int \frac{d^d k}{(2\pi)^d}\frac{1}{k^2+m^2}$$
$$\left\{\frac{1}{(\mathbf{k}+\mathbf{q})^2+m^2} - \frac{1}{\mathbf{k}^2+m^2}\right\} \qquad (104)$$

The divergence of $\Gamma_{0,2}$ containing two operators $\tfrac{1}{2}\varphi^2$, which does not appear as subdivergence in any diagram, is not totally removed by the Z_2 renormalization. Note that it occurs even for the free field, i.e. to order zero. An additional subtraction is needed for its compensation,

$$\Gamma_{0,2}(\mathbf{p},-\mathbf{p}) = Z_2^2\left\{\Gamma_{0,2}^R(\mathbf{p},-\mathbf{p}) + A\right\}$$
$$\Gamma_{0,2}^R(\mathbf{0},\mathbf{0}) = 0 \qquad (105)$$

where the normalization implies that the corresponding bare term is included in the specific heat.

Finally, the quantity $\Gamma_{0,1} = \langle \tfrac{1}{2}\varphi^2(\mathbf{0})\rangle$ diverges quadratically in four dimensions. This constant may be used in the computation of the internal energy, but will not play a role in the following. Insertions of operators of higher degree will be studied later.

5.3 The renormalization group

5.3.1 *Renormalization flow*

For several reasons, the previous discussion might superficially look irrelevant in the study of critical phenomena. First, we were initially interested in the bare theory. Indeed, the coefficients which are directly related to the temperature are those of the original Lagrangian. Secondly, there is no reason to assume that the coupling constant g_0 is small. In dimension d smaller than four, this is rather a large quantity, due to its positive dimension $4-d$ in terms of the original scale $1/a$. A perturbative expansion is thus not suitable, except if we succeed in extracting information

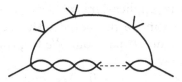

Fig. 4 Example of a diagram with a chain of "bubbles" leading to infrared divergences in a massless theory.

after resummation. Finally, we have discussed the short-range singularities, although we are interested in the infrared domain where the correlation length is infinite. For $d < 4$, the infrared divergences at $m = 0$ are more and more severe as the perturbative order increases. For instance, if p represents the momentum scale, the 4-point function behaves as $p^{-(4-d)}$. Hence its iteration n times will produce infrared divergences if $n(4-d) > d$ in diagrams which are ultraviolet convergent (figure 4).

To control infrared divergences, we will keep a nonzero mass m for $d < 4$. The critical theory appears thus in a domain $p/m \gg 1$ and $p/\Lambda \ll 1$. From the point of view of the renormalized theory, the scale Λ has been absorbed in a redefinition of T_c by the use of the correlation length m^{-1} as a fundamental parameter and by a rescaling of the field (the magnetization). The second condition $p/\Lambda \ll 1$ is then automatically fulfilled. The first condition shows that there is a relation between the critical behaviour and the short-distance analysis of the renormalized theory in a scale $1/m$. In order to obtain the scale θ of the deviations of the temperature from T_c, the relation between m and m_0 has to be elucidated. These indications suggest how the problems raised above will be solved. One will choose as an expansion parameter the dimensionless renormalized coupling constant g, which will turn out to remain finite whatever the value of $g_0 \geq 0$. For $d < 4$, one keeps a finite mass in the symmetric (high temperature) phase and discusses the corresponding ultraviolet behaviour. On the other hand, there are no infrared difficulties when $d = 4$. In this case, it is possible to study directly the massless theory, which is related to the $g \to 0$ limit. In this sense, we obtain an infrared asymptotically free theory in four dimensions. It will then be possible to proceed to a double expansion in g and $\varepsilon = 4 - d$.

Consequently, the applications of renormalization theory to critical phenomena can be presented either as a study at fixed $d < 4$ (2 or 3), or as an ε-expansion. We begin with a survey from the first point of view.

Let us assume $d < 4$, $m > 0$. We shall see that the critical theory corresponds to the limit where the bare coupling goes to infinity, $g_0 \to \infty$. The essential hypothesis is that the correlation functions, renormalized following the scheme valid at $d = 4$ (renormalizations of the mass, the coupling constant and the wavefunction) remain finite as $g_0 \to \infty$. This hypothesis seems to be verified when d is close to 4. We know from quantum mechanics that scattering amplitudes approach the Born terms at high energy. Similarly, the super-renormalizability of the φ^4 theory suggests that the Green functions tend towards the leading terms of their perturbative expansion when the momenta increase indefinitely. In particular, the 4-point vertex function tends towards $g_0 \to \infty$. The hypothesis amounts to saying that the value at fixed momentum remains finite although the short-distance behaviour exhibits ultraviolet divergences.

To investigate this question, let us consider a change of scale in the correlation functions. This can be done by varying the mass m at fixed g_0 and Λ. For a bare vertex function, we obtain

$$m \frac{\partial}{\partial m} \Gamma_n(\mathbf{p}; m_0, g_0, \Lambda) \Big|_{g_0, \Lambda}$$

$$= \left(m \frac{\partial}{\partial m} m_0^2 \Big|_{g_0, \Lambda} \right) \Gamma_{n,1}(\mathbf{p}, \mathbf{q} = \mathbf{0}; m_0, g_0, \Lambda) \quad (106)$$

The bare and renormalized functions are related through

$$\Gamma_n(\mathbf{p}; m_0, g_0, \Lambda) = Z_1^{-n/2} \Gamma_n^R(\mathbf{p}; m, g)$$
$$\Gamma_{n,1}(\mathbf{p}, \mathbf{q}; m_0, g_0, \Lambda) = Z_1^{-n/2} Z_2 \Gamma_{n,1}^R(\mathbf{p}, \mathbf{q}; m, g)$$

We introduce the dimensionless quantities

$$\beta = m \frac{\partial g}{\partial m} \Big|_{g_0, \Lambda}$$

$$\gamma_1 = Z_1^{-1} m \frac{\partial Z_1}{\partial m} \Big|_{g_0, \Lambda} \qquad (107)$$

$$\gamma_2 = Z_2^{-1} m \frac{\partial Z_2}{\partial m} \Big|_{g_0, \Lambda}$$

and express equation (106) in terms of renormalized functions as

$$\left\{ m\frac{\partial}{\partial m} + \beta\frac{\partial}{\partial g} - \tfrac{1}{2}n\gamma_1 \right\} \Gamma_n^R(\mathbf{p}; m, g)$$

$$= Z_2 \left(m\left.\frac{\partial m_0^2}{\partial m}\right|_{g_0, \Lambda} \right) \Gamma_{n,1}^R(\mathbf{p}, \mathbf{0}; m, g)$$

Note that β, γ_1 and γ_2 (which does not yet occur) are *a priori* functions of g and m/Λ, but do not depend on n and on the momenta. In particular, if we choose $n = 2$ and $\mathbf{p} = \mathbf{0}$, we have

$$\Gamma_2^R(\mathbf{p} = \mathbf{0}; m, g) = m^2 \quad \text{and} \quad \Gamma_{2,1}^R(\mathbf{p} = \mathbf{0}, \mathbf{q} = \mathbf{0}; m, g) = 1$$

Therefore,

$$(2 - \gamma_1)m^2 = Z_2 m\left.\frac{\partial m_0^2}{\partial m}\right|_{g_0, \Lambda} \tag{108}$$

Inserting this expression in the r.h.s. of the above relation, any explicit reference to the bare theory disappears, and thus the coefficients β and γ_1 either depend only on g or are infinite. This last possibility can be excluded by expressing these coefficients in terms of the renormalized vertex functions for particular values of the momenta. Hence one obtains the set of equations

$$\left\{ m\frac{\partial}{\partial m} + \beta(g)\frac{\partial}{\partial g} - \tfrac{1}{2}n\gamma_1(g) \right\} \Gamma_n^R(\mathbf{p}; m, g)$$

$$= (2 - \gamma_1(g))m^2\Gamma_{n,1}^R(\mathbf{p}, \mathbf{0}; m, g) \tag{109}$$

Equivalent forms of such relations were first derived in the context of quantum electrodynamics, in a discussion involving the arbitrariness of renormalization conditions. The subject was brought back to life by Callan and Symanzik in an analysis of anomalies pertaining to scale invariance. The quantities β, γ_1 and γ_2, which partly depend on the normalization conditions, have finite perturbative expansions in g, and equation (109) relates only renormalized quantities. It remains valid at $d = 4$.

The derivation of these equations is based in an essential way on the multiplicative renormalization of the fields and the fact that the dimensionful cutoff factor Λ disappears in the renormalized theory expressed in terms of m and g. The multiplicative renormalization is responsible for the factor $\gamma_1(g)$, related to the anomalous dimension of the field φ, while $\gamma_2(g)$ is similarly related to the anomalous dimension of $\frac{1}{2}\varphi^2$. The interesting phenomenon is the appearance of the factor $\beta(g)$. It expresses the fact that a change in scale is accompanied by a flow of the coupling. This phenomenon would obviously generalize to a multidimensional case of several coupling constants. The *critical points* where $\beta(g_c) = 0$ play the crucial role of attractive or repulsive points. This is not completely unexpected for $d < 4$, since the dimensionful initial coupling constant would have to change with scale. The fact might seem more surprising in dimension four since both g and g_0 are dimensionless. However, it then appears as a relic of the existence of a cutoff Λ in the underlying regularized theory. Let us point out also that the right-hand side in equation (109) shows that these equations are the first in a hierarchy involving more insertions of $\frac{1}{2}\varphi^2$

$$\left\{ m\frac{\partial}{\partial m} + \beta(g)\frac{\partial}{\partial g} - \tfrac{1}{2}n\gamma_1(g) + s\gamma_2(g) \right\} \Gamma_{n,s}^{R}(\{\mathbf{p}\}, \{\mathbf{q}\}; m, g)$$

$$= (2 - \gamma_1(g))m^2\Gamma_{n,s+1}^{R}(\{\mathbf{p}\}, \{\mathbf{q}, \mathbf{0}\}; m, g) \quad (110)$$

Derive equations (110).

This system might seem of little utility, except in cases where there are regions in momentum space where the right-hand side is negligible. Let us consider for instance the relations (109). Although the momentum configuration of the r.h.s. is in a sense exceptional (the insertion of φ^2 is at zero momentum), $\Gamma_{n,1}^{R}(\mathbf{p}, \mathbf{0})$ is of order $p^{-2}\Gamma_n^{R}(\mathbf{p})$, up to powers of $\ln p$, for large $\mathbf{p}$, order by order in a perturbative expansion. This is an example of a general theorem obtained by Weinberg. If these logarithms do not sum up to compensate the factor p^{-2}, this r.h.s. can be neglected in the region of large momenta. The coherence of this procedure can be checked on the next equation of the hierarchy (110). Finally, near $d = 4$, one can expect that the deviations from canonical dimensions are at least of order $(4 - d)$.

(1) Show that equation (108) can be re-expressed as a relation between Z_2 and the other renormalization constants, in the form

$$Z_2 = Z_1 \frac{\partial}{\partial m_0^2} \left(\frac{m^2}{Z_1} \right)_{g_0, \Lambda} \qquad (111)$$

(2) In the limit $\Lambda \to \infty$, g_0/m^{4-d} is a function of g only (if g and m are finite), so that $\beta(g)$ is obtained for $d < 4$ by the relation

$$\left\{ 4 - d + \beta(g) \frac{d}{dg} \right\} \frac{g_0}{m^{4-d}} = 0 \qquad (112)$$

As g_0/m^{4-d} tends towards g for $g \to 0$, it follows that

$$\beta(g) \underset{g \to 0}{\sim} -(4 - d)g + \mathcal{O}(g^2) \qquad (113)$$

As an illustration, let us present the one-loop computation of the coefficients β, γ_1, γ_2. First, let us summarize the expressions of g_0, Z_1, Z_2 and m_0^2 in both cases $d < 4$ and $d = 4$. We also include the symmetry factors for and n-vector model with a $(\Phi^2)^2$ interaction.

$$d < 4 \begin{cases} g_0 = m^{4-d} \left\{ g + \dfrac{n+8}{6} \dfrac{\Gamma(2 - d/2)}{(4\pi)^{d/2}} g^2 + \cdots \right\} \\[2mm] Z_1 = 1 + \cdots \\[2mm] m_0^2 = m^2 \left\{ 1 - \dfrac{n+2}{6} \dfrac{\Gamma(1 - d/2)}{(4\pi)^{d/2}} g + \cdots \right\} \\[2mm] Z_2 = 1 - \dfrac{n+2}{6} \dfrac{\Gamma(2 - d/2)}{(4\pi)^{d/2}} g + \cdots \end{cases} \qquad (114a)$$

$$d = 4 \begin{cases} g_0 = g + \dfrac{n+8}{6} \dfrac{\ln \frac{\Lambda^2}{m^2} - \ln 2 - 1 - \gamma}{(4\pi)^2} g^2 + \cdots \\[2mm] Z_1 = 1 + \cdots \\[2mm] m_0^2 = m^2 \left\{ 1 - \dfrac{n+2}{6} \dfrac{\frac{\Lambda^2}{m^2} - 1 - \ln \frac{\Lambda^2}{m^2}}{(4\pi)^2} g + \cdots \right\} \\[2mm] Z_2 = 1 - \dfrac{n+2}{6} \dfrac{\ln \frac{\Lambda^2}{m^2} - \ln 2 - 1 - \gamma}{(4\pi)^2} g + \cdots \end{cases} \qquad (114b)$$

The series are inverted and lead to

$$
d < 4 \begin{cases}
g = \dfrac{g_0}{m^{4-d}} - \dfrac{n+8}{6}\dfrac{\Gamma(2-d/2)}{(4\pi)^{d/2}}\left(\dfrac{g_0}{m^{4-d}}\right)^2 + \cdots \\[2ex]
m\left.\dfrac{\partial g}{\partial m}\right|_{g_0,\Lambda} = -(4-d)\dfrac{g_0}{m^{4-d}} \\[2ex]
\qquad + \dfrac{n+8}{3}\dfrac{(4-d)\Gamma(2-d/2)}{(4\pi)^2}\left(\dfrac{g_0}{m^{4-d}}\right)^2 + \cdots \\[2ex]
\beta(g) = -(4-d)g + \dfrac{n+8}{3}\dfrac{\Gamma(3-d/2)}{(4\pi)^{d/2}}g^2 + \cdots
\end{cases}
\tag{115a}
$$

$$
d = 4 \begin{cases}
g = g_0 - \dfrac{n+8}{6}\dfrac{\ln\frac{\Lambda^2}{m^2} - \ln 2 - 1 - \gamma}{(4\pi)^2}g_0^2 + \cdots \\[2ex]
m\left.\dfrac{\partial g}{\partial m}\right|_{g_0,\Lambda} = \dfrac{n+8}{3}\dfrac{g_0^2}{(4\pi)^2} + \cdots \\[2ex]
\beta(g) = \dfrac{n+8}{3}\dfrac{g^2}{(4\pi)^2} + \cdots
\end{cases}
\tag{115b}
$$

Note that the limit $d \to 4$ is regular term by term and yields a function $\beta(g)$ perturbatively finite, as a consequence of the renormalizability of the theory.

For $d \le 4$, one gets

$$
\gamma_1(g) = \mathcal{O}(g^2)
\tag{116}
$$

$$
\gamma_2(g) = \left(\frac{n+2}{3}\right)\frac{\Gamma(3-d/2)}{(4\pi)^{d/2}}g + \cdots
$$

$$
= \left(\frac{n+2}{3}\right)\frac{g}{(4\pi)^2} \qquad (d=4)
\tag{117}
$$

Compute $\beta(g)$ using equation (113) and check (116) and (117).

For large Λ and fixed g_0, the variation of g is related to the last free parameter, i.e. the temperature or equivalently, the correlation length. As this length increases, m decreases and g varies according to the sign of $-\beta(g)$. For small g, we have just seen that β is negative for $d < 4$. Recall that the reasoning is valid in the symmetric phase ($T > T_c$). Therefore, as T decreases towards T_c, m decreases and g increases. If g is bounded, this growth has to stop and $\beta(g)$ should vanish for some positive value g_c, as suggested by the one-loop formula. Moreover, as the

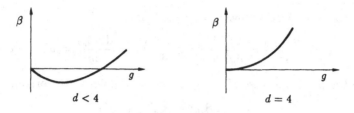

Fig. 5   Behaviour of the function $\beta(g)$.

coefficients of the expansion have a finite limit when $d \to 4$, and if the perturbative expansion is asymptotically related to a well-defined function, g_c is of order $4 - d$ for $d \to 4$. This hypothetical behaviour of $\beta(g)$ is displayed in figure 5. Of course, we only consider the stable perturbative region $g > 0$, although the point $g = 0$ does not seem to be singular at this level.

Using the terms explicitly given in (115), one obtains

$$\frac{g_c}{(4\pi)^{d/2}} = \left(\frac{3}{n+8}\right)\frac{4-d}{\Gamma(3-d/2)} + \cdots \qquad (118)$$

behaving as $(4-d)3/(n+8)$ as $d \to 4$. In three dimensions (although this calculation might seem unrealistic in this case!), we find

$$\frac{g_c}{16\pi} = \frac{3}{n+8} + \cdots \qquad (119)$$

Covariance of the flow in coupling space   The notion of a fixed point is independent of a regular reparametrization in coupling space. For the φ^4 model, there is only one coupling constant. Given an invertible transformation $g' = g'(g)$, the equality of vector fields

$$\beta(g)\frac{\partial}{\partial g} = \beta'(g')\frac{\partial}{\partial g'} \qquad (120a)$$

yields the transformation law

$$\beta'(g') = \beta(g)\frac{\mathrm{d}g'}{\mathrm{d}g} \qquad (120b)$$

If the transformation is monotonic – as is required if it is to be invertible – there is a one-to-one correspondence between the zeroes of β and β' as well as between the sign of the functions. Let $g_c \leftrightarrow g'_c$ be a fixed point where both functions vanish. The slope is a reparametrization invariant, according to

$$\omega' = \frac{\mathrm{d}\beta'(g_c')}{\mathrm{d}g'} = \frac{\mathrm{d}g}{\mathrm{d}g'}\frac{\mathrm{d}}{\mathrm{d}g}\left(\beta(g)\frac{\mathrm{d}g'}{\mathrm{d}g}\right)_{g_c} = \frac{\mathrm{d}\beta(g_c)}{\mathrm{d}g} = \omega \qquad (121)$$

In the particular case of a degenerate fixed point at the origin,

$$\beta(g) = \beta_3 g^3 + \beta_5 g^5 + \mathcal{O}(g^7)$$
$$g' = g + ag^3 + \cdots \qquad (122a)$$

we observe that in

$$\beta'(g') = \beta_3 g'^3 + \beta_5 g'^5 + \mathcal{O}(g'^7) \qquad (122b)$$

the first two coefficients are reparametrization invariants. Generalize these properties to the case of a multidimensional coupling constant space.

When $d < 4$, as the coupling varies with m according to (107), the quantities Z_1, Z_2 and $g_0/m^{4-d}g$ can be computed by integrating the flow equations (107) and (112) using their values 1 for $g = 0$,

$$Z_1 = \exp \int_0^g \mathrm{d}g' \frac{\gamma_1(g')}{\beta(g')}$$

$$Z_2 = \exp \int_0^g \mathrm{d}g' \frac{\gamma_2(g')}{\beta(g')} \qquad (123)$$

$$\frac{g_0}{m^{4-d}} = g \exp - \int_0^g \mathrm{d}g' \left(\frac{4-d}{\beta(g')} + \frac{1}{g'}\right)$$

Looking at the last relation at fixed g_0 and $m \to 0$, one sees that g has to increase up to the first zero of $\beta(g)$ where g_0/m^{4-d} diverges. This can be reinterpreted at fixed m by saying that, as g runs from 0 to g_c, g_0 varies from 0 to infinity.

Suppose that g_c is a simple zero of $\beta(g)$ with a nonvanishing slope ω, as suggested in figure 5,

$$\omega = \beta'(g_c) \qquad (124)$$

Close to this point, one gets

$$\frac{g_0}{m^{4-d}} \cong A\left(1 - \frac{g}{g_c}\right)^{-(4-d)/\omega}$$

$$A = g_c \exp - \int_0^{g_c} \mathrm{d}g' \left(\frac{4-d}{\beta(g')} + \frac{1}{g'} - \frac{4-d}{\omega(g' - g_c)}\right) \qquad (125a)$$

or equivalently,

$$g = g_c \left[1 - \left(\frac{Am^{4-d}}{g_0} \right)^{\omega/(4-d)} \right] \qquad (125b)$$

This relation involves fractional powers of g_0, showing its nonperturbative nature. As g_c behaves as $4 - d$, this expression becomes meaningless when $d \to 4$ where the interval in which g is allowed to vary shrinks to zero. The inescapable and rather astounding conclusion is that the renormalized φ^4 theory obtained from a cutoff regularized model becomes free in four dimensions! On second thought, this ties up very nicely with the idea that four is an upper critical dimension. To investigate this question in a meaningful way, one can consider the case where the cutoff factor is large, but not infinite. It leaves a small interval for the variation of g of order Λ^{-1}. Alternative methods, in particular numerical ones, seem to confirm the triviality of the four-dimensional ϕ^4 theory.

> Show that, if the slope ω at the fixed point vanishes, i.e. for a degenerate double zero with $\beta''(g_c) \neq 0$, and $\beta(g) \sim -\frac{1}{2}\omega'(g - g_c)^2$, then
>
> $$g_c - g \sim \frac{2(4-d)}{\omega' \ln \left(g_0/(A'm^{4-d}) \right)} \qquad (126)$$
>
> What is the corresponding expression for higher order zeroes? One notes that, as g_0 goes to infinity, g approaches g_c slower and slower.

The coefficients $\gamma_1(g)$ and $\gamma_2(g)$ have no reason to be singular at the critical point. In a generic situation $\omega \neq 0$, we get the following behaviours

$$
\begin{aligned}
Z_1 &\sim C_1 m^{\gamma_1(g_c)} \left[1 + \bar{C}_1 \left(m g_0^{1/(d-4)} \right)^{\omega} + \cdots \right] \\
Z_2 &\sim C_2 m^{\gamma_2(g_c)} \left[1 + \bar{C}_2 \left(m g_0^{1/(d-4)} \right)^{\omega} + \cdots \right]
\end{aligned}
\qquad (127)
$$

The critical theory is characterized by scaling laws in powers of the correlation length. We shall now relate them to the standard exponents.

5.3.2   *Critical exponents*

The present discussion is relative to the symmetric high temperature phase where

$$m_0^2 \sim a + b\theta \qquad\qquad \theta = \frac{T - T_c}{T_c} > 0 \qquad (128)$$

Recall the definition of the exponents ν and γ

$$m = \xi^{-1} \sim \theta^\nu$$
$$\chi = G_2(\mathbf{p} = \mathbf{0}) = \frac{Z_1}{m^2} \sim \theta^{-\gamma} \qquad (129)$$

and the relation

$$Z_2 = \left.\frac{\partial m^2}{\partial m_0^2}\right|_{g_0,\Lambda} (1 - \tfrac{1}{2}\gamma_1(g))$$

where $1 - \tfrac{1}{2}\gamma_1(g_c)$ generically does not vanish.

Analyze the meaning of $\gamma(g_c) = 2$.

When $m \to 0$, the l.h.s. of this last equality behaves as $m^{\gamma_2(g_c)} \sim \theta^{\nu\gamma_2(g_c)}$ according to equation (127), while the r.h.s. is proportional to $dm^2/d\theta \sim \theta^{2\nu-1}$. Hence $\gamma_2(g_c)$ is related to ν through

$$\boxed{\nu = \frac{1}{2 - \gamma_2(g_c)}} \qquad (130)$$

Similarly from equations (127, 129), one gets

$$\boxed{\frac{\gamma}{\nu} = 2 - \gamma_1(g_c)} \qquad (131)$$

Using equation (4.11), this relation is equivalent to

$$\boxed{\eta = \gamma_1(g_c)} \qquad (132)$$

which relates the exponent η to the abnormal dimension of the field φ. The two quantities $\gamma_1(g_c)$ and $\gamma_2(g_c)$ will yield all exponents, provided that the critical theory satisfies the standard scaling relations. To prove this fact, we will have to look at the behaviour of correlation functions. Observe also that, from equations (127), the exponent ω characterizes the deviations from critical behaviour. The coefficients of the terms in $m^\omega \sim \theta^{\omega\nu}$ depend on the details of microscopic parameters.

Let us apply now the results of the one-loop calculations to the computation of critical exponents. One gets

$$\beta(g) = -(4-d)g(1-g/g_c) \qquad g_c = \frac{4-d}{n+8}\frac{3(4\pi)^{d/2}}{\Gamma(3-d/2)} \qquad (133)$$

and therefore

$$\boxed{\omega = 4 - d + \cdots} \qquad (134b)$$

At the one-loop order, $\gamma_1(g)$ vanishes. Thus

$$\boxed{\eta = 0 + \cdots} \qquad (134c)$$

Finally from (117)

$$\gamma_2(g_c) = \frac{g_c}{(4\pi)^{d/2}}\frac{n+2}{3}\Gamma(3-d/2) = (4-d)\frac{n+2}{n+8} + \cdots$$

yielding

$$\boxed{\begin{aligned} \nu &= \frac{1}{2-(4-d)(n+2)/(n+8)+\cdots} \\ \gamma &= \frac{2+\cdots}{2-(4-d)(n+2)/(n+8)+\cdots} \end{aligned}} \qquad (134d)$$

These one loop results yield the exact $n \to \infty$ limit (chapter 3)

$$\eta = 0 \qquad \tfrac{1}{2}\gamma = \nu = 1/(d-2) \qquad (135)$$

Explain why.

The predictions of the renormalization group are already non-trivial at the one-loop level. Deviations from the mean field predictions are of order $\varepsilon = 4 - d$. Close to four dimensions, this procedure is coherent. Indeed, g_c is of order ε and this fact justifies the renormalized perturbative expansion used to compute the coefficients $\beta(g)$, $\gamma_1(g)$, $\gamma_2(g)$,... of the renormalization group equations. The whole scheme assumes that these equations have a domain of validity independent from the perturbative method. Many efforts are presently being made to justify this point.

It is remarkable that setting $d = 3$ in the one-loop expressions for the exponents yields values very close to the correct ones. For the Ising model ($n = 1$) for instance, one gets

$$\nu = \tfrac{1}{2}\gamma = 0.6 \qquad \eta = 0 \qquad (d = 3,\ n = 1) \qquad (136)$$

to be compared with the values given in table I which appears later in this chapter.

5.3.3   *From the Gaussian ultraviolet fixed point to the infrared critical point in dimension less than four*

The proof of equations (123) implicitly assumes a trivial behaviour as $g \to 0$. More generally, in dimension less than four, we expect a trivial short-distance behaviour of the massive φ^4 model, which corresponds to a super-renormalizable theory, i.e. it has a free field behaviour, corrected by the first nonvanishing perturbative corrections. If this is the case, the bare functions tend toward their tree level expressions (Born amplitudes) for large momenta. Hence,

$$\lim_{\lambda \to \infty} \lambda^{-2} \mathbf{p}^{-2} \Gamma_2^{\mathrm{R}}(\lambda \mathbf{p}; m, g) = Z_1$$

$$\lim_{\lambda \to \infty} \Gamma_4^{\mathrm{R}}(\{\lambda \mathbf{p}\}; m, g) = m^{4-d} \frac{g_0}{m^{4-d}} Z_1^2 \qquad (137)$$

$$\lim_{\lambda \to \infty} \Gamma_{2,1}^{\mathrm{R}}(\{\lambda \mathbf{p}\}, \lambda \mathbf{q}; m, g) = Z_1 Z_2^{-1}$$

In the last two formulae, all points in configuration space should become coincident in the limit. Hence, one imposes the following technical restriction that none of the partial sums of momenta vanishes.

The limits in (137) state that fluctuations do not modify the dominant perturbative behaviour. This can be checked order by order in the renormalized theory for $d < 4$.

Verify this assertion at the one-loop order. For instance, for $n = 1$,

$$\Gamma_2^{\mathrm{R}} = m^2 + \mathbf{p}^2 + \cdots$$

$$\Gamma_4^{\mathrm{R}} = m^{4-d} g - \tfrac{1}{2} g^2 m^{4-d} \sum_{i=2}^{4} \int \frac{\mathrm{d}^d \mathbf{q}}{(2\pi)^d}$$

$$\frac{1}{\mathbf{q}^2 + m^2} \left(\frac{1}{(\mathbf{p}_1 + \mathbf{p}_i - \mathbf{q})^2 + m^2} - \frac{1}{\mathbf{q}^2 + m^2} \right) + \cdots$$

$$\Gamma_{2,1}^{\mathrm{R}} = 1 - \tfrac{1}{2} m^{4-d} g \int \frac{\mathrm{d}^d \mathbf{q}}{(2\pi)^d}$$

$$\frac{1}{\mathbf{q}^2 + m^2} \left(\frac{1}{(\mathbf{p}_1 + \mathbf{p}_2 - \mathbf{q})^2 + m^2} - \frac{1}{\mathbf{q}^2 + m^2} \right) + \cdots$$

Hence, from (137),

$$Z_1 = 1 + \cdots$$

$$\frac{g_0}{m^{4-d}} = g + \tfrac{3}{2}g^2 \int \frac{d^d\mathbf{q}}{(2\pi)^d} \frac{1}{(\mathbf{q}^2+1)^2} + \cdots = g + \tfrac{3}{2}g^2\frac{\Gamma(2-d/2)}{(4\pi)^{d/2}} + \cdots$$

$$Z_2 = 1 - g \int \frac{d^d\mathbf{q}}{(2\pi)^d} \frac{1}{(\mathbf{q}^2+1)^2} + \cdots = 1 - \tfrac{1}{2}g\frac{\Gamma(2-d/2)}{(4\pi)^{d/2}} + \cdots$$

in agreement with the expressions given in (114).

We can confirm the trivial ultraviolet behaviour by studying the variation of g with respect to the scale $m \ll \Lambda$ at fixed g_0 and Λ. Since $dg/\beta(g) = d\ln m$, one has

$$\frac{m_2}{m_1} = \exp \int_{g_1}^{g_2} \frac{dg'}{\beta(g')}$$

As β is negative, the effective coupling constant $g_2 \equiv g(m_2)$ vanishes exponentially in the region $\Lambda \gg m_2 \gg m_1$

$$\frac{g_2}{g_1} \sim \left(\frac{m_1}{m_2}\right)^{4-d} \qquad \frac{m_2}{m_1} \gg 1 \qquad (138)$$

The situation seems to be paradoxically very different in four dimensions, where $\beta(g)$ is positive and of order g^2 ($\beta(g) \sim 3g^2/(4\pi)^2 + \cdots$ for $n=1$). Apparently, the origin is a repulsive ultraviolet fixed point. As m increases, g also increases and one enters an unknown nonperturbative domain. It is conceivable that $\beta(g)$ might have a maximum, then vanish again at a nontrivial ultraviolet fixed point. This unlikely possibility seems excluded by numerical simulations, and thus we assume that it does not occur. Note that the region $g > 0$ for $d = 4$ corresponds to the domain $g > g_c$ for $d < 4$. We just saw that, at fixed m and Λ and for $d < 4$, the bare coupling constant covers the whole interval $[0, \infty]$ when the renormalized coupling varies in $[0, g_c]$. Therefore, the region $g > g_c$ is not accessible without an analytical continuation which reveals pathologies in the correlation functions, as e.g. imaginary parts. This is physically unacceptable and signals instabilities. One is tempted to make the same analysis in dimension four for the whole domain $g > 0$ (figure 6).

Formally, the terms of the perturbative renormalized expansion are nevertheless well-defined for $d = 4$. The question is to know whether these series are asymptotic to meaningful correlation

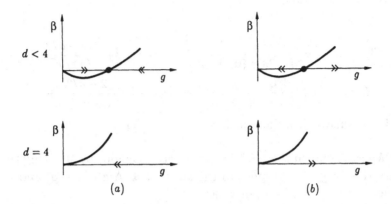

Fig. 6   Infrared (*a*) and ultraviolet (*b*) behaviours for $d < 4$ and $d = 4$.

functions. From the preceding discussion, this does not seem to be the case, but, up to now, the justification seems to be circular, since it requires that regular functions such as $\beta(g),...$ are evaluated perturbatively. The computation of such quantities within an ill-defined model seems paradoxical. However, in the framework of dimensional continuation, it is conceivable to consider these quantities as regular limits of well-defined functions for $d < 4$.

As $m \to \infty$, (after taking first the limit $\Lambda \to \infty$) the theory becomes trivial. This is not the case when m decreases. In this case, g tends towards the nontrivial attractive infrared fixed point. Let us study the expressions resulting from the renormalization group equations.

5.3.4   Correlation functions at the critical point

We fix the dimension to some value less than four. Standard dimensional analysis yields

$$\Gamma_n^R(\lambda\mathbf{p}; \lambda m, g) = \lambda^{d-\frac{1}{2}n(d-2)}\Gamma_n^R(\mathbf{p}; m, g) \qquad (139a)$$

and allows derivatives in m at fixed $\mathbf{p}$ to be changed into derivatives in $\mathbf{p}$ at fixed m. Using the notation

$$d_n^\Gamma = d - \tfrac{1}{2}n(d - 2) \qquad (139b)$$

we can rewrite the Callan–Symanzik equations (109) as

$$\left\{-\lambda\frac{\partial}{\partial\lambda} + \beta(g)\frac{\partial}{\partial g} + (d_n^\Gamma - \tfrac{1}{2}n\gamma_1(g))\right\}\Gamma_n^R(\lambda\mathbf{p}; m, g) =$$
$$(2 - \gamma_1(g))m^2\Gamma_{n,1}^R(\lambda\mathbf{p}, \mathbf{0}; m, g) \tag{140}$$

which makes obvious the interpretation of $\tfrac{1}{2}\gamma_1(g)$ as the abnormal dimension of the field φ

$$d_n^\Gamma - \tfrac{1}{2}n\gamma_1(g) = d - n\left(\tfrac{1}{2}(d-2) + \tfrac{1}{2}\gamma_1(g)\right) \tag{141}$$

Renormalized connected Green functions, here written without the index c for simplicity, obey similar equations. Setting

$$d_n^G = d - \tfrac{1}{2}n(d+2) \tag{142}$$

they read

$$\left\{-\lambda\frac{\partial}{\partial\lambda} + \beta(g)\frac{\partial}{\partial g} + (d_n^G + \tfrac{1}{2}n\gamma_1(g))\right\}G_n^R(\lambda\mathbf{p}; m, g) =$$
$$(\gamma_1(g) - 2)m^2 G_{n,1}^R(\lambda\mathbf{p}, \mathbf{0}; m, g) \tag{143}$$

Prove equation (143).

Let us introduce a coupling constant $\bar{g}(\lambda, g)$ depending on the scale λ, as the solution of the first order differential equation

$$\lambda\frac{\mathrm{d}\bar{g}(\lambda, g)}{\mathrm{d}\lambda} = \beta[\bar{g}(\lambda, g)] \qquad \bar{g}(1, g) = g \tag{144}$$

This function is given implicitly by

$$\ln\lambda = \int_g^{\bar{g}(\lambda,g)} \frac{\mathrm{d}g'}{\beta(g')} \tag{145}$$

Recall also that, according to equations (123),

$$Z_1(g) = \exp\int_0^g \mathrm{d}g'\frac{\gamma_1(g')}{\beta(g')} \tag{146}$$

Finally, for short, we introduce the notation

$$\tilde{G}_n^R(\mathbf{p}; m, g) = (2 - \gamma_1(g))m^2 G_{n,1}^R(\mathbf{p}, \mathbf{0}; m, g) \tag{147}$$

A solution of (143) can thus be written as

$$\lambda_2^{-d_n^G} Z_1^{n/2}(g) G_n^R(\lambda_2 \mathbf{p}; m, g)$$
$$- \lambda_1^{-d_n^G} Z_1^{n/2}(\bar{g}[\frac{\lambda_2}{\lambda_1}, g]) G_n^R(\lambda_1 \mathbf{p}; m, \bar{g}[\frac{\lambda_2}{\lambda_1}, g]) =$$
$$= \int_{\lambda_1}^{\lambda_2} \frac{d\lambda}{\lambda} \lambda^{-d_n^G} Z_1^{n/2}(\bar{g}[\frac{\lambda_2}{\lambda}, g]) \tilde{G}_n^R(\lambda \mathbf{p}; m, \bar{g}[\frac{\lambda_2}{\lambda}, g])$$

$$(148)$$

Using the function $\bar{g}(\lambda, g)$, show that the equations (143) transform into first order differential equations and yield (148).

In order to clarify the interpretation of equation (148), let us assume the r.h.s. negligible. Hence, with $\lambda_1 = 1$ and $\lambda_2 = \lambda$,

$$G_n^R(\lambda \mathbf{p}; m, g) \simeq \lambda^{d_n^G} \left(\frac{Z_1(\bar{g}[\lambda, g])}{Z_1(g)} \right)^{n/2} G_n^R(\mathbf{p}; m, \bar{g}[\lambda, g]) \quad (149)$$

This states that changing the scale of the momenta is equivalent to the multiplication by a factor depending on the dilatation parameter accompanied by a change in the coupling constant. At the fixed point of the transformation, the Green functions are homogeneous. Of course, this result presupposes that the r.h.s. is absent. Note that the structure of (149) is reminiscent of the relation between bare and renormalized functions.

Returning to equation (148), let us consider the ultraviolet limit where λ_1 tends to zero, $\bar{g}(\lambda_2/\lambda_1, g)$ goes to zero and $Z_1(\bar{g}(\lambda_2/\lambda_1), g)$ to 1. The Green function is finite at zero momenta since the mass does not vanish. Finally, for $n \geq 1$,

$$\lambda_1^{-d_n^G} = \lambda_1^{\frac{1}{2}n(d+2)-d}$$

also vanishes (this is the reason why we use correlation functions instead of vertex functions, which would require additional subtractions for $n = 2, 4$). Hence,

$$G_n^R(\lambda \mathbf{p}; m, g) = \lambda^{d_n^G} \int_0^\lambda \frac{d\mu}{\mu} \mu^{-d_n^G} \left(\frac{Z_1(\bar{g}[\frac{\lambda}{\mu}, g])}{Z_1(g)} \right)^{n/2} \tilde{G}_n^R(\mu \mathbf{p}; m, \bar{g}(\frac{\lambda}{\mu}, g))$$

$$(150)$$

Let us now study the critical limit $m_0^2 \to m_{0c}^2$, equivalent to $m^2 \to 0$ at fixed g_0, or also to $g \to g_c$ at fixed momenta. Moreover, this is also equivalent to letting m go to zero or the momenta to infinity, since $G_n(\lambda\mathbf{p}; m, g) = m^{d_n^G} G_n(\lambda\mathbf{p}/m, 1, g)$. We want to control the behaviour of the functions $\tilde{G}$ defined in (147) in the large momentum region. Up to a finite renormalization, the perturbative expansion of these functions looks similar to that for correlation functions, except that in turn each propagator is squared. Therefore, the high momentum behaviour is improved order by order. Let us assume that this perturbative property remains valid after resummation, and, more precisely, that

$$\lim_{\lambda \to \infty} \tilde{G}_n^{\mathrm{R}}(\lambda\mathbf{p}) \big/ G_n^{\mathrm{R}}(\lambda\mathbf{p}) = 0 \qquad (151)$$

Due to equation (112), one has

$$\lim_{\lambda \to \infty} \lambda^{4-d} \bar{g}[\lambda, g] = g_0/m^{4-d} \qquad (152)$$

Combining (151) and (152), one checks that, at fixed g such that $0 \le g \le g_c$, the limit $\lambda \to \infty$ of the integral in (150) reproduces the perturbative expansion. This was expected because the ultraviolet behaviour of the model is trivial.

This is no longer the case for $g = g_c$. In this limit,

$$\bar{g}[x, g_c] = g_c$$

$$\lim_{g \to g_c} \frac{Z_1(\bar{g}[x, g])}{Z_1(g)} = \lim_{g \to g_c} \exp \int_g^{\bar{g}} \mathrm{d}g' \frac{\gamma_1(g')}{\beta(g')} = \exp \gamma_1(g_c) \ln x \qquad (153)$$

Hence,

$$G_n^{\mathrm{R}}(\lambda\mathbf{p}; m, g_c) = \lambda^{d_n^G + \frac{1}{2}n\gamma_1(g_c)} \int_0^\lambda \frac{\mathrm{d}\mu}{\mu^{1+d_n^G+\frac{1}{2}n\gamma_1(g_c)}} \tilde{G}_n^{\mathrm{R}}(\mu\mathbf{p}; m, g_c) \qquad (154)$$

As $\lambda \to \infty$, the hypothesis (151) implies that the integral is convergent, so that the behaviour of the correlation functions at g_c in the region $|\mathbf{p}| \gg m$ is a pure power law

$$\lim_{\lambda \to \infty} G_n^{\mathrm{R}}(\lambda\mathbf{p}; m, g_c) = \lambda^{d_n^G + \frac{1}{2}n\gamma_1(g_c)} \int_0^\infty \frac{\mathrm{d}\mu}{\mu^{1+d_n^G+\frac{1}{2}n\gamma_1(g_c)}} \tilde{G}_n^{\mathrm{R}}(\mu\mathbf{p}; m, g_c) \qquad (155)$$

This discontinuity in the ultraviolet behaviour arises from the fact that g_c is a repulsive ultraviolet fixed point. The scaling behaviour (155) corresponds of course to the massless theory. The underlying regularity hypothesis is that the integral is convergent at the upper limit, i.e.

$$\lim_{\lambda \to \infty} \lambda^{-d_n^G - \frac{1}{2} n \gamma_1(g_c)} \tilde{G}_n^{\mathrm{R}}(\lambda \mathbf{p}, m, g_c) = 0$$

This implies that at criticality the scaling behaviour of correlation functions can be deduced consistently from the low frequency modes of the theory. This is an aspect of universality, i.e. the independence on the details of the bare theory at short distance.

Deduce equation (155) directly from the differential equation (143), using the fact that $\beta(g_c) = 0$.

These results can now be inserted in the *bare* correlation functions expressed in terms of the *renormalized* mass m instead of the bare one m_0

$$G_n(\mathbf{p}; \frac{m}{\lambda}, g_0)_\Lambda = \int_0^\lambda \frac{\mathrm{d}\mu}{\mu^{1+d_n^G}} \left(Z_1(\bar{g}[\mu^{-1}, g]) \right)^{n/2} \tilde{G}_n^{\mathrm{R}}(\mu \mathbf{p}; m, \bar{g}[\mu^{-1}, g])$$

(156)

where $g = g(m, g_0, \Lambda)$, $g_\lambda = g(m/\lambda, g_0, \Lambda)$, and where we used the relation $\bar{g}(\lambda/\mu, g_\lambda) = \bar{g}(1/\mu, g)$.

When $\lambda \to \infty$, $m/\lambda \to 0$, the integral may be divergent in the region of large μ. If one uses a perturbative reasoning, one encounters the difficulties discussed at the beginning of this section. Indeed, $\bar{g}$ and Z_1 are polynomials in μ^{4-d} and this generates divergences at a fixed finite order when $\mu \to \infty$. The renormalization group technique avoids this difficulty. Indeed, $\bar{g}(\mu^{-1}, g)$ tends toward g_c and one obtains finite expressions for the massless theory

$$G_n(\lambda \mathbf{p}; 0, g_0)_\Lambda = \lambda^{d_n^G} \int_0^\infty \frac{\mathrm{d}\mu}{\mu^{1+d_n^G}} \left(Z_1(\bar{g}[\frac{\lambda}{\mu}, g]) \right)^{n/2} \tilde{G}_n^{\mathrm{R}}(\mu \mathbf{p}; m, \bar{g}[\frac{\lambda}{\mu}, g])$$

(157)

This expression illustrates the difference between the large- and short-distance behaviour. If λ becomes very large (but always such that $|\lambda \mathbf{p}| \ll \Lambda$), one recovers the perturbative expansion in g_0, with singularities arising at the same orders as those of the infrared perturbative divergences. Conversely, as $\lambda \to 0$, $\bar{g}(\lambda/\mu, g)$ tends towards g_c for any finite value of μ, and, from (157),

$$\lim_{\lambda \to 0} G_n(\lambda \mathbf{p}; 0, g_0)_\Lambda$$

$$= \lambda^{d_n^G + \frac{1}{2} n \gamma_1(g_c)} \left(Z_1(g) \exp \int_g^{g_c} dg' \frac{\gamma_1(g') - \gamma_1(g_c)}{\beta(g')} \right)^{n/2}$$

$$\times \int_0^\infty \frac{d\mu}{\mu^{1 + d_n^G + \frac{1}{2} n \gamma_1(g_c)}} \tilde{G}_n^R(\mu \mathbf{p}; m, g_c) \tag{158}$$

As a consequence, up to a renormalization factor, the infrared behaviour of the massless bare theory (at T_c) is identical to the ultraviolet behaviour of the massive renormalized theory at $g = g_c$.

With $n = 2$ and $d_2^G = -2$ in (158), one obtains the small momentum behaviour of the 2-point correlation function

$$G_2(\mathbf{p}) \quad \sim \quad \frac{1}{p^{2 - \gamma_1(g_c)}} \tag{159}$$

Equivalently, the large-distance behaviour in configuration space reads $1/|\mathbf{x}|^{d-2+\gamma_1(g_c)}$. We recover the result (132) of the preceding subsection, $\eta = \gamma_1(g_c)$.

(1) Corrections to the dominant scaling behaviour. Check, using an expansion of (157) beyond the dominant term, that the correction of order $g_\lambda - g_c$ for $\lambda \to 0$ is (cf. (133))

$$G_n(\lambda \mathbf{p}; 0, g_0)_\Lambda \underset{\lambda \to 0}{\sim} \mathrm{Cst} \lambda^{d_n^G + \frac{1}{2} n \gamma_1(g_c)} \left(1 + \mathcal{O}(\lambda^\omega) \right) \tag{160}$$

where $\omega = \beta'(g_c)$ is the invariant slope of the β function at the critical point. We shall return later to the corrections to the scaling laws due to irrelevant terms in the Lagrangian.

(2) Nonanalytic behaviour of the critical theory with respect to the bare coupling constant. In dimension less than four, the mass introduced in the renormalized theory plays the role of a regulator of infrared perturbative divergences and of a scaling factor independent from the cutoff $\Lambda \gg m$. Using the homogeneity of $\tilde{G}_n^R$, equation (158) is rewritten as

$$\lim_{\lambda \to 0} G_n(\lambda \mathbf{p}; 0, g_0)_\Lambda = \lambda^{d_n^G + \frac{1}{2} n \gamma_1(g_c)} m^{-\frac{1}{2} n \gamma_1(g_c)} \times$$

$$\left\{ \left(\frac{g_c}{g} \right)^{\frac{\gamma_1(g_c)}{4-d}} e^{\gamma_1(g_c) \int_0^g dg' \left(\frac{1}{\beta(g')} + \frac{1}{(4-d)g'} \right)} e^{\int_0^{g_c} dg' \left(\frac{\gamma_1(g') - \gamma_1(g_c)}{\beta(g')} - \frac{\gamma_1(g_c)}{(4-d)g'} \right)} \right\}^{n/2}$$

$$\times \int_0^\infty \frac{d\mu}{\mu^{1 + d_n^G + \frac{1}{2} n \gamma_1(g_c)}} \tilde{G}_n^R(\mu \mathbf{p}; m = 1, g_c) \tag{161}$$

The dependence on g (function of g_0) is only contained within the braces. The last integral is a constant, the second term is a regular function as $g \to 0$. There just remains

$$\left[\left(\frac{g_c}{g}\right)^{\gamma_1(g_c)/(4-d)} m^{-\gamma_1(g_c)}\right]^{n/2} \sim \left(\frac{g_c}{g_0}\right)^{\frac{1}{2}n\gamma_1(g_c)/(4-d)}$$

which exhibits a non analytic behaviour in g_0. This is of course once more related to the infrared singularities in the massless perturbative theory.

5.3.5   *Expansion near the critical point*

Critical scaling laws require the introduction of an arbitrary scale, essential to restore the standard dimensional analysis. We used previously the correlation length near T_c. However, this is not very practical when one explores the neighborhood of T_c, and in particular the ordered region $\theta < 0$. The procedure would be simplified by starting directly from the massless theory. This is of course impossible for $d < 4$ due to infrared divergences, but can be implemented in a double expansion in g and $\varepsilon = 4 - d$ following the original suggestion of Wilson and Fisher. In order to carry out this program, we introduce an arbitrary scale μ in the normalization conditions, according to

$$\Gamma_2^R(\mathbf{p}^2; g, \mu)\big|_{\mathbf{p}^2=0} = 0$$

$$\frac{\partial \Gamma_2^R}{\partial \mathbf{p}^2}(\mathbf{p}^2; g, \mu)\bigg|_{\mathbf{p}^2=\mu^2} = 1 \qquad (162)$$

$$\Gamma_4^R(\mathbf{p}_s; g, \mu) = \mu^{4-d} g$$

These conditions now replace equations (88). The symmetric point $\mathbf{p}_s$ satisfies

$$\mathbf{p}_i \cdot \mathbf{p}_j = \tfrac{1}{3}\mu^2(4\delta_{ij} - 1) \qquad (163)$$

The choice of nonvanishing momenta in the normalization of the four-point function and the derivative of the two-point function is such as to avoid the occurence of infrared divergences. As we shall see later, when the massless theory is well under control, the mass can be introduced semiperturbatively, so that the small θ domain can be explored. The method is based on the fact that the functions $\beta(g, \varepsilon)$ and $\gamma_i(g, \varepsilon)$ which determine the critical behaviour can be entirely computed within the massless theory.

The renormalization group equations express now the invariance with respect to variations of the normalization scale μ, according to the original point of view of Gell-Mann and Low. Indeed the regularized bare theory does not refer to μ. Hence, differentiating the relation $\Gamma_n(\mathbf{p}; g_0, m_0^c)|_{\Lambda} = Z_1^{-\frac{1}{2}n} \Gamma_n^R(\mathbf{p}; g, \mu)$ with respect to μ, one obtains the homogeneous equation

$$\left\{ \mu \frac{\partial}{\partial \mu} + \beta(g) \frac{\partial}{\partial g} - \tfrac{1}{2} n \gamma_1(g) \right\} \Gamma_n^R(\mathbf{p}; g, \mu) = 0 \qquad (164)$$

with

$$\beta(g) = \mu \left. \frac{\partial g}{\partial \mu} \right|_{g_0, \Lambda}$$

$$\gamma_1(g) = Z_1^{-1} \mu \left. \frac{\partial Z_1}{\partial \mu} \right|_{g_0, \Lambda} \qquad (165)$$

Although these quantities might differ from those defined in (107), we use the same notation. The analysis of equation (164) leads however to similar conclusions. There is a infrared stable fixed point of order $g_c \sim \varepsilon$ and the critical exponent η is equal to $\gamma_1(g_c)$. Equations (164) can also be generalized to functions with $\frac{1}{2}\varphi^2$ insertions, which are normalized according to

$$\Gamma_{n,s}(\mathbf{p}, \mathbf{q}; g_0, \Lambda) = Z_1^{-n/2} Z_2^s \left[\Gamma_{n,s}^R(\mathbf{p}, \mathbf{q}; g, \mu) + A \delta_{n,0} \delta_{s,2} \right]$$

$$\Gamma_{2,1}^R(\mathbf{p}_1, \mathbf{p}_2, \mathbf{q} = -\mathbf{p}_1 - \mathbf{p}_2; g, \mu) \Big|_{\mathbf{p}_1^2 = \mathbf{p}_2^2 = -3\mathbf{p}_1 \cdot \mathbf{p}_2 = \mu^2} = 1 \quad (166)$$

$$\Gamma_{0,2}^R(\mathbf{q}; g, \mu) \Big|_{\mathbf{q}^2 = \frac{4}{3}\mu^2} = 0$$

and obey

$$\left\{ \mu \frac{\partial}{\partial \mu} + \beta(g) \frac{\partial}{\partial g} - \tfrac{1}{2} n \gamma_1(g) + s \gamma_2(g) \right\} \Gamma_{n,s}^R = \delta_{n,0} \delta_{s,2} B(g) \quad (167)$$

with

$$\gamma_2(g) = Z_2^{-1} \mu \left. \frac{\partial Z_2}{\partial \mu} \right|_{g_0, \Lambda} \qquad (168)$$

and

$$B(g) = \left\{ \mu \frac{\partial}{\partial \mu} + 2\gamma_1(g) \right\} A \qquad (169)$$

This last quantity has dimension $d - 4$. Excepting the case $n = 0$, $s = 2$ (which corresponds to the specific heat), the standard dimensional analysis, which yields

$$\Gamma^{R}_{n,s}(\lambda\mathbf{p}, \lambda\mathbf{q}; g, \lambda\mu) = \lambda^{d-\frac{1}{2}n(d-2)-2s}\Gamma^{R}_{n,s}(\mathbf{p}, \mathbf{q}; g, \mu) \qquad (170)$$

and the relation (167) taken at $g = g_{\mathrm{c}}$ where $\beta(g_{\mathrm{c}}) = 0$, lead to the small momentum behaviour

$$\Gamma^{R}_{n,s}(\lambda\mathbf{p}, \lambda\mathbf{q}; g, \mu) \underset{\lambda\to 0}{\sim} \lambda^{d-\frac{1}{2}n(d-2+\eta)-s/\nu} \qquad (171)$$

Hence, according to (130) and (132),

$$\begin{aligned} \gamma_1(g_{\mathrm{c}}) &= \eta \\ \gamma_2(g_{\mathrm{c}}) &= 2 - \nu^{-1} \end{aligned} \qquad (172)$$

For $n = 0$, $s = 2$, the inhomogeneous term yields an additional constant which may be dominant

$$\Gamma^{R}_{0,2}(\lambda\mathbf{q}; g, \mu) \underset{\lambda\to 0}{\sim} a_1 + a_2\lambda^{d-2/\nu} \qquad (173)$$

The massless symmetric theory is thus constructed close to four dimensions in a $4-d$ expansion. The neighborhood of T_{c} will then be studied by resummation of a series in the quantity $m_0^2 - m_{0\mathrm{c}}^2$ proportional to the deviation θ from criticality, with θ coupled to the operator $\frac{1}{2}\varphi^2$. For the bare functions, one has a expansion

$$\Gamma_n(\mathbf{p}, m_0^2 - m_{0\mathrm{c}}^2) = \sum_{s=0}^{\infty} \frac{(m_0^2 - m_{0\mathrm{c}}^2)^s}{s!}\Gamma_{n,s}(\mathbf{p}, \mathbf{q} = 0; m_{0\mathrm{c}}^2)$$

Due to the relation

$$(m_0^2 - m_{0\mathrm{c}}^2)\tfrac{1}{2}\varphi^2 = \theta\frac{Z_1}{Z_2}\left(\tfrac{1}{2}\varphi^2\right)_{\mathrm{R}} \qquad (174)$$

which can be interpreted as $Z_2(m_0^2 - m_{0\mathrm{c}}^2) = \theta$, the renormalized quantities satisfy

$$\Gamma^{R}_n(\mathbf{p}; \theta) = \sum_{s=0}^{\infty} \frac{\theta^s}{s!}\Gamma^{R}_{n,s}(\mathbf{p}, \mathbf{q} = 0, \theta = 0) \qquad (175)$$

Because of infrared divergences, this expansion is, however, only valid if we use a space dependent quantity $\theta(\mathbf{x})$, constant at finite distance, and vanishing sufficiently fast at infinity. It should more properly be written as

$$\Gamma^{R}_n(\mathbf{p}; \theta(\mathbf{x})) = \sum_{s=0}^{\infty} \frac{1}{s!}\int\prod_{k=1}^{s}\left(\frac{\mathrm{d}^d\mathbf{q}_k}{(2\pi)^d}\tilde{\theta}(\mathbf{q}_k)\right)\Gamma^{R}_{n,s}(\mathbf{p}, \{\mathbf{q}_k\}; \theta = 0)$$
$$(176)$$

From (167), one gets

$$\left\{\mu\frac{\partial}{\partial\mu} + \beta(g)\frac{\partial}{\partial g} - \tfrac{1}{2}n\gamma_1(g) + \gamma_2(g)\int\frac{d^dq}{(2\pi)^d}\tilde\theta(\mathbf{q})\frac{\delta}{\delta\tilde\theta(\mathbf{q})}\right\}$$
$$\times\,\Gamma_n^R(\mathbf{p};\theta(\mathbf{x}),g,\mu) = 0$$

In the limit of uniform θ, which corresponds to the introduction of a massive propagator, one finally obtains

$$\left\{\mu\frac{\partial}{\partial\mu} + \beta(g)\frac{\partial}{\partial g} - \tfrac{1}{2}n\gamma_1(g) + \gamma_2(g)\theta\frac{\partial}{\partial\theta}\right\}\Gamma_n^R(\mathbf{p};\theta,g,\mu) = 0$$
(177)

These equations are to be compared to relations (110). They differ only by finite renormalizations, but the crucial point is that they are homogeneous. The derivation shows that the coefficients β, γ_1, γ_2 for the massive theory are the same as those computed in the simpler massless theory through a double expansion in g and ε.

From equations (177), we can now extract all scaling relations. Let us consider first the symmetric phase $\theta > 0$ where the Green functions with an odd number of arguments identically vanish. With the notation

$$\mu(\lambda) = \lambda\mu$$
$$\lambda\frac{dg(\lambda)}{d\lambda} = \beta(g(\lambda)) \qquad\qquad g(1) = g \qquad\qquad (178)$$
$$\lambda\frac{d\ln\theta(\lambda)}{d\lambda} = \gamma_2(g(\lambda)) \qquad\qquad \theta(1) = \theta$$

the homogeneous equations (177) are solved in the form

$$\Gamma_n^R(\mathbf{p};\theta,g,\mu) = \exp\left\{-\tfrac{1}{2}n\int_1^\lambda\frac{d\lambda'}{\lambda'}\gamma_1(g(\lambda'))\right\}\Gamma_n^R(\mathbf{p};\theta(\lambda),g(\lambda),\lambda\mu)$$
(179)

Let us choose λ_0 such that $\theta(\lambda_0)$ has a fixed value of the form

$$\theta(\lambda_0) = \lambda_0^2\mu^2 \qquad\qquad (180)$$

which defines λ_0 implicitly through

$$\ln(\theta/\mu^2) = \int_g^{g(\lambda_0)} dg'\,(2 - \gamma_2(g'))\,/\beta(g')$$

With this choice, equation (179) can be rewritten as

$$\Gamma_n^R(\mathbf{p};\theta,g,\mu) = Z_1^{n/2}(m/\mu,g(\lambda_0))\Gamma_n^R(\mathbf{p};m,g(\lambda_0)) \qquad (181)$$

where we use the notations

$$m = \mu \exp \int_g^{g(\lambda_0)} \frac{dg'}{\beta(g')} \tag{182a}$$

$$Z_1(m/\mu, g(\lambda_0)) = \exp - \int_g^{g(\lambda_0)} dg' \, \gamma_1(g')/\beta(g') \tag{182b}$$

Again these quantities differ from the previous ones at most by finite renormalizations. Let us now choose the normalization scale μ in the ultraviolet domain, i.e. such that

$$\theta^{1/2}, |\mathbf{p}| \ll \mu \ll \Lambda \tag{183}$$

In this way, μ appears as an additional arbitrary cutoff factor, but, in contradistinction with the bare theory in which we have to neglect terms of order $1/\Lambda$, this new formulation is now minimal and exact. As $\theta(\lambda_0)$ is fixed whereas μ tends to infinity, λ_0 tends to zero, and $g(\lambda_0)$ approaches the infrared attractive fixed point g_c. According to (172) and (182),

$$\frac{\theta}{\mu^2} \to \left(\frac{m}{\mu}\right)^{1/\nu}$$

$$Z_1 \to \left(\frac{m}{\mu}\right)^{-\eta} \simeq \left(\frac{\theta}{\mu^2}\right)^{-\eta\nu} \tag{184}$$

In the expressions for the vertex functions at finite $\mathbf{p}/m$, $g(\lambda_0)$ can be replaced in the r.h.s. of (181) by g_c, and, using the dimension

$$d_n^\Gamma = d - \frac{1}{2}n(d-2)$$

one finally gets

$$\Gamma_n^R(\mathbf{p}; \theta, g, \mu) = \mu^{d-\frac{1}{2}n(d-2)} \left(\frac{\theta}{\mu^2}\right)^{\nu[d-\frac{1}{2}n(d-2+\eta)]} f_n\left(\frac{\mathbf{p}}{\mu}\left(\frac{\mu^2}{\theta}\right)^\nu\right) \tag{185}$$

which is the scale invariant form in the critical domain. The correlation length ξ proportional to $\theta^{-\nu}$ appears naturally in the reduced function f_n. For $\theta > 0$, Γ_n^R has a finite limit at zero momenta behaving as $\theta^{\nu[d-\frac{1}{2}n(d-2+\eta)]}$. In particular, the inverse susceptibility corresponds to the case $n = 2$ and hence the critical exponent γ satisfies

$$\boxed{\gamma = \nu(2-\eta)} \tag{186}$$

The Γ_n^R's have been constructed in order to be finite at $\theta = 0$. The prefactor with a fractional power of θ in (185) should thus be compensated by the behaviour of the reduced functions f_n when their arguments tend to infinity ($\theta^\nu \ll |\mathbf{p}| \ll \mu$); hence

$$\theta^\nu \ll \lambda |\mathbf{p}| \ll \mu \qquad \Gamma_n^R(\lambda\mathbf{p}; \theta, g, \mu) \sim \lambda^{d-\frac{1}{2}n(d-2+\eta)} \quad (187)$$

which relates the ultraviolet behaviour of the massive theory to the scaling behaviour of the massless one.

Similarly, for the functions with $\frac{1}{2}\varphi^2$ insertions, one finds (except in the cases $n = 0$, $s = 1, 2$)

$$\left\{ \mu\frac{\partial}{\partial\mu} + \beta(g)\frac{\partial}{\partial g} - \frac{1}{2}n\gamma_1(g) + \gamma_2(g)(s + \theta\frac{\partial}{\partial\theta}) \right\} \Gamma_{n,s}^R(\mathbf{p}, \mathbf{q}; \theta, g, \mu) = 0 \tag{188}$$

which leads to

$$\Gamma_{n,s}^R(\mathbf{p}, \mathbf{q}; \theta, g, \mu) = \mu^{d-\frac{1}{2}n(d-2)-2s} \left(\frac{\theta}{\mu^2}\right)^{-s+\nu[d-\frac{1}{2}n(d-2+\eta)]}$$

$$\times f_{n,s}\left(\frac{\mathbf{p}}{\mu}\left(\frac{\mu^2}{\theta}\right)^\nu, \frac{\mathbf{q}}{\mu}\left(\frac{\mu^2}{\theta}\right)^\nu\right) \tag{189}$$

in the region $\theta \ll \mu^2$, $|\mathbf{p}|, |\mathbf{q}| \ll \mu$. Note that the prefactor can also be interpreted as

$$m^{-2s}\left(\frac{m}{\mu}\right)^{s\gamma_2(g_c)} \sim \mu^{-2s}\left(\frac{\mu^2}{\theta}\right)^s$$

For insertions at zero momentum $\mathbf{q} = \mathbf{0}$, there is an additional divergent factor $1/\theta$ which was to be expected since this insertion is equivalent to a derivative with respect to θ. As $\theta \to 0$, the same mechanism of compensations yields the scaling behaviour (171).

An additional term should be introduced in the r.h.s. of equation (188) for the function $\Gamma_{0,2}$; writing it as $\mu^{d-4}b(g)$, one obtains

$$\Gamma_{0,2}^R(\mathbf{q}; \theta, g, \mu) \sim \mu^{d-4}\left\{ \left(\frac{\theta}{\mu^2}\right)^{-(2-\nu d)} f_{0,2}\left(\frac{\mathbf{q}}{\mu}\left(\frac{\mu^2}{\theta}\right)^\nu\right) - \frac{\nu}{2-\nu d}b(g_c) \right\} \tag{190}$$

for $\theta \ll \mu^2$, $|\mathbf{q}| \ll \mu$. As $\mathbf{q} \to 0$, we obtain the specific heat singularity in $\theta^{-\alpha}$ with

$$\boxed{\alpha = 2 - \nu d} \tag{191}$$

The additive constant in (190) is the regular part of the specific heat, and ensures the coherence of the results when α is negative. Finally, normalizing the free energy to zero at $\theta = 0$ in the absence of field, and noting that $\Gamma_{0,2}^{R}(\theta) = d^2\Gamma^{R}(\theta)/d\theta^2$, one obtains its expression in the critical region for $\theta > 0$ as

$$\Gamma^{R}(\theta) \sim \mu^d \left[f_{0,2}(0) \frac{1}{(2-\alpha)(1-\alpha)} \left(\frac{\theta}{\mu^2} \right)^{2-\alpha} \right.$$

$$\left. - \frac{\nu}{2\alpha} b(g_c) \left(\frac{\theta}{\mu^2} \right)^2 + C \frac{\theta}{\mu^2} \right] \quad (192)$$

The constant C is related to $\Gamma_{0,1}^{R}$.

5.3.6   Scaling laws below the critical temperature

If the argument φ is given a uniform value M, the functional $\Gamma(\varphi)$ becomes proportional to the volume, with a coeffecient given by the effective potential $V(M)$

$$\varphi(\mathbf{x}) \rightarrow M$$

$$\Gamma^{R}(\varphi) \rightarrow \int d^d\mathbf{x}\, V(M) \quad (193)$$

$$V(M, \theta, g, \mu) = \sum_{n=0}^{\infty} \frac{M^n}{n!} \Gamma_n^{R}(\mathbf{p} = \mathbf{0}; \theta, g, \mu)$$

The corresponding uniform external field, denoted H, equals

$$H = \frac{\partial V}{\partial M} = \sum_{n=1}^{\infty} \frac{M^n}{n!} \Gamma_{n+1}^{R}(\mathbf{p} = \mathbf{0}; \theta, g, \mu) \quad (194)$$

The generalization to the case of multicomponent fields does not present difficulties. We restrict ourselves for the time being, to the scalar case. Let us establish the form of the equation of state in the scaling region. We do not assume a given sign for θ. The name equation of state refers to the interpretation of the model in terms of a liquid–vapor equilibrium near the critical point. Dimensional analysis yields

$$[V] = d \quad [M] = \tfrac{1}{2}(d-2) \quad [H] = \tfrac{1}{2}(d+2) = 3 - \tfrac{1}{2}(4-d) \quad (195)$$

Repeating the derivation of the renormalization flow equations, we obtain

$$\left\{ \mu\frac{\partial}{\partial\mu} + \beta(g)\frac{\partial}{\partial g} - \tfrac{1}{2}\gamma_1(g)\left[1 + M\frac{\partial}{\partial M}\right] + \gamma_2(g)\theta\frac{\partial}{\partial\theta} \right\}$$
$$\times H(M,\theta,g,\mu) = 0 \quad (196)$$

Using the notation (178) and the dimensions (195), this yields

$$H(M,\theta,g,\mu) = \mu^{\frac{1}{2}(d+2)} H\left(\frac{M}{\mu^{\frac{1}{2}(d-2)}}, \frac{\theta}{\mu^2}, g, 1 \right)$$
$$= (\lambda\mu)^{\frac{1}{2}(d+2)} \exp\left\{ -\tfrac{1}{2}\int_g^{g(\lambda)} dg'\frac{\gamma_1(g')}{\beta(g')} \right\}$$
$$\times H\left(\frac{M(\lambda)}{(\lambda\mu)^{\frac{1}{2}(d-2)}}, \frac{\theta(\lambda)}{\lambda^2\mu^2}, g(\lambda), 1 \right) \quad (197)$$

with $M(\lambda)$ given by

$$\ln\frac{M(\lambda)}{M} = -\tfrac{1}{2}\int_g^{g(\lambda)} dg'\frac{\gamma_1(g')}{\beta(g')} \quad (198)$$

In order to eliminate $M(\lambda)$, we choose λ in such a way that

$$\frac{M(\lambda)}{(\lambda\mu)^{\frac{1}{2}(d-2)}} = 1 \quad (199)$$

In other words, λ is determined by the condition

$$\ln\frac{M}{\mu^{\frac{1}{2}(d-2)}} = \tfrac{1}{2}\int_g^{g(\lambda)} dg'\frac{d-2+\gamma_1(g')}{\beta(g')} \quad (200)$$

In the critical domain, $M/\mu^{\frac{1}{2}(d-2)} \ll 1$. Hence (200) is satisfied if $g(\lambda) \to g_c$ and if λ tends to zero according to

$$\lambda \sim \left(\frac{M}{\mu^{\frac{1}{2}(d-2)}} \right)^{2/(d-2+\eta)} = \left(\frac{M}{\mu^{\frac{1}{2}(d-2)}} \right)^{\nu/\beta} \quad (201)$$

where the critical exponent β is defined by

$$\boxed{\beta = \tfrac{1}{2}\nu(d-2+\eta)} \quad (202)$$

Hence, as $g(\lambda) \to g_c$,

$$\frac{\theta(\lambda)}{\lambda^2\mu^2} \sim \frac{\theta}{\mu^2}\lambda^{\gamma_2(g_c)-2} = \frac{\theta}{\mu^2}\lambda^{-1/\nu} = \frac{\theta}{\mu^2}\left(\frac{\mu^{\frac{1}{2}(d-2)}}{M} \right)^{2/(\nu(d-2+\eta))}$$
$$\quad (203)$$

Inserting relation (199) into (197), one gets

$$H(M, \theta, g, \mu) \sim \mu^{\frac{1}{2}(d+2)} \left(\frac{M}{\mu^{\frac{1}{2}(d-2)}} \right)^{(d+2-\eta)/(d-2+\eta)}$$

$$f \left[\frac{\theta}{\mu^2} \left(\frac{\mu^{\frac{1}{2}(d-2)}}{M} \right)^{2/(\nu(d-2+\eta))} \right] \qquad (204)$$

Choosing now for definiteness $\mu = 1$,

$$\frac{H}{M^\delta} = f(\frac{\theta}{M^{1/\beta}}) \qquad (205)$$

where

$$\boxed{\delta = \frac{d+2-\eta}{d-2+\eta}} \qquad (206)$$

We have therefore recovered all the relations between the exponents which were discussed in the previous chapter.

We know that H is analytic and odd with respect to M when θ is positive. This forces the following large x behaviour of $f(x)$ in equation (205)

$$f(x) \underset{x \to \infty}{\sim} \sum_{p=0}^{\infty} f_p x^{\beta\delta - \beta(2p+1)} = \sum_{p=0}^{\infty} f_p x^{\gamma - 2p\beta} \qquad (207)$$

since $\beta(\delta - 1) = \nu(2 - \eta) = \gamma$. For small θ and finite M, H remains analytic in θ as long as M does not vanish. Hence $f(x)$ is regular at the origin.

The spontaneous magnetization can be extracted from equation (205) by letting H approach zero, assuming that the equation $f(x) = 0$ has a negative solution $-x_0$ (θ is of course negative). Hence

$$M_{\text{sp}} = \mu^{\frac{1}{2}(d-2)} \left(-x_0 \frac{\theta}{\mu^2} \right)^\beta \qquad (208)$$

This gives the usual interpretation of the critical exponent β.

In the scalar case considered so far, Goldstone's theorem does not apply. For a given magnetization, the correlation functions in the low temperature phase

$$\Gamma_n^R(\mathbf{p}_1,\ldots,\mathbf{p}_n;\theta,M,g,\mu) =$$

$$= \int \prod_1^n (e^{i\mathbf{p}_k\cdot\mathbf{x}_k}d^d\mathbf{x}_k)\,\frac{\delta^n}{\delta\varphi(\mathbf{x}_1)\cdots\delta\varphi(\mathbf{x}_n)}\Gamma^R(\varphi;\theta,g,\mu)\bigg|_{\varphi(\mathbf{x})=M}$$

$$= \sum_{k=0}^{\infty}\frac{M^k}{k!}\Gamma_{n+k}^R(\mathbf{p}_1,\ldots,\mathbf{p}_n,0,\ldots,0;g,\theta,\mu)$$

$$\tag{209}$$

satisfy

$$\left\{\mu\frac{\partial}{\partial\mu}+\beta(g)\frac{\partial}{\partial g}-\frac{1}{2}\gamma_1(g)\left[n+M\frac{\partial}{\partial M}\right]+\gamma_2(g)\theta\frac{\partial}{\partial\theta}\right\}$$
$$\times\,\Gamma_n^R(\mathbf{p};\theta,M,g,\mu)=0 \quad (210)$$

Hence

$$\Gamma_n^R(\mathbf{p};\theta,M,g,\mu) = (\lambda\mu)^{d-\frac{1}{2}n(d-2)}\left[\exp-\frac{1}{2}n\int_g^{g(\lambda)}dg'\frac{\gamma_1(g')}{\beta(g')}\right]$$

$$\times\,\Gamma_n^R(\frac{\mathbf{p}}{\lambda\mu};\frac{\theta(\lambda)}{\lambda^2\mu^2},\frac{M(\lambda)}{(\lambda\mu)^{\frac{1}{2}(d-2)}},g(\lambda),1)$$

$$\tag{211}$$

Let us use the same condition (199) to determine λ, and pick $M \ll \mu^2$ in the critical domain (which forces $\lambda \to 0$ and $g(\lambda) \to g_c$). One gets

$$\Gamma_n^R(\mathbf{p};\theta,M,g,\mu) \sim \mu^{d-\frac{1}{2}n(d-2)}\left(\frac{M}{\mu^{\frac{1}{2}(d-2)}}\right)^{[2d/(d-2+\eta)]-n}$$

$$\times\,f_n\left(\frac{\mathbf{p}}{\mu}\left(\frac{\mu^{\frac{1}{2}(d-2)}}{M}\right)^{\nu/\beta},\frac{\theta}{\mu^2}\left(\frac{\mu^{\frac{1}{2}(d-2)}}{M}\right)^{1/\beta}\right)$$

$$|\mathbf{p}| \ll \mu \qquad\quad \theta \ll \mu^2 \qquad\quad M \ll \mu^{\frac{1}{2}(d-2)} \quad (212)$$

Note that Γ_n^R is regular in M for $\theta > 0$ and in θ for $M \neq 0$.

The vertex functions for $M \neq 0$ are the Legendre transforms of the connected Green functions which, in the present case, contain the subtraction of the 1-point function, i.e. the magnetization M. For $\theta < 0$, $M \to M_{sp}$ when the external field vanishes. With $\mu = 1$, the functions behave in the critical domain as

$$\Gamma_n^R(\mathbf{p};\theta,M_{sp}) \sim (-\theta)^{\nu[d-\frac{1}{2}n(d-2+\eta)]}f_n((-\theta)^{-\nu}\mathbf{p}) \tag{213}$$

This is the same structure as in equation (185) for $\theta > 0$, but n is not constrained to be even. In particular, for $\theta < 0$, the correlation length diverges as

$$\xi \sim (-\theta)^{-\nu'} \qquad \text{with} \quad \nu' = \nu \qquad (214)$$

The thermal exponent ν takes the same value below as well as above the critical temperature. Similarly, for $n = 2$ and $\mathbf{p} \to 0$, the susceptibility diverges as

$$\chi \sim (-\theta)^{-\gamma'} \qquad \text{with} \quad \gamma' = \gamma = \nu(2 - \eta) \qquad (215)$$

with the same exponent as above T_c. Generalizing these results to the derivatives with respect to θ, the specific heat divergence is found to be

$$\Gamma^R_{0,2}(\mathbf{q} = 0; \theta, M_{\text{sp}}) \sim (-\theta)^{-\alpha'} \qquad \text{with} \quad \alpha' = \alpha = 2 - \nu d \qquad (216)$$

Again, one finds the same exponent above and below the critical temperature.

In the case of a multicomponent field, one should distinguish transverse and longitudinal components with respect to the external field or the magnetization. If there exists a symmetry group for, e.g. $O(n)$, the functionals $F(\mathbf{H})$ and $\Gamma(\mathbf{M})$ are of course invariant.

Let $H = |\mathbf{H}|$ and $M = |\mathbf{M}|$ denote the magnitudes of a uniform external field and the corresponding magnetization. The effective potential depends only on H and therefore $\mathbf{M}$ is parallel to the applied field

$$H_a = \frac{M_a}{M}\frac{dV}{dM} \qquad\qquad H = \frac{dV}{dM} \qquad (217)$$

The susceptibility and its inverse are now matrices.

$$\chi^{-1}_{ab} = \frac{\partial H_a}{\partial M_b} = \frac{M_a M_b}{M^2}\chi_l^{-1} + \left(\delta_{ab} - \frac{M_a M_b}{M^2}\right)\chi_t^{-1}$$

$$\chi_l^{-1} = \frac{d^2 V}{dM^2} \qquad (218)$$

$$\chi_t^{-1} = \frac{1}{M}\frac{dV}{dM} = \frac{H}{M}$$

Below the critical point $(\theta < 0)$, $M \to M_{\text{sp}}$ and χ_t^{-1} vanishes when $H \to 0$. This is Goldstone's theorem, asserting that the transverse susceptibility is infinite in the ordered phase of a system with a continuous symmetry, due to the existence of zero modes. On the other hand, the longitudinal susceptibility has the same behaviour as in one-component systems.

5.4   Corrections to scaling laws

The discussion of correlations is not limited to the dominant scaling behaviour. In particle physics, after an analytic continuation in the Minkovskian region, the Green functions also describe bound states, scattering amplitudes, etc. We do not discuss these problems here. It is nevertheless interesting, in the framework of statistical mechanics, to compute the leading corrections to scaling. We examine successively deviations from the critical point, the role of marginal operators and the contributions of irrelevant terms.

5.4.1   Deviation from the critical point in dimension lower than four

Looking at the renormalization flow equations, such as e.g. (210), in the framework of the φ^4 model, where irrelevant operators are neglected, one sees that the correlation functions are homogeneous at $g = g_c$, whatever the auxiliary normalization parameter μ is. Let us consider a small deviation of g from g_c. The effect can be absorbed in part in a change of the normalization of correlations at g_c, the remaining part representing the essential effect.

In order to perform this splitting, let

$$M(g) = M \exp \tfrac{1}{2} \int_{g_c}^{g} \mathrm{d}g' \frac{\gamma_1(g') - \gamma_1(g_c)}{\beta(g')}$$

$$\theta(g) = \theta \exp - \int_{g_c}^{g} \mathrm{d}g' \frac{\gamma_2(g') - \gamma_2(g_c)}{\beta(g')} \tag{219}$$

Hence, dropping the index for renormalized functions,

$$\Gamma_n(\mathbf{p}; M, \theta, g, \mu) = \Gamma_n(\mathbf{p}; M(g), \theta(g), g_c, \mu)$$

$$\times \exp \left\{ \tfrac{1}{2} n \int_{g_c}^{g} \mathrm{d}g' \frac{\gamma_1(g') - \gamma_1(g_c)}{\beta(g')} \right\} D_n(\mathbf{p}; M(g), \theta(g), g, \mu) \tag{220}$$

with

$$\left\{\mu\frac{\partial}{\partial\mu} - \tfrac{1}{2}\gamma_1(g_c)(n + M\frac{\partial}{\partial M}) + \gamma_2(g_c)\theta\frac{\partial}{\partial\theta}\right\}\Gamma_n(\mathbf{p}; M, \theta, g_c, \mu) = 0 \tag{221}$$

$$\begin{cases} \left\{\mu\dfrac{\partial}{\partial\mu} + \beta(g)\dfrac{\partial}{\partial g} - \tfrac{1}{2}\gamma_1(g_c)M\dfrac{\partial}{\partial M} + \gamma_2(g_c)\theta\dfrac{\partial}{\partial\theta}\right\} \\ \qquad\times D_n(\mathbf{p}; M, \theta, g, \mu) = 0 \\ D_n(\mathbf{p}; M, \theta, g_c, \mu) = 1 \end{cases} \tag{222}$$

The parametrization (220) has been chosen in such a way that, in the equation for D_n, the only reference to g appears in the term $\beta(g)\partial/\partial g$. Moreover, n does not appear explicitly in this equation. The function D_n is normalized to unity at the critical point and is dimensionless, so that

$$D_n \equiv D_n\left(\frac{\mathbf{p}}{\mu}; \frac{M}{\mu^{\frac{1}{2}(d-2)}}, \frac{\theta}{\mu^2}, g, 1\right) \tag{223}$$

The solution of equation (222) can thus be written as

$$\begin{aligned} D_n(\mathbf{p}; M, \theta, g, \mu) &= D_n(\mathbf{p}; M\lambda^{-\frac{1}{2}\gamma_1(g_c)}, \theta\lambda^{\gamma_2(g_c)}, g(\lambda), \lambda\mu) \\ &= D_n\left(\frac{\mathbf{p}}{\lambda}; M\lambda^{-\frac{1}{2}\gamma_1(g_c)-\frac{1}{2}(d-2)}, \theta\lambda^{\gamma_2(g_c)-2}, g(\lambda), \mu\right) \end{aligned} \tag{224}$$

where

$$\ln\lambda = \int_g^{g(\lambda)} \frac{dg'}{\beta(g')} \tag{225}$$

Let us now choose λ such that

$$1 = |\theta|\,\lambda^{\gamma_2(g_c)-2} = |\theta|\,\lambda^{-1/\nu} \tag{226}$$

The previous relation reads

$$D_n(\mathbf{p}; \theta, M, g, \mu = 1) = D_n\left(\frac{\mathbf{p}}{|\theta|^\nu}; \frac{M}{|\theta|^\beta}, g(|\theta|^\nu), 1\right) \tag{227}$$

$$\nu\ln|\theta| = \int_g^{g(|\theta|^\nu)} \frac{dg'}{\beta(g')}$$

We approximate the function $\beta(g)$ by a linear term near g_c

$$\beta(g) \simeq \omega(g - g_c) + \mathcal{O}((g - g_c)^2) \tag{228}$$

If g is close enough to g_c and θ small, $g(|\theta|^\nu)$ is also close to g_c. Indeed, from equations (227) and (228),

$$|\theta|^{\nu\omega} = \frac{g_c - g(|\theta|^\nu)}{g_c - g} + \cdots \tag{229}$$

and hence the correction factor D_n behaves as

$$D_n(\mathbf{p}; \theta, M, g, 1) \simeq 1 + (g - g_c)|\theta|^{\nu\omega} d_n \left(\frac{\mathbf{p}}{|\theta|^\nu}, \frac{M}{|\theta|^\beta}\right) \tag{230}$$

The exponent

$$\boxed{\omega = \beta'(g_c) > 0} \tag{231}$$

characterizes therefore the deviations from $g = g_c$ independently of which of the Green functions is considered. As indicated in equation (220), this correction is accompanied by a change in the normalization of the correlation functions.

In equation (230), we have expressed the deviations from criticality in terms of the temperature. It is also possible to use the momenta or the magnetization. The corresponding corrections behave as

$$\begin{array}{l} \text{for } |\theta| \text{ as } |\theta|^{\omega\nu} \\ \text{for } |\mathbf{p}| \text{ as } |\mathbf{p}|^\omega \\ \text{for } M \text{ as } M^{\omega\nu/\beta} \end{array} \tag{232}$$

To order $\varepsilon = 4 - d$, $\beta = \varepsilon(-1 + g/g_c)$ where g_c is of order ε. Hence

$$\omega = \varepsilon + \mathcal{O}(\varepsilon^2) \tag{233}$$

As $d \to 4$, the corrections are confluent with the dominant term.

5.4.2   *Logarithmic corrections in dimension four*

In the marginal dimension, the stable infrared fixed point g_c coincides with the ultraviolet fixed point at the origin, and the scale invariant theory is the Gaussian model, up to logarithmic corrections. These corrections are thus important for characterizing the critical behaviour. The equation of state (197) now reads

$$H(M,\theta,g,\mu) = (\lambda\mu)^3 \left[\exp -\tfrac{1}{2} \int_g^{g(\lambda)} dg' \frac{\gamma_1(g')}{\beta(g')} \right] H\left(1, \frac{\theta(\lambda)}{\lambda^2\mu^2}, g(\lambda), 1\right)$$

(234)

with

$$\lambda = \frac{M}{\mu} \exp -\tfrac{1}{2} \int_g^{g(\lambda)} dg' \frac{\gamma_1(g')}{\beta(g')}$$

(235)

and, of course, $g(\lambda)$ and $\theta(\lambda)$ are defined through

$$\ln \lambda = \int_g^{g(\lambda)} \frac{dg'}{\beta(g')}$$

$$\ln \frac{\theta(\lambda)}{\theta} = \int_g^{g(\lambda)} dg' \frac{\gamma_2(g')}{\beta(g')}$$

(236)

In the critical domain $M \ll \mu^2$, λ is vanishingly small. The function $\beta(g)$ is given by (cf. (115))

$$\beta(g) = \frac{n+8}{3} \frac{g^2}{(4\pi)^2} + \cdots$$

(237)

therefore

$$\frac{g(\lambda)}{(4\pi)^2} \sim \frac{3}{n+8} \frac{1}{(-\ln \lambda)}$$

(238)

This expression is typical of asymptotic infrared freedom. In the exponential (235), $\gamma_1(g')$ is of order g'^2 and the integral is finite, i.e. M is of order λ. Finally, as

$$\gamma_2(g) = \frac{n+2}{8} \frac{g}{(4\pi)^2} + \cdots$$

(239)

$$\frac{\theta(\lambda)}{\theta} \sim \mathrm{Cst}\,(-\ln \lambda)^{-(n+2)/(n+8)} \sim \mathrm{Cst}\left(\ln \frac{\mu}{M}\right)^{-(n+2)/(n+8)}$$

(240)

Inserting this in (234), one gets

$$H(M,\theta,g,\mu=1) \simeq M^3$$

$$\times H\left(1, \mathrm{Cst}\frac{\theta}{M^2}(-\ln M)^{-(n+2)/(n+8)}, \frac{3}{n+8}\frac{(4\pi)^2}{(-\ln M)}, 1\right)$$

(241)

Since $g(\lambda)$ is vanishingly small, the dependence of the r.h.s. on this variable can be computed perturbatively. To lowest order, we have

$$V(M) = \tfrac{1}{2}\theta M^2 + \frac{g}{4!}M^4 + \cdots$$
$$H = \theta M + \tfrac{1}{6}gM^3 + \cdots$$

Therefore relation (241) becomes (with A denoting a constant)

$$H(M,\theta,g,\mu=1)$$
$$\simeq M^3 \left\{ A\frac{\theta}{M^2}\left(-\ln M\right)^{-(n+2)/(n+8)} - \frac{(4\pi)^2}{2(n+8)\ln M}\right\}$$
$$\times \left\{1 + \mathcal{O}\left(\frac{1}{\ln M}\right)\right\} \tag{242}$$

where the correction arises from the term in $g(\lambda)$.

At the critical temperature, this relation yields

$$\boxed{H \simeq \frac{M^3}{-\ln M}} \tag{243}$$

The critical exponent δ takes its mean field value $\delta = 3$, but there is a logarithmic correction. This result can be obtained heuristically, by replacing g by its running value $g(\lambda) \simeq g(M) \sim -1/\ln M$ in the mean field expression $\tfrac{1}{6}gM^3$.

The spontaneous magnetization is obtained for vanishing H and $T < T_c$, as

$$\boxed{M \simeq \sqrt{-\theta}\,(-\ln-\theta)^{3/(n+8)}} \tag{244}$$

with logarithmic corrections (depending on n) to the mean field behaviour characterized by an exponent $\beta = \tfrac{1}{2}$.

In zero external field and for $\theta \neq 0$, the susceptibility (the longitudinal one for $n \geq 2$ and $\theta < 0$) deduced from (242) reads

$$\boxed{\chi = |\theta|^{-1}\,(-\ln|\theta|)^{(n+2)/(n+8)}} \tag{245}$$

with γ equal to -1 up to logarithmic corrections.

To study the specific heat, let us look at the equation for the dimensionless function $\Gamma_{0,2}^{\mathrm{R}}$

$$\left\{\mu\frac{\partial}{\partial\mu} + \beta(g)\frac{\partial}{\partial g} + \gamma_2(g)\left[2 + \theta\frac{\partial}{\partial\theta}\right]\right\}\Gamma_{0,2}^{\mathrm{R}} = b(g) \tag{246}$$

where $b(g)$ approaches a finite constant b_0 as g vanishes. Upon integration,

$$\Gamma_{0,2}^R(\mathbf{q}; \theta, g, \mu) = \left[\exp 2 \int_g^{g(\lambda)} dg' \frac{\gamma_2(g')}{\beta(g')} \right] \Gamma_{0,2}^R(\mathbf{q}; \theta(\lambda), g(\lambda), \lambda\mu)$$
$$- \int_g^{g(\lambda)} \frac{dg'}{\beta(g')} b(g') \exp 2 \int_g^{g'} dg'' \frac{\gamma_2(g'')}{\beta(g'')}$$

(247)

At zero momentum, using the fact that $\Gamma_{0,2}^R$ is dimensionless, we have

$$\Gamma_{0,2}^R(0; \theta, g, \mu = 1) = \left[\exp 2 \int_g^{g(\lambda)} dg' \frac{\gamma_2(g')}{\beta(g')} \right] \Gamma_{0,2}^R \left(0; \frac{\theta(\lambda)}{\lambda^2}, g(\lambda), 1 \right)$$
$$- \int_g^{g(\lambda)} \frac{dg'}{\beta(g')} b(g') \exp 2 \int_g^{g'} dg'' \frac{\gamma_2(g'')}{\beta(g'')}$$

(248)

Let us choose λ by requiring that $|\theta(\lambda)| = \lambda^2$, thus

$$\lambda^2 \simeq |\theta| \left(\frac{g(\lambda)}{g} \right)^{(n+2)/(n+8)}$$

(249)

Since

$$g(\lambda) \simeq \frac{\text{Cst}}{\ln \lambda}$$

(250)

the dominant contribution in (248) comes from the inhomogeneous term and one obtains

$$\boxed{\Gamma_{0,2}^R \simeq \frac{\text{Cst}}{(-\ln|\theta|)^{(n-4)/(n+8)}} + \text{Cst}}$$

(251)

with again a logarithmic correction.

These results agree in the limit $n \to \infty$ with those obtained in chapter 3. Note a change in the behaviour of the specific heat for $n = 4$ with a divergence for $n < 4$. For instance, when $n = 1$, the specific heat diverges as $(-\ln|\theta|)^{1/3}$.

Let us again emphasize the paradox of the four-dimensional continuous φ^4 theory. The renormalized theory is free, since the interval $[0, g_c]$ shrinks to zero. The preceding results are meaningful only if one keeps a large, but not infinite, cutoff Λ, so that the interval is not strictly 0, but of order $1/\ln\Lambda$.

5.4.3 *Irrelevant operators*

We finally discuss the effect of keeping local terms with higher powers of the field or its derivatives in the Lagrangian. The presence of such terms cannot be avoided in any consistent regularization scheme. Close to four dimensions, these terms have coefficients with a negative power of Λ, and therefore were neglected in a first approach. It is crucial to the proof of universality to make sure that, through renormalization effects, they do not affect the leading scaling behaviour. We do not discuss here dimensions low enough for some of these operators to become marginal or relevant, as is a φ^6 term in dimension three. This corresponds to a richer phase diagram, with a possible occurrence of first order transitions and tricritical phenomena (appendix A).

We shall content ourselves here with estimating perturbatively the effect of irrelevant operators, taking into account renormalization effects. We examine an operator $\mathcal{O}_i$ with n_1 fields and n_2 derivatives (symbolically $\partial^{n_2}\varphi^{n_1}$). Its coefficient in the Lagrangian will be

$$g(\mathcal{O}_i) \sim \Lambda^{-\delta_i}$$
$$\delta_i = \tfrac{1}{2}n_1(d-2) + n_2 - d \tag{252}$$

and its canonical degree is δ_i. For instance,

$$\delta_{\varphi^6} = 2d - 6$$
$$\delta_{\varphi^2(\partial\varphi)^2} = d - 2 \tag{253}$$

At first sight, operators with $\delta_i > 0$ are negligible. However, their insertion in Green functions lead to additional divergences, and their effect should therefore be analyzed more carefully. For instance, we have seen the important effect of the marginal φ^4 term in four dimensions. Concerning operators with negative degree, we have already analyzed the φ^2 case which generates logarithmic divergences in dimension four.

Consider now operators with nonnegative degree in dimension four. The results will then be continued using an ε-expansion. The insertion of φ^4 corresponds to a shift of the coupling constant from its fixed point value. Vertex functions Γ_{2,φ^4} and Γ_{4,φ^4} are respectively quadratically and logarithmically divergent. The logarithmic divergence of the 4-point function can be treated by a multiplicative renormalization, but the 2-point function needs new subtractions. More precisely,

$$\Gamma_{2,\varphi^4}(\mathbf{p}_1,\mathbf{p}_2,\mathbf{q})_\Lambda = a_1(\Lambda)+a_2(\Lambda)(\mathbf{p}_1^2+\mathbf{p}_2^2)+a_3(\Lambda)\mathbf{p}_1\cdot\mathbf{p}_2 \quad +finite \tag{254}$$

The coefficient a_1 is quadratic, a_2 and a_3 are logarithmic in Λ. As Γ_{2,φ^2} also has a logarithmic divergence, it is possible to construct a linear combination in φ^2 and φ^4 which eliminates the singular part of a_1. Similarly, using the operators $\varphi(\Delta\varphi)$ and $(\partial\varphi)^2$, the divergences of a_2 and a_3 can also be suppressed. The idea is thus to form linear combinations of operators with equal or lower dimensions in order to obtain finite renormalized quantities. We have also to give precise normalization conditions. For the example of operators of degree 0, the three possible candidates are

$$\mathcal{O}_1 = \varphi^4 \qquad\qquad \mathcal{O}_2 = \varphi(\Delta\varphi) \qquad\qquad \mathcal{O}_3 = (\partial\varphi)^2$$

Their insertion leads to renormalized vertex functions related to the bare ones through

$$\Gamma_{n,\mathcal{O}_i}^{R} = Z_1^{\frac{1}{2}n}\left[\sum_j z_{ij}\Gamma_{n,\mathcal{O}_j} + z_i\Gamma_{n,\frac{1}{2}\varphi^2}\right] \tag{255}$$

in which twelve new *a priori* divergent constants z_{ij} and z_i have been introduced. Hence twelve conditions are necessary, as e.g.

$$\Gamma_{4,\varphi^4}^{R}(0) = 4! \qquad\qquad \Gamma_{2,\varphi^4}^{R}(\mathbf{p}) = \mathcal{O}(\mathbf{p}^4)$$
$$\Gamma_{4,\varphi(\Delta\varphi)}^{R}(0) = 0 \qquad\qquad \Gamma_{2,\varphi(\Delta\varphi)}^{R}(\mathbf{p}) = \mathbf{p}_1^2 + \mathbf{p}_2^2 + \mathcal{O}(\mathbf{p}^4) \tag{256}$$
$$\Gamma_{4,(\partial\varphi)^2}^{R}(0) = 0 \qquad\qquad \Gamma_{2,(\partial\varphi)^2}^{R}(\mathbf{p}) = 2\mathbf{p}_1\cdot\mathbf{p}_2 + \mathcal{O}(\mathbf{p}^4)$$

Note that each function $\Gamma_{2,\mathcal{O}_i}^{R}$ satisfies three conditions at $\mathbf{p} = \mathbf{0}$.

This example can now be generalized. The insertion of a renormalized operator is equivalent, in the bare theory, to the insertion of a linear combination of all operators of equal or lower dimension. Their number might be reduced if one takes into account symmetries or relations arising from the equations of motion. Indeed, if the regularized theory admits some symmetry, it is clear that we can only combine quantities with the same behaviour. Note also that the Green functions obey relations which result from the structure of the Lagrangian (as a consequence of the so-called *equations of motion*).

To first order, the introduction in the action of additional local terms requires that one estimates the corresponding insertions at zero momentum in the correlation functions. The renormalized operator $\mathcal{O}_i$ is coupled to other bare $\mathcal{O}_j$'s and in particular to $\frac{1}{2}\varphi^2$ which is of degree -2. Therefore, its insertion displaces the critical temperature. This effect can be compensated by subtracting from $\mathcal{O}_i$ a term proportional to $\frac{1}{2}\varphi^2$,

$$\hat{O}_i = \mathcal{O}_i - \tfrac{1}{2}a_i\varphi^2 \qquad (257)$$

in which a_i is to be chosen in such a way that the expansion of the renormalized operator does not contain any contribution of $\frac{1}{2}\varphi^2$.

A definite advantage of this procedure is to suppress additional infrared divergences which would correspond to a perturbative shift of the critical temperature. Furthermore, it is convenient to include in $\mathcal{O}_i$ and $\hat{O}_i$ the dimensionful factor $\Lambda^{-\delta_i}$ which ensures that the corresponding coupling is dimensionless. As we have seen, this modification also affects the coupling constant g_0 (coupled to φ^4) for $d < 4$.

Using these conventions, the zero momentum insertions in the critical theory read

$$\Gamma^R_{n,\hat{O}_i}(\mathbf{p}; g, \mu) = Z_1^{n/2}(g_0, \Lambda/\mu) \sum_j (\Lambda/\mu)^{\delta_j} z_{ij}(g_0, \Lambda/\mu)\Gamma_{n,\hat{O}_j}(\mathbf{p}; g_0, \Lambda)$$

$$(258)$$

where the sum on j runs over all operators of degree less than or equal to δ_i except $\frac{1}{2}\varphi^2$. In dimension four, the quantities z_{ij} only contain logarithmic divergences.

The corresponding Callan–Symanzik equations for the bare vertex functions are

$$\left\{ \left[\Lambda\frac{\partial}{\partial\Lambda} + \beta(g_0)\frac{\partial}{\partial g_0} - \tfrac{1}{2}n\gamma_1(g_0) \right] \delta_{ij} - \gamma_{ij}(g_0) \right\} \Gamma_{n,\hat{O}_i}(\mathbf{p}; g_0, \Lambda) = 0$$

$$\gamma_{ij}(g_0) = \{z^{-1}\}_{ik} \left[\delta_k + \Lambda\frac{\partial}{\partial\Lambda} \right] z_{kj} \qquad (259)$$

For small momenta, the integration of this equation generates additional powers of $\mathbf{p}/\Lambda$ with respect to functions without insertions.

The matrix γ_{ij} has a block triangular structure. Let η_a be its eigenvalues. One finds as a result of equation (259) that

$$\Gamma_{n,\hat{O}_i}(\lambda\mathbf{p}; g_0, \Lambda) \underset{\lambda\to 0}{\sim} \lambda^{d-\frac{1}{2}n(d-2+\eta)} \sum_a C_{ia}\lambda^{\eta_a} \qquad (260)$$

According to equation (252), the eigenvalues behave as

$$\eta_a = \delta_a + \mathcal{O}(4-d) \tag{261}$$

The block triangular structure of z_{ij} shows that, when the degree δ_i of the operator increases, one obtains for its eigenvalues the previous values η_a corresponding to lower degrees together with new eigenvalues, close to the degree of the inserted operator. As we have compensated $\frac{1}{2}\varphi^2$, the lowest eigenvalues, which are the dominant ones in (260), correspond to the operators φ^4 and $(\partial\varphi)^2$ (note that, at zero momentum, the two operators $\varphi(-\triangle\varphi)$ and $(\partial\varphi)^2$ are equivalent). These operators are marginal in four dimensions. Their effect is to change the coupling constant and to renormalize the field. The corresponding eigenvalue is precisely ω (see (231)) and the correction behaves as $(|\mathbf{p}|/\Lambda)^\omega$.

To summarize, the effect of higher local terms in the Lagrangian is to correct the scaling laws by subdominant corrections. The leading one corresponds to the exponent ω which can be directly computed within the framework of the φ^4 theory. Hence, the truncation of the Lagrangian is *a posteriori* justified.

5.5   Numerical results

It is not possible to collect here the many perturbative results about critical quantities, nor their detailed derivation. The interested reader can find a partial list of references at the end of this chapter. We only reproduce here some series. Their numerical treatment is essential from the practical point of view. Important progress was made when it was realized that the series, although divergent, are nevertheless asymptotic, and that their behaviour at large order can be analyzed using saddle point methods. The initial idea is due to Dyson in the context of electrodynamics. The large order behaviour of perturbative series in field theory has been studied extensively by Lipatov and others. This information allows one to make the best use of rather short series. The stability of the estimates (i.e. the theoretical uncertainties) are discussed in the original papers.

We first give some results on the ε-expansion. Then we also present some series in fixed dimension three.

5.5.1 *ε- expansion of critical exponents*

The computation scheme described in section 3 has been developed and applied by Wilson, Brézin, Le Guillou, Zinn-Justin, Nickel and others to compute the following series

$$\eta = \gamma_1(g_c) = \frac{n+2}{2(n+8)^2}\varepsilon^2 + \frac{n+2}{8(n+8)^4}[24(3n+14) - (n+8)^2]\varepsilon^3$$

$$+ \frac{n+2}{2(n+8)^6}\left[-\frac{5}{16}n^4 - \frac{115}{8}n^3 + \frac{281}{4}n^2 + 1120n\right.$$

$$+ \quad 2884 - 24(n+8)(5n+22)\zeta(3)\Bigg] \varepsilon^4 + \mathcal{O}(\varepsilon^5) \qquad (262)$$

$$2 - \nu^{-1} = \gamma_2(g_c) = \frac{n+2}{n+8}\varepsilon + \frac{n+2}{2(n+8)^3}(13n+44)\varepsilon^2$$

$$+ \frac{n+2}{(n+8)^4}\left[36\frac{(3n+14)(n+3)}{n+8}\right.$$

$$- \frac{3n^2 + 388n + 848}{8} - 12(5n+22)\zeta(3)\Bigg] \varepsilon^3 + \mathcal{O}(\varepsilon^4)$$

$$(263)$$

with

$$\zeta(3) = \sum_{j=1}^{\infty}\frac{1}{j^3} \quad = 1.20206... \qquad (264)$$

From the relation $\gamma = \nu(2 - \eta)$, one gets

$$\gamma = 1 + \frac{n+2}{2(n+8)}\varepsilon + \frac{(n+2)(n^2 + 22n + 52)}{4(n+8)^3}\varepsilon^2$$

$$+ \frac{n+2}{(n+8)^5}\left[\frac{n^4}{8} + \frac{11n^3}{3} + 83n^2 + 312n\right. \qquad (265)$$

$$+ \quad 388 - 6(n+8)(5n+22)\zeta(3)]\,\varepsilon^3 + \mathcal{O}(\varepsilon^4)$$

The exponent ω governing the corrections to the dominant scaling behaviour, defined by equation (231), is given by

$$\omega = \beta'(g_c) = \varepsilon - \frac{3(3n+14)}{(n+8)^2}\varepsilon^2 + \frac{1}{(n+8)^3}\left[\frac{33n^2}{4} + \frac{461n}{2}\right.$$

$$+ 740 + 24(5n+22)\zeta(3) - 18\frac{(3n+14)^2}{n+8}\Bigg]\varepsilon^3 + \mathcal{O}(\varepsilon^4)$$

$$(266)$$

Table I. Critical exponents of the three-dimensional Ising model

$\mathcal{O}(\varepsilon^k)$		Padé–Borel	High temp. series	$d = 3$ renorm. group
η	0.019 0.037 0.029	0.04^a 0.0333 ± 0.0001^b	0.041 ± 0.01	0.031 ± 0.001
γ	1.167 1.244 1.19	1.242^a 1.235 ± 0.004^b	1.250 ± 0.003^c 1.245 ± 0.003^d	1.241 ± 0.002
ν	0.589 0.634 0.606	0.632^a 0.628 ± 0.02^b	0.638 ± 0.002^c	0.630 ± 0.0015

Note: The values listed $\mathcal{O}(\varepsilon)$ are for $k = 1, 2, 3$ and should be understood $\mathcal{O}(\varepsilon^{+1})$ for the exponent η.

[a] J. Zinn-Justin and J.-C. Le Guillou, *Phys. Rev.* **B21** 3976 (1980).
[b] G. Parisi, *J. Stat. Phys.* **23**, 49 (1980).
[c] C. Domb, *Phase Transitions and Critical Phenomena*, vol. 3, C. Domb and M.S. Green eds (Academic Press, London).
[d] J. Zinn-Justin, *J. Physique* **40** 969 (1979).

If the corrections are expressed in terms of the deviation θ to the critical temperature, the corresponding exponent is (cf. (232))

$$\omega\nu = \tfrac{1}{2}\varepsilon + \frac{n^2 - 8n - 68}{4(n+8)^2}\varepsilon^2 + \frac{1}{8(n+8)^3}\left[(n+2)(n+3)(n+20)\right.$$
$$- 72\frac{(3n+14)^2}{n+8} + 15n^2 + 802n + 2792$$
$$\left. + 96(5n+22)\zeta(3)\right]\varepsilon^3 + \mathcal{O}(\varepsilon^4) \tag{267}$$

The most naive interpretation is to insert the corresponding value of ε and to evaluate the successive contributions. Table I displays the successive approximations for the Ising model $n = 1$ in three dimensions. At second order, the values are in good agreement with the results from high temperature expansions, but the disagreement increases at the next order. The fourth column displays more accurate values based on a Padé–Borel resummation method, which uses large order estimates. Finally, the last two columns show the results of high temperature expansions and similar results of the renormalization group analysis obtained in dimension three.

5.5.2 *Equation of state*

The equation of state can also be obtained in an ε-expansion. Let us compute the effective potential $V(\mathbf{M})$ for an n-vector model as the generating functional of irreducible diagrams. Up to one-loop order,

$$V(\mathbf{M}) = \tfrac{1}{2}\theta\mathbf{M}^2 + \frac{g(\mathbf{M}^2)^2}{4!}$$

$$+ \tfrac{1}{2}\int \frac{d^d\mathbf{p}}{(2\pi)^d}\,\mathrm{Tr}\ln[(\mathbf{p}^2 + \theta)\delta_{ij} + \tfrac{1}{6}g(\mathbf{M}^2\delta_{ij} + 2M_iM_j)]$$

$$- \text{subtractions}$$

(268)

The trace is relative to internal indices. Subtractions contain an $\mathbf{M}$-independent polynomial $P_2(\theta)$ of degree 2 in θ, which corresponds to the divergences of $\Gamma_{0,0}$, $\Gamma_{0,1}$, $\Gamma_{0,2}$. Further subtractions correspond to the renormalization of $\Gamma_{2,0}$, $\Gamma_{4,0}$, $\Gamma_{2,1}$ in the massless theory. In particular, the functions $\Gamma_{4,0}$ and $\Gamma_{2,1}$ are normalized at the symmetric point $\mathbf{p}_i \cdot \mathbf{p}_j = -\mathbf{p}_i^2/3$ and we choose $\mu^2 = 4\mathbf{p}_i^2/3$ as a mass scale. Hence

$$V(\mathbf{M}) = \tfrac{1}{2}\theta\mathbf{M}^2 + g\frac{(\mathbf{M}^2)^2}{4!}$$

$$+ \tfrac{1}{2}\int \frac{d^d\mathbf{p}}{(2\pi)^d}\,\mathrm{Tr}\left\{\ln\left[\delta_{ij} + \frac{1}{\mathbf{p}^2}(\theta + \tfrac{1}{6}g\mathbf{M}^2)\delta_{ij} + \frac{g}{3\mathbf{p}^2}M_iM_j\right]\right.$$

$$- \frac{1}{\mathbf{p}^2}(\theta + \tfrac{1}{6}g\mathbf{M}^2)\delta_{ij} - \frac{g}{\mathbf{p}^2}M_iM_j$$

$$+ \frac{1}{2\mathbf{p}^2(\mathbf{p}+\mathbf{p}_i+\mathbf{p}_j)^2}\left[(\theta + \tfrac{1}{6}g\mathbf{M}^2)\delta_{ij} + \tfrac{2}{3}g(\theta + \tfrac{1}{9}g\mathbf{M}^2)M_iM_j\right]\right\}$$

$$+ P_2(\theta)$$

(269)

The transverse and longitudinal parts are treated independently. Taking into account the quartic subtractions which avoid infrared divergences, one gets as $d \to 4$

$$V = \tfrac{1}{2}\theta\mathbf{M}^2 + g\frac{(\mathbf{M}^2)^2}{4!} +$$

$$\frac{1}{8(4\pi)^2}\left\{(\theta + \tfrac{1}{2}g\mathbf{M}^2)^2[2\ln(\theta + \tfrac{1}{2}g\mathbf{M}^2) + 1]\right.$$

$$\left. + (n-1)(\theta + \tfrac{1}{6}g\mathbf{M}^2)[2\ln(\theta + \tfrac{1}{6}g\mathbf{M}^2) + 1]\right\} + P_2(\theta)$$

(270)

We substitute for g its critical value (118),

$$g_c = (4\pi)^2 \frac{3}{n+8}\varepsilon + \cdots \tag{271}$$

and get

$$H = \frac{\partial V}{\partial M} = M\left\{\theta + \tfrac{1}{6}y + \frac{3\varepsilon}{2(n+8)}\left[(\theta + \tfrac{1}{2}y)\ln(\theta + \tfrac{1}{2}y)\right.\right.$$
$$\left.\left. + 1 + \tfrac{1}{3}(n-1)(\theta + \tfrac{1}{6}y)(\ln(\theta + \tfrac{1}{6}y) + 1)\right] + \cdots\right\} \tag{272}$$

with $y = g_c M^2$. This expression can be identified with the expansion of

$$H = M^\delta f\left(\frac{\theta}{M^{1/\beta}}\right) \tag{273}$$

To order ε,

$$\eta = 0 + \cdots$$
$$\nu = \tfrac{1}{2} + \frac{\varepsilon}{4}\frac{n+2}{n+8} + \cdots$$
$$\beta = \tfrac{1}{2}\nu(d-2+\eta) = \tfrac{1}{2} - \frac{3\varepsilon}{2(n+8)} \tag{274}$$
$$\delta = \frac{d+2-\eta}{d-2+\eta} = 3 + \varepsilon + \cdots$$

If we insert these values in (273) and compare with (272), changing the scale of M and θ so that $f(0) = 1$ and $f(-1) = 0$, we get

$$H = M^\delta\left(1 + x + \frac{\varepsilon}{2(n+8)}[(n-1)(x+1)\ln(x+1)\right.$$
$$\left. + 3(x+3)\ln(x+3) + 6x\ln 2 - 9(x+1)\ln 3] + \cdots\right) \tag{275}$$

with

$$x = \frac{\theta}{M^{1/\beta}}$$

which gives the function $f(x)$ to order ε. Further computations at higher orders can be found in the references.

5.5.3   Amplitude ratios

It is also possible to compute other universal quantities describing the critical behaviour which are independent of the arbitrary scale of the temperature and the external field. For instance, consider the specific heat $A_\pm(\pm\theta)^{-\alpha}$ above and below the critical temperature. The ratio of amplitudes A_+/A_- is such a universal quantity. This ratio is also the one of singular parts in the free energy in the absence of an external field. These quantities are contained in the expression of the function $f(x)$ in the equation of state (275).

Taking into account that $2-\alpha = \beta(1+\delta)$, the effective potential is of the form

$$V(M,\theta) = \theta^{2-\alpha}\varphi(x) \qquad\qquad x = \frac{\theta}{M^{1/\beta}} \qquad (276)$$

The free energy is a regular function of θ in presence of a field, i.e. $x^{2-\alpha}\varphi(x)$ is analytic at the origin. As

$$\varphi'(x) = -x^{\alpha-3}f(x) \qquad (277)$$

with the same function f as occured in equations (273) and (275), an integration is sufficient to determine φ. We take a convention of a phase $e^{i\pi}$ for x negative, and determine A_+/A_- as the ratio of the limits of V as $M \to 0$ for $\theta > 0$ and as $M \to M_{\mathrm{sp}} = (-\theta)^\beta$ for $\theta < 0$. Hence

$$\frac{A_+}{A_-} = \frac{\varphi(+\infty)}{e^{i\pi(2-\alpha)}\varphi(-1)} \qquad (278)$$

Integrating equation (277) assuming $0 < \alpha < 1$, one obtains

$$\varphi(x) = \frac{\beta}{\alpha-2}\left\{\int_0^x dy\, y^{\alpha-2}[f'(y) - f'(0)] + f'(0)\frac{x^{\alpha-1}}{\alpha-1} - x^{\alpha-2}f(x)\right\}$$
$$0 < \alpha < 1 \qquad (279)$$

For an alternative derivation, we may return to equation (269) and substitute g_c as well as the corresponding scales for M and θ. In the range $0 < \alpha < 1$, we obtain

$$\frac{A_+}{A_-} = \frac{\int_0^\infty dx\, x^{\alpha-2}[f'(x) - f'(0)]}{\int_{-1}^0 dx(-x)^{\alpha-2}[f'(0) - f'(x)] + f'(0)/(1-\alpha)}$$
$$0 < \alpha < 1 \qquad (280)$$

Similar expressions exist if α lies in other intervals.

Using the result (275) as well as $\alpha = \frac{1}{2}\varepsilon(4-n)/(n+8) + \cdots$, one finds that the leading term in the ratio A_+/A_- is

$$\frac{A_+}{A_-} = \frac{1}{4}n + \mathcal{O}(\varepsilon) \tag{281}$$

An elaborate computation up to order ε^2 yields

$$\frac{A_+}{A_-} = \frac{1}{4}n2^\alpha \left[1 + \varepsilon + \frac{\varepsilon^2}{2(n+8)^2}[3n^2 + 26n + 100 \tag{282}\right.$$
$$\left. + \zeta(2)(4-n)(n-1) - 6\zeta(3)(5n+22) + \frac{9}{2}(4-n)I] + \mathcal{O}(\varepsilon^3)]$$

with

$$I = \int_0^1 dx \frac{\ln x(1-x)}{1 - x(1-x)} = -2.349...$$

In the mean field approximation, the specific heat is discontinuous at the critical temperature and the above ratio vanishes. As $\varepsilon \to 0$, that is in four dimensions, the ratio is relative to the coefficient of the singularity in $(\ln|\theta|)^{(4-n)/(n+8)}$ and its value is $\frac{1}{4}n$. In three dimensions, taking $\nu = 0.63$ and therefore $\alpha = 0.11$ for $n = 1$, equation (282) predicts the value

$$d = 3 \qquad n = 1 \qquad \frac{A_+}{A_-} = 0.54 \tag{283}$$

For the Ising model, the susceptibility χ and the correlation length are finite below the critical temperature, since there are no zero modes. It is then possible to define a similar ratio χ_+/χ_- for the susceptibility, written as

$$\chi(\theta) = \chi_\pm |\theta|^{-\gamma} \tag{284}$$

From the equation of state, one derives

$$\frac{\chi_+}{\chi_-} = \lim_{x \to \infty} f'(-1) \frac{x^\gamma}{\beta\delta f(x) - x f'(x)} \tag{285}$$

which is meaningful only if $f(x)/x^\gamma$ behaves as a power series in $x^{-2\beta} = M^2/\theta^{2\beta}$ for large x. Using

$$\beta\delta = \frac{3}{2} + \cdots$$
$$\gamma = 1 + \frac{1}{6}\varepsilon + \cdots \tag{286}$$

and the expression (275) for $f(x)$, one obtains

$$\frac{\chi_+}{\chi_-} = 2 + \varepsilon \left(1 + \tfrac{1}{3}\ln 2\right) + \mathcal{O}(\varepsilon^2) \qquad (287)$$

which can be rewritten as

$$\frac{\chi_+}{\chi_-} = 2^{\gamma-1}\frac{\gamma}{\beta} + \mathcal{O}(\varepsilon^2) \qquad (288)$$

The mean field value 2 is obtained in the limit $\varepsilon \to 0$.

5.5.4   Three-dimensional results

The direct treatment of the critical theory in fixed dimension less than four yields more accurate results, although no small parameter analogous to ε is available. We have discussed above the computational scheme. Let us display the results obtained by Baker, Nickel, Green and Meiron for $d = 3$ and $n = 1$. They use a coupling constant normalized such that the first two terms of the β function have coefficients -1 and 1. This means

$$\bar{g} = \frac{3g}{16\pi} \qquad \bar{\beta}(\bar{g}) = \frac{3}{16\pi}\beta(g)$$

The series are

$$\bar{\beta}(\bar{g}) = -\bar{g} + \bar{g}^2 - 0.4224965707\bar{g}^3 + 0.3510695978\bar{g}^4 \qquad (289)$$
$$- 0.3765268283\bar{g}^5 + 0.49554751\bar{g}^6 - 0.749689\bar{g}^7 + \cdots$$

$$\gamma_1(\bar{g}) = 0.0109739369\bar{g}^2 + 0.0009142223\bar{g}^3 \qquad (290)$$
$$+ 0.0017962229\bar{g}^4 - 0.00065370\bar{g}^5 + 0.0013878\bar{g}^6 + \cdots$$

$$\gamma_2(\bar{g}) = \tfrac{1}{3}\bar{g} - 0.0631001372\bar{g}^2 + 0.0452244754\bar{g}^3 \qquad (291)$$
$$- 0.0377233460\bar{g}^4 + 0.04374666\bar{g}^5 - 0.0589754\bar{g}^6 + \cdots$$

This extends considerably the meager expressions given in (115)–(118). We do not elaborate here on the methods needed to extract the information contained in these series, in particular to find $\bar{g}_{\mathrm{c}}$, and which also make an essential use of the asymptotic behaviour. Table II displays the results obtained for the critical exponents.

Appendix 5.A   Multicritical points

We have presented a detailed study of renormalization applied to the standard φ^4 critical point. It was observed that irrelevant

Table II. Critical exponents of n-vector models in dimension three

	$n = 0$	$n = 1$	$n = 2$	$n = 3$
$\bar{g}_c$	1.421 ± 0.004	1.416 ± 0.0015	1.406 ± 0.005	1.392 ± 0.009
ω	0.794 ± 0.006	0.788 ± 0.003	0.78 ± 0.01	0.78 ± 0.02
$\eta + \nu^{-1} - 2$	-0.274 ± 0.01	-0.382 ± 0.005	-0.474 ± 0.008	-0.550 ± 0.012
ν	0.588 ± 0.001	0.630 ± 0.002	0.669 ± 0.003	0.705 ± 0.005
γ	1.161 ± 0.003	1.241 ± 0.004	1.316 ± 0.009	1.39 ± 0.01
α	0.236 ± 0.004	0.110 ± 0.008	-0.007 ± 0.009	-0.115 ± 0.015
β	0.302 ± 0.004	0.324 ± 0.006	0.346 ± 0.009	0.362 ± 0.012
$\frac{1}{2}(3\nu + \gamma)$	1.462 ± 0.004	1.566 ± 0.006	1.662 ± 0.009	1.753 ± 0.012
η	0.026 ± 0.014	0.031 ± 0.011	0.032 ± 0.015	0.031 ± 0.022
δ	4.85 ± 0.08	4.82 ± 0.06	4.81 ± 0.08	4.82 ± 0.12

Note: From Baker, Nickel, Green and Meiron, *Phys. Rev. Lett.* **36**, 1351 (1976), *Phys. Rev.* **B17**, 1365 (1978).

operators did not affect the leading scaling behaviour, giving a foundation to the property of universality. This holds close to the upper marginal dimension for two effective macroscopic control parameters (such as the external field and the temperature). There are circumstances when this is no longer the case, as exemplified by the tricritical singularity, occurring for instance in $He^3 - He^4$ mixtures or for antiferromagnets in the presence of a strong external field (metamagnets). Here one deals with a three-dimensional phase diagram. We give as a complement a brief introduction to this subject, present some sample calculations, and quote in the notes relevant work giving both a more complete physical picture as well as more details on renormalization.

For simplicity assume a symmetric situation with a one component field φ (i.e. no explicit symmetry breaking term is present). To discuss a tricritical singularity we retain in a mean field approach terms up to order φ^6 in a Landau Lagrangian. Neglecting fluctuations, the free energy per unit volume is expressed as a potential term

$$V(\varphi) = \tfrac{1}{2}t\varphi^2 + \tfrac{1}{4}s\varphi^4 + \tfrac{1}{6}g\varphi^6 \qquad (A.1)$$

One requires g to be positive to insure that $V(\varphi)$ is bounded from below. In the $s-t$ plane, let us plot the various possible situations following the minima of $V(\varphi)$ which should satisfy

$$\varphi\{t + s\varphi^2 + g\varphi^4\} = 0 \qquad (A.2)$$

The origin is always an extremum but there exist other possibilities if $s^2 - 4tg > 0$ (dotted curve on figure 7, which will be the boundary of a metastable zone).

If t is negative, the absolute minimum is for

$$\varphi_+^2 = \frac{-s + \sqrt{s^2 - 4tg}}{2g}$$

in which case

$$V(\varphi_+) - V(0) = \tfrac{1}{3}\varphi_+^2 \left[t + s\tfrac{1}{4}\varphi_+^2\right] = \tfrac{1}{12}\varphi_+^2 \left[3t - g\varphi_+^4\right]$$

is negative, indicating a broken symmetry with a spontaneous magnetization φ_+.

If both s and t are positive, the only real root of $(A.2)$ is $\varphi = 0$ and, as one crosses the positive s axis, one has a typical second order mean field transition controlled by t and the term $g\varphi^6$ is

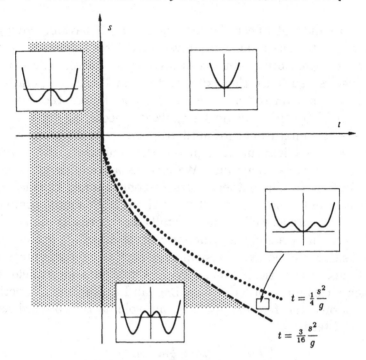

Fig. 7 Phase diagram for a tricritical point. Heavy line: second order transition; broken line: first order transition; dotted curve: limit of a metastability region. At the tricritical point, the two transition lines merge smoothly. In each region a typical behaviour of the potential is shown. The region to the left of the transition lines is the ordered one.

irrelevant. If however t is positive but s negative and $4tg < s^2$, one has two extra extrema $\varphi_\pm$. The sign of $V(\varphi_\pm) - V(0)$ is the sign of

$$8gt + s\left(-s \pm \sqrt{s^2 - 4gt}\right) = \pm s\sqrt{s^2 - 4gt} - (s^2 - 8gt)$$

Obviously $V(\varphi_-) - V(0)$ is positive but $V(\varphi_+) - V(0)$ is negative i.e. φ_+ represents a spontaneous magnetization only up to the curve $s\sqrt{s^2 - 4gt} = s^2 - 8gt$, i.e. the arc of parabola

$$s^2 = \tfrac{16}{3}gt \qquad\qquad (A.3)$$

its negative s-branch describing a line of first order transitions.

Thus, as shown on figure 7, a line of second order transitions merges smoothly at the tricritical point ($s = t = 0$) into a line of first order transitions. The discontinuous jump in magnetization along the first order line vanishes linearly as $\varphi_+ = \frac{3}{4}(-s)/g$ if one uses s (rather than t) as a parameter along the curve. Mean field exponents are readily defined in the vicinity of the tricritical point.

Such a mechanism can be generalized to multicritical points of order k with a multidimensional phase space $((k-1)$ dimensions in the absence of a symmetry breaking field). Physical examples are not so readily available but this extension is interesting in view of the two-dimensional discussion to be given in chapter 9 (volume 2). At multicriticality the effective potential reduces to

$$V_{\text{crit}} = g\frac{\varphi^{2k}}{(2k)!} \qquad (A.4)$$

where for convenience the coupling constant has been rescaled by a numerical factor $[(2k-1)!]^{-1}$. The upper critical dimension, such that g becomes dimensionless, is according to the familiar criterion

$$d_c = \frac{2k}{k-1} \qquad (A.5)$$

Thus $d_c = 4$ for $k = 2$, $d_c = 3$ for $k = 3$, the two important physical instances, while d_c tends to 2 as k grows indefinitely.

Let us sketch the renormalization of the corresponding massless theory in an ε-expansion such that $d = d_c - \varepsilon$. We have in mind logarithmic corrections for $d = k = 3$ (this is left as an exercise for the reader) or the approach to two dimensions in an ε expansion for $k \to \infty$, the rationale being that exact results are known in this case. It will be seen that this comparison is unfortunately not convincing, although it seems to offer an interesting testing ground for the concept of dimensional continuation.

The cutoff-dependent bare Lagrangian is written in terms of a renormalized field φ

$$\varphi^{\text{bare}} = Z^{1/2}\varphi \qquad (A.6)$$

and renormalized coupling g (the bare one is denoted g_0), as

$$\mathcal{L} = \tfrac{1}{2}(\partial\varphi)^2 + \frac{g}{(2k)!}\mu^{\varepsilon(k-1)}\varphi^{2k} + (Z-1)\tfrac{1}{2}\partial\varphi^2$$

$$+ \frac{g}{(2k)!}\mu^{\varepsilon(k-1)}(Z^k Z_g - 1)\varphi^{2k} \qquad (A.7)$$

$$+ \,[\text{counterterms for } \varphi^2, \ldots, \varphi^{2k-2}]$$

The extra counterterms are designed to cancel the divergent parts in the irreducible functions $\Gamma^{(2)}$, ..., $\Gamma^{(2k-2)}$ normalized to zero at zero momentum in Fourier transformed space. The logarithmic divergent function $\Gamma^{(2k)}$ (at $\varepsilon = 0$) is normalized at a symmetric point scaled by μ, thus defining $Z^k Z_g$. More precisely the normalization conditions read

$$\Gamma^{(2)}(0) = \cdots = \Gamma^{(2k-2)}(0) = 0$$

$$\left.\frac{\partial}{\partial p^2}\Gamma^{(2)}(p, -p)\right|_{p^2=\mu^2} = 1 \qquad (A.8)$$

$$\Gamma^{(2k)}(p_1, \ldots, p_{2k})_{S_\mu} = g\mu^{\varepsilon(k-1)}$$

The symmetric point S_μ for $2k$ momenta satisfying $\sum_1^{2k} p_j = 0$ is chosen in such a way that $p_j^2 = \mu^2(2k - 1)/k^2$ and for $i \neq j$ $p_i \cdot p_j = -\mu^2/k^2$, with the result that the sum of any k distinct momenta has a square equal to μ^2. The factor $\mu^{\varepsilon(k-1)}$ insures that the renormalized coupling constant g is dimensionless. The bare functions

$$\Gamma^{(2\ell)}_{\text{bare}} = Z^{-\ell}\Gamma^{(2\ell)} \qquad (A.9)$$

are of course μ -independent.

The irreducible functions $\Gamma^{(2)}$ and $\Gamma^{(2k)}$ are the only ones required to obtain the renormalization factors Z and Z_g. To lowest order, as the reader will check, the only logarithmic singularities (at $\varepsilon = 0$) are accounted for by the graphs depicted in figure 8 (dimensional regularization suppresses in particular tadpole insertions). Explicitly

$$\Gamma^{(2)}(p, -p) = Zp^2 - \qquad (A.10)$$

$$- \frac{g^2}{(2k-1)!}\mu^{2\varepsilon(k-1)}\int \prod_1^{2k-2} \frac{d^d q_j}{(2\pi)^d} \frac{1}{\prod_j q_j^2 \left(p - \sum_j q_j\right)^2}$$

$$\Gamma^{(2k)}(\{p\}) = g\mu^{\varepsilon(k-1)}Z_g Z^k - \qquad (A.11)$$

$$- \frac{g^2}{k!}\mu^{2\varepsilon(k-1)}\sum\int \prod_1^{k-1} \frac{d^d q_j}{(2\pi)^d} \frac{1}{\prod_j q_j^2 \left(\sum \tilde{p} - \sum q_j\right)^2}$$

On the right-hand side in equation $(A.11)$ the sum runs over the $(2k)!/2(k!)^2$ ways of splitting the $2k$ external momenta into two groups of k momenta as $\{p\} = \{\tilde{p}\} \cup \{\tilde{\tilde{p}}\}$.

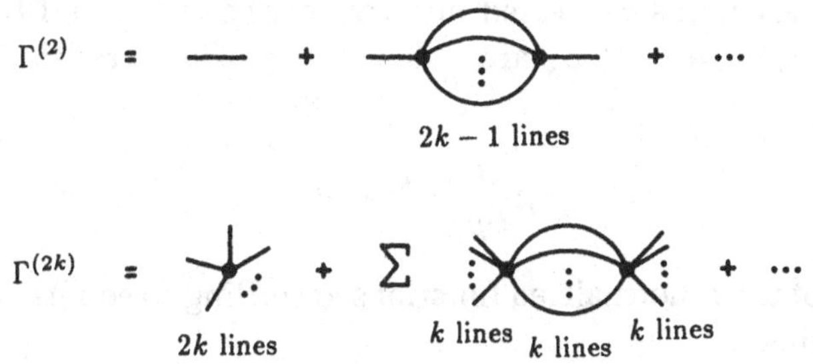

Fig. 8   Diagrams contributing to the leading singular behaviour in the renormalization factors Z and Z_g.

We use dimensional regularization in the form

$$\int \frac{d^d q}{(2\pi)^d} \frac{1}{[q^2]^\alpha} \frac{1}{[(p-q)^2]^\beta} = \frac{1}{(4\pi)^{d/2}} \frac{\Gamma(\tfrac{1}{2}d - \alpha)\Gamma(\tfrac{1}{2}d - \beta)}{\Gamma(d - \alpha - \beta)}$$
$$\times \frac{\Gamma(\alpha + \beta - \tfrac{1}{2}d)}{\Gamma(\alpha)\Gamma(\beta)} \frac{1}{(p^2)^{\alpha+\beta-d/2}}$$
$$(A.12)$$

and recursively

$$\int \prod_1^{m-1} \frac{d^d q_j}{(2\pi)^d} \frac{1}{\prod_j [q_j^2]^{\alpha_j} \left[(p - \sum q_j)\right]^{\alpha_m}} =$$
$$= \frac{1}{(4\pi)^{\frac{1}{2}(m-1)d}} \frac{\Gamma(\tfrac{1}{2}d - \alpha_1) \cdots \Gamma(\tfrac{1}{2}d - \alpha_m)}{\Gamma\left(m\tfrac{d}{2} - \alpha_1 - \cdots - \alpha_m\right)}$$
$$\times \frac{\Gamma(\alpha_1 + \cdots + \alpha_m - \tfrac{1}{2}(m-1)d)}{\Gamma(\alpha_1) \cdots \Gamma(\alpha_m)} \frac{1}{(p^2)^{\alpha_1 + \cdots + \alpha_m - \tfrac{1}{2}(m-1)d}}$$
$$(A.13)$$

In equation $(A.10)$ we need the case $m = 2k - 1$, in $(A.11)$ the case $m = k$ with the α's all equal to unity. Setting the derivative $\partial \Gamma^{(2)}/\partial p^2$ at $p^2 = \mu^2$ equal to unity yields in the limit $\varepsilon \to 0$

$$Z = 1 - \frac{g^2}{(4\pi)^{2k}} \frac{1}{(2k-1)!} \frac{\Gamma\left(1/(k-1)\right)^{2k-1}}{\Gamma\left((2k-1)/(k-1)\right)} \frac{1}{(k-1)\varepsilon} + O(g^3)$$
$$(A.14)$$

Similarly from the normalization $(A.8)$ of the $2k$-point function

$$Z_g = 1 + \frac{g}{(4\pi)^k} \frac{(2k)!}{(k!)^3} \frac{\Gamma\left(1/(k-1)\right)^k}{\Gamma\left(k/(k-1)\right)} \frac{1}{(k-1)\varepsilon} + O(g^2) \quad (A.15)$$

The bare functions expressed in terms of the cutoff and the bare coupling $g_0 = g\mu^{\varepsilon(k-1)}Z_g$ are independent of μ. Translating the equations

$$\mu\frac{\partial}{\partial\mu}\Gamma^{(2k)}_{\text{bare}} = 0 \qquad (A.16)$$

in terms of the renormalized functions according to equation $(A.9)$, and defining

$$\beta(g) = \mu\left.\frac{\partial}{\partial\mu}g\right|_{g_0,\Lambda}$$
$$\gamma(g) = \mu\left.\frac{\partial}{\partial\mu}\ln Z\right|_{g_0,\Lambda} \qquad (A.17)$$

one finds the Callan–Symanzik equations in the form

$$\left[\mu\frac{\partial}{\partial\mu} + \beta(g)\frac{\partial}{\partial g} - \ell\gamma(g)\right]\Gamma^{(2\ell)} = 0 \qquad (A.18)$$

with

$$\beta(g) = -\varepsilon(k-1)g + \frac{g^2}{(4\pi)^{2k}}\frac{(2k)!}{(k!)^3}\frac{\Gamma\left(1/(k-1)\right)^k}{\Gamma\left(k/(k-1)\right)} + O(g^3)$$
$$\gamma(g) = 2\frac{g^2}{(4\pi)^{2k}}\frac{1}{(2k-1)!}\frac{\Gamma\left(1/(k-1)\right)^{2k-1}}{\Gamma\left((2k-1)/(k-1)\right)} + O(g^3)$$

$$(A.19)$$

The β-function exhibits an infrared stable fixed point g_c of order ε. Upon integration of the renormalization flow equations, one obtains

$$\Gamma^{(2\ell)}\left(\{p\}, g_2, \mu\exp\int_{g_1}^{g_2}\frac{du}{\beta(u)}\right)$$
$$= \exp\left\{\ell\int_{g_1}^{g_2}\frac{\gamma(u)du}{\beta(u)}\right\}\Gamma^{(2\ell)}\left(\{p\}, g_1, \mu\right) \quad (A.20)$$

Combining with a rescaling of p and μ, this reads

$$\Gamma^{(2\ell)}\left(\{\lambda p\}, g_2, \lambda\mu\exp\int_{g_1}^{g_2}\frac{du}{\beta(u)}\right) =$$
$$= \lambda^{d-\ell(d-2)}\exp\left\{\ell\int_{g_1}^{g_2}\frac{\gamma(u)du}{\beta(u)}\right\}\Gamma^{(2\ell)}\left(\{p\}, g_1, \mu\right)$$

$$(A.21)$$

Choose g_1 as a function of g_2 and λ through

$$\lambda \exp \int_{g_2}^{g_1} \frac{du}{\beta(u)} = 1 \qquad (A.22)$$

and let $\lambda \to 0$, which drives g_1 to the fixed point g_c. We get therefore the scaling laws

$$\lim_{\lambda \to 0} \Gamma^{(2\ell)} (\{\lambda p\}, g_2, \mu) = \lambda^{d - \ell[d - 2 + \gamma(g_c)]} \Gamma (\{p\}, g_c, \mu) \qquad (A.23)$$

The anomalous dimension of the field φ is therefore $\frac{1}{2}\gamma(g_c)$ and the exponent $\eta^{[k]}$ defined as in the φ^4 theory as giving the departure of the two-point correlation from a $r^{-(d-2)}$ behaviour (the appended index $[k]$ recalls the order of multicriticality) is given by

$$\eta^{[k]} = \gamma(g_c) \qquad (A.24)$$

To lowest order in ε

$$\frac{g_c}{(4\pi)^k} = \varepsilon(k-1) \frac{(k!)^3}{(2k)!} \frac{\Gamma(k/(k-1))}{\Gamma(1/(k-1))^k} + \cdots \qquad (A.25)$$

and

$$\eta^{[k]} = 2 \left[\frac{g_c}{(4\pi)^k} \right]^2 \frac{1}{(2k-1)!} \frac{\Gamma(1/(k-1))^{2k-1}}{\Gamma((2k-1)/(k-1))} + \cdots \qquad (A.26)$$

reads

$$\eta^{[k]} = 4 \left[\varepsilon(k-1) \right]^2 \left[\frac{k!^2}{(2k)!} \right]^3 + \cdots \qquad (A.27)$$

a result which agrees to order ε^2 with equation (262) in the case $k = 2$ when the number n of field components is set equal to one. In both cases, $\eta^{[2]} = \frac{1}{54}\varepsilon^2 + \cdots$. One could of course compute other exponents such as the anomalous dimensions of the fields φ^2, φ^3, ...

As will be shown in chapter 9 (volume 2), when $d = 2$, the anomalous dimension of the field φ for the tricritical model ($k = 3$) is believed to be 3/40 leading to $\eta^{[3]} = 3/20$. If we estimate the same exponent using the leading term only, given in $(A.27)$, taking $\varepsilon = 1$, $k = 3$, we find $\eta^{[3]}_{\text{app}} = \frac{1}{500}$. The disagreement by a factor 75 is quite striking, and increases exponentially with the order of multicriticality. It is suggested that in two dimensions

$$d = 2 \qquad \eta^{[k]} = \frac{3}{(k+1)(k+2)} \qquad (A.28)$$

while the estimate from equation $(A.27)$ obtained by taking $(k-1)\varepsilon = 2$, reads

$$d = 2 \qquad \eta^{[k]}_{\mathrm{app}} = 16 \left[\frac{k!^2}{(2k)!} \right]^3 \qquad (A.29)$$

The reason for this discrepancy may be understood from the fact that the true expansion parameter is not the small number ε, but $\varepsilon(k-1)$ which takes the finite value two (the dimension of the bare coupling) when one attempts to reach two dimensions starting from the upper critical dimension. Thus the ε-expansion is useless under these circumstances, unless one attempts a resummation based on the computation of several other terms in the expansion.

Notes

The renormalization group methods as an effective tool in the computation of critical properties are due to K.G. Wilson. Some relevant works have already been quoted in the previous chapter. Among early contributions are those by C. Di Castro and G. Jona Lasinio, *Phys. Lett.* **29A**, 322 (1969); K.G. Wilson and M.E. Fisher, *Phys. Rev. Lett.* **28**, 240 (1972); K.G. Wilson, *Phys. Rev. Lett.* **28**, 548 (1972). Rather than giving an exhaustive list of articles which would require a full volume, we refer the reader to the following original reviews, K.G. Wilson and J. Kogut, *Physics Report* **12C**, 75 (1974); M.E. Fisher, *Rev. Mod. Phys.* **46**, 597 (1974); K.G. Wilson, *Rev. Mod. Phys.* **47**, 773 (1975); E. Brezin, J.C. Le Guillou and J. Zinn-Justin in volume VI of the Domb and Green series (1976); S.K. Ma, *Modern Theory of Critical Phenomena*, Benjamin, New York (1976); G. Parisi's contribution to the 1973 Cargèse school which appeared in *J. Stat. Phys.*, **23**, 49 (1980). The book by D.J. Amit, *Field Theory, the Renormalization Group and Critical Phenomena*, 2nd edition, World Scientific, Singapore (1984) reflects also in part the influence of the Saclay group of our friends E. Brézin, J.-C. Le Guillou and J. Zinn-Justin, who have been, together with their collaborators, among the most active in developing a systematic field theoretic approach. We have based most of our presentation on their impressive work as well as G. Parisi's contribution.

Forthcoming books by G. Parisi and by J. Zinn-Justin should give an exhaustive and authoritative view.

The three-dimensional computations quoted in the text are from G.A. Baker, B.G. Nickel, M.S. Green and D.I. Meiron, *Phys. Rev. Lett.* **36**, 1351 (1976); G.A. Baker, B.G. Nickel and D.I. Meiron, *Phys. Rev.* **B17**, 1365 (1978); J. Zinn-Justin and J.C. Le Guillou, *Phys. Rev.* **B21**, 3976 (1980). No attempt has been made to incorporate the latest progress in this domain.

A lattice model for the tricritical singularity is due to J. Blume, V.J. Emery, R.B. Griffiths, *Phys. Rev.* **A4**, 1071 (1971). For the general scaling theory, see E.K. Riedel and F.J. Wegner, *Phys. Rev.* **B7**, 248 (1973), R.B. Griffiths, *Phys. Rev.* **B7**, 545 (1973), E. Abrahams, M.J. Stephen and J.P. Straley, *Phys. Rev.* **B12**, 256 (1975) and M. Wohrer, Thesis, Saclay (1976). The computations in appendix A are due to E. Brézin and M. Bauer.

Some of the works dealing with the continuous aspects in statistical mechanics are based on earlier investigations in quantum field theory. The renormalization group appears in E.C.G. Stueckelberg and A. Peterman, *Helv. Phys. Acta* **26**, 499 (1953); M. Gell-Mann and F.E. Low, *Phys. Rev.* **95**, 1300 (1954). A thorough discussion is found in N.N. Bogoliubov and D.V. Shirkov, *Introduction to the Theory of Quantized Fields*, Interscience, New York (1959). The modern approach with its emphasis on broken scale invariance is due to K. Symanzik, *Comm. Math. Phys.* **18**, 227 (1970) and C.G. Callan, *Phys. Rev.* **D2**, 1541 (1970).

This short summary of some essential references does not do justice to the impressive amount of contributions to the subject and to the impact of renormalization group ideas in many different areas.

6

LATTICE GAUGE FIELDS

Up to now, continuous field theory has appeared as a tool in the study of critical phenomena. Conversely, techniques from statistical mechanics can be useful in field theory. In 1973, Wilson proposed a lattice analog of the Yang–Mills gauge model. Its major aim was to explain the confinement of quarks in quantum chromodynamics. The lattice implementation of a local symmetry yields a transparent geometric interpretation of the gauge potential degrees of freedom, the latter being replaced by group elements assigned to links. Strong coupling expansions predict a linearly rising potential energy between static sources. Complex phase diagrams emerge when gauge fields are coupled to matter fields, and new phenomena appear, such as the absence of local order parameters. The discretization of fermions leads also to interesting relations with topology. This chapter is devoted to the theoretical developments of these ideas.

6.1 Generalities

6.1.1 Presentation

Schematic as they are, statistical models have directly a physical background at any temperature. Lattices may represent the crystalline structure of solids. They play an important role at short distance, but become irrelevant in the critical region, except as a regulator for the field theory describing the approach to critical points. The opposite point of view can also be considered. A lattice is artificially introduced as a regulator for a continuous field theory. The lattice system has no physical meaning, but can be studied at any "temperature", so that one can get information about its critical region, hopefully described by the initial continuous theory.

In particle physics, an important class of models can be built using the idea of minimal interactions with a gauge field, enabling the conservation of local internal symmetries (Yang–Mills, 1956). One of their striking properties is to be asymptotically free in the ultraviolet regime in four dimensions, at least for nonabelian continuous gauge groups. At short distances, the effective coupling between fields – describing "quarks" and "gluons", assumed to be the fundamental constituents of hadrons – decreases indefinitely, and this fact allows a perturbative treatment. Hence theoretical predictions can be compared to experimental data in this regime. *Quantum chromodynamics*, based on an internal *color* $SU(3)$ symmetry, provides results in remarkable agreement with experimental observations. Examples are deep inelastic scattering of leptons on nuclear targets, hadronic jet production in very high energy collisions, or the spectroscopy of bound states involving heavy quarks. Yang–Mills fields are also used in the unified electroweak model of Glashow, Weinberg and Salam, which is now considered, together with chromodynamics, as the standard model of particle physics.

However, the large-distance behaviour of chromodynamics cannot be reached in the perturbative approach. This method does not explain the permanent confinement of quarks, which cannot be experimentally isolated as elementary particles, at least under present laboratory conditions. The determination of the hadronic mass spectrum – bound states of quarks and gluons – is also a typical strong coupling problem which can be studied by perturbative field theory only within a phenomenological framework. The discretization of space-time using a lattice, has provided a partial answer to these questions. Lattice gauge field systems are studied using statistical techniques; high temperature series yield strong coupling expansions. Mean field approximation methods, as well as numerical simulations, also provide interesting results. The study of the critical region should now be refined for strong interaction physics. Ideas developed in gauge field theory for particle physics have applications in disordered or amorphous systems. Finally, there is some connection between gauge fields and general relativity (which can also be discretized following Regge (1960, see chapter 11 in volume 2)).

The fundamental concept is gauge invariance. Let us consider transformations $g(\mathbf{x})$, belonging to some group G, and depending

on the space point $\mathbf{x}$. "Charged"matter fields $\varphi(\mathbf{x})$ are transformed under a finite, generally linear representation of $g(\mathbf{x})$ at
the same point $\varphi(\mathbf{x}) \rightarrow g(\mathbf{x})\varphi(\mathbf{x})$. This is a shorthand notation
for the action of the group elements on the fields. In a continuous
model, derivative terms of the Lagrangian, which are responsible
for propagation, induce variations $\partial g(\mathbf{x})$. For continuous groups,
if one wishes to construct a local gauge invariant model, these
terms should be compensated by the introduction of a new gauge
field $A_\mu(\mathbf{x})$, taking its values in the Lie algebra of G. This leads to
the concept of minimal coupling. In a discretized model, derivative
terms are replaced by finite differences, so that the action contains
bilocal terms, for instance the nearest neighbour interaction

$$S_{\text{mat}} = \beta_{\text{m}} \sum_{(i,j)} \varphi_i \cdot \varphi_j \tag{1}$$

The above notation has been shortened for convenience. Fields
$\varphi_i = \{\varphi_i^\alpha, \alpha = 1, 2, \ldots, d_r\}$ are transformed through the linear
representation r of the symmetry group $G : \varphi_i^\alpha \rightarrow (g\varphi_i)^\alpha = \mathcal{D}_{\alpha\beta}^r(g)\varphi_i^\beta$. The dot product $\varphi_i \cdot \varphi_j$ is the quadratic or sesquilinear
form which is preserved by the group. Up to now, we have mostly
studied models invariant under the orthogonal group $SO(n)$; we
are also interested in this chapter by the unitary group $SU(n)$,
which leaves invariant the sesquilinear form $\sum_\alpha \varphi_i^{\alpha*}\varphi_j^\alpha$. Simpler
groups $(Z_2, Z_n, U(1), \ldots)$ will also be considered.

    The action in equation (1) is only globally invariant. In order
to make it locally invariant, one introduces gauge fields $U_{ij} \in G$
associated to any ordered pair of nearest neighbours (Wilson,
1974). Thus, in going from a continuous to a discretized form, the
novelty is the use of the group instead of the Lie algebra. This has
the virtue of making local invariance explicit and simple. It also
affords the opportunity to use discrete gauge groups which have no
direct interpretation in the continuous limit. The modified action

$$S_{\text{mat}} = \beta_{\text{m}} \sum_{(i,j)} \varphi_i U_{ij} \varphi_j = \beta_{\text{m}} \sum_{(i,j)} \varphi_i^{\alpha*} \mathcal{D}_{\alpha\beta}^r(U_{ij})\varphi_j^\beta \tag{2}$$

is invariant if the gauge fields transform as

$$U_{ij} \rightarrow g_i U_{ij} g_j^{-1} \tag{3}$$

with the constraint

$$U_{ij} \equiv U_{ji}^{-1} \tag{4}$$

The field U_{ij} is not associated to a lattice site, but to an *oriented* lattice link ij. The action has to be completed with an additional term S_J depending only on these new fields. Let us discuss the general form of such a term, which should of course be locally invariant. The product of fields along a closed curve $C = i_1 i_2 \cdots i_n i_1$ drawn on the lattice

$$U_C = U_{i_1 i_2} U_{i_2 i_3} \cdots U_{i_n i_1} \tag{5}$$

transforms as

$$U_C \rightarrow g_{i_1} U_C g_{i_1}^{-1} \tag{6}$$

i.e. it remains in the same conjugacy class of the group. Let us recall briefly that the group G is partitioned into classes by the following relation. Two elements h and h' are equivalent (conjugate) if there exists $g \in G$ such that $gh = h'g$. A function $\chi(U_C)$ depending only on the class of U_C is gauge invariant, and can therefore be used in the action S_J. The cyclic invariance property of such a function implies that $\chi(U_C)$ is moreover independent from the origin chosen on the closed curve C. Finally, χ should also be a real function (so is the action), and hence the orientation of C has no influence.

Among all possible choices, the simplest curve C (the shortest one) is the boundary of a *plaquette*, an elementary square on the hypercubical lattice. The path forth and back along a link leads to a trivial term, due to the constraint (4). From group theory, any class function can be decomposed on the basis of characters χ_r which are the traces of the corresponding matrices in the irreducible representation r : $\chi(U_C) = \sum_r c_r \chi_r(U_C)$. The "natural" choice of the fundamental representation for the matrix groups $SU(n)$ and $SO(n)$ leads to the Wilson's action

$$S_J = \beta_J \sum_p \frac{1}{n} \mathrm{Re\,Tr}\, U_p \tag{7}$$

with $U_p = U_{ij} U_{jk} U_{kl} U_{li}$ for the plaquette $p \equiv ijkl$. The dimension of the fundamental representation of G in which the trace Tr is taken is n.

Geometrical actions   Another choice follows from the following considerations. When β_J is large, U_p fluctuates around unity. In this case (which, as will be seen later, corresponds to the continuous limit), the most important terms are quadratic in the fluctuations of the field. A geometrical choice uses the geodesic distance from U_p to unity to measure these fluctuations. The metric is induced on G by the invariant quadratic form on the Lie algebra. For $G = U(1)$, $U_p = \exp(i\theta_p)$ and the action $\beta_J \sum_p \theta_p^2$ seems natural. However, it is multivalued. This defect may be corrected using a generalization of an idea due to Villain for the abelian case. For any Lie group, the *heat kernel* Boltzmann factor is the response to a local heating at the unit element on the group manifold after a time $t \approx 1/\beta_J$. This manifold is a circle for the group $U(1)$, a 3-sphere for $SU(2)$,... Hence, the Boltzmann factor is the solution of the heat equation

$$\Delta_g f(g,t) = \frac{\partial f(g,t)}{\partial t}, \qquad \text{with} \quad f(g,0) = \delta(g,1) \qquad (8)$$

where Δ_g is the Laplace–Beltrami operator on the group. The time is proportional to the inverse of β_J. This leads to the generalized Villain action

$$\exp S_J = \prod_p \exp S_J(p),$$

$$\exp S_J(p) = \sum_r d_r \chi_r(U_p) \exp\left(-\frac{C_r^{(2)}}{n\beta_J}\right) \qquad (9)$$

where $C_r^{(2)}$ is the value of the quadratic Casimir operator in the representation r, i.e. the corresponding eigenvalue of $-\Delta_g$. Summation runs over all inequivalent irreducible representations of dimension d_r (harmonic analysis on the group). In the $SU(2)$ case, if U is a generic 2×2 matrix conjugate to the diagonal matrix $(e^{i\theta/2}, e^{-i\theta/2})$, the plaquette contribution is the function

$$\exp S_J(p) = \sum_{2j=0}^{\infty} (2j+1)\frac{\sin(j + \frac{1}{2})\theta}{\sin\frac{1}{2}\theta} \exp{-\frac{j(j+1)}{2\beta_J}} \qquad (10)$$

Any other reasonable action is asymptotically equivalent to this one in the large β_J limit. The corresponding Boltzmann weight avoids in particular spurious singularities which may appear in some special limits such as the $n \to \infty$ limit of Wilson's action.

6.1.2   *The continuous limit*

Before embarking on a detailed analysis of gauge systems at any coupling β_J, we consider first its formal continuous limit. For the time being, we assume that fluctuations around this limit are small and without essential qualitative effects. The reasoning is only

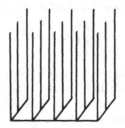

Fig. 1   Maximal tree on a cubic lattice, defining the axial gauge

valid for continuous gauge groups. As in the previous chapters, we develop the action in local terms depending only on fields and their derivatives at a point. The resulting local, regularized Lagrangian is then studied using renormalization group techniques. This procedure, however, needs some modifications to take into account specific properties of the gauge field U_{ij}.

The first step is to characterize the configurations $\{U_{ij}^0\}$ for which the action is maximal and around which the perturbative expansion will be done. They satisfy

$$U_p^0 = U_{ij}^0 U_{jk}^0 U_{kl}^0 U_{li}^0 = 1 \tag{11}$$

for any plaquette (all characters are simultaneously maximal for the group identity). Let us search a gauge transformed configuration $U_{ij}^{0'} = g_i^{-1} U_{ij}^0 g_j$, such that $U_{ij}^{0'} = 1$ on a maximal tree of the lattice, i.e. on a maximal set of links without any loop. Figure 1 displays such a set, and more precisely defines the so-called *axial gauge*, where all gauge fields are set equal to unity in a "*timelike*" direction. The reader can easily convince himself that the gauge transformation g_i is defined from site to site. It then follows that $U_{ij}^{0'}$ is equal to unity everywhere due to relation (11). Hence the only configurations maximizing the action are pure gauge configurations

$$U_{ij}^0 = g_i g_j^{-1} \tag{12}$$

Generalizing the previous property, one might think that any plaquette configuration $\{U_p\}$ is sufficient to define a link configuration $\{U_{ij}\}$, up to a gauge transformation. Show that this is the case if the gauge group is abelian and if the six plaquette variables

U_p bounding any tridimensional cube satisfy the constraint

$$\prod_{p \in \partial c} U_p = 1 \tag{13}$$

This constraint disappears in two dimensions, and the result remains valid for nonabelian groups. In this case, the variables $\{U_p\}$ are independent and do not interact in the action. As a consequence, a two-dimensional pure gauge model is trivial.

Around the classical solution (12), fields are parametrized using a local gauge field $A_\mu(\mathbf{x})$, taking its values in the Lie algebra of the group G, in the form

$$U_{ij} = g_i \exp(\mathrm{i}a A_\mu(\mathbf{x})) g_j^{-1} \tag{14}$$

with $\mathbf{x}_j = \mathbf{x}_i + a\hat{\mu}$, $\mathbf{x} = \frac{1}{2}(\mathbf{x}_i + \mathbf{x}_j)$, a being the lattice spacing. This parametrization cannot be distinguished from another possible one using the path ordered integral $P \exp\{\mathrm{i} \int_{t=0}^{1} \mathrm{d}\mathbf{x}^\mu(t) A_\mu(\mathbf{x}(t))\}$, with $\mathbf{x}(t) = (1-t)\mathbf{x}_i + t\mathbf{x}_j$, if one neglects higher order corrections in a. When evaluating the plaquette term U_p, the generally noncommutative products are computed using Baker–Campbell–Hausdorff formula

$$\mathrm{e}^A \mathrm{e}^B = \mathrm{e}^{A+B+\frac{1}{2}[A,B]+\cdots}$$

Recall that the general relation

$$\exp C = \exp A \exp B,$$

$$C = A + B + \tfrac{1}{2}[A,B] + \tfrac{1}{12}[A,[A,B]] + \tfrac{1}{12}[B,[B,A]]$$
$$- \tfrac{1}{48}[A,[B,[A,B]]] + \cdots$$

is obtained by the following algorithm. The product

$$C = \ln\left[\left(1 + A + \frac{A^2}{2!} + \cdots\right)\left(1 + B + \frac{B^2}{2!} + \cdots\right)\right]$$

is expanded as a sum of homogeneous monomials. If $z_1 z_2 \cdots z_p$ is such a monomial of degree p, it is replaced by the expression $(1/p)[z_1, [z_2, [\cdots [z_{p-1}, z_p]] \cdots]$ belonging to the Lie algebra generated by A and B.

The expression in terms of the fields at the center of the plaquette is obtained after some straightforward algebra

$$U_p = g_i \exp[\mathrm{i}a^2 F_{\mu\nu} + \mathcal{O}(a^4)] g_j^{-1} \tag{15}$$

with

$$F_{\mu\nu} = \partial_\mu A_\nu - \partial_\nu A_\mu + \mathrm{i}[A_\mu, A_\nu] \tag{16}$$

Finally, the exponential (15) is expanded as a power series in a, and the sum over plaquettes in the action is replaced by an integral (with corrections of order a, which can be computed using the Euler–MacLaurin formula). Up to an additive constant, which can be removed by a redefinition of the partition function, the final result reads

$$S_{\mathrm{J}} = \frac{\beta_{\mathrm{J}} g_0^2 a^{4-d}}{2n} \int d^d x \left(-\frac{1}{2g_0^2} \,\mathrm{Tr}(F_{\mu\nu}^2) + \mathcal{O}(a^2) \right) \tag{17}$$

The leading term in this formula coincides with the continuous action postulated by Yang and Mills when the multiplicative term in front of the integral is set equal to unity by an appropriate choice of g_0. Requiring that the coefficient in front of the integral be unity yields a relation between the inverse temperature β_{J} of the gauge model and the bare coupling constant g_0 in the continuous theory of the form

$$\beta_{\mathrm{J}} = \frac{2n}{g_0^2} a^{d-4} \tag{18}$$

Naive dimensional analysis leads to $[A_\mu/g_0] = \frac{1}{2}(d-2)$, $[g_0^2] = 4 - d$. Hence the continuous theory is renormalizable in four dimensions provided one can show that local invariance can be preserved under renormalization. The higher order terms in a, which have not been explicitly written in equation (17), involve higher derivatives, and their canonical dimension is greater than four. It is therefore expected that, in the large distance limit, their effect is limited to a finite renormalization, without qualitative change in the critical behaviour.

To obtain the continuous limit, one chooses a physical length ξ, computed from correlation functions, and one keeps it fixed (by tuning the bare coupling constant g_0) while the lattice spacing shrinks indefinitely. In lattice units, ξ/a diverges. In other words, one approaches a critical point $g_0 \to g_{0\mathrm{c}}$. This process follows the renormalization group equation

$$a\frac{d\xi}{da} \equiv \left(a\frac{\partial}{\partial a} - \beta(g_0)\frac{\partial}{\partial g_0} \right) \xi = 0 \tag{19}$$

with

$$\beta(g_0) = -a \left. \frac{\partial g_0}{\partial a} \right|_{\xi} \tag{20}$$

The notation $\beta(g_0)$ for the Callan–Symanzik function is the traditional one and should not be confused with the constants β_J, β_m appearing as coefficients in the action. In integral form, equation (20) yields

$$\xi = a \exp \int^{g_0} \frac{dg}{\beta(g)} \tag{21}$$

Hence g_{0c} is a zero of the function $\beta(g)$, with a negative slope (if it is a simple zero). Relation (21) indicates how g_0 approaches g_{0c} when a vanishes. A direct computation of $\beta(g)$ could be performed using the lattice regularization. In chapter 5, we mentioned however that the slope of $\beta(g)$ at g_{0c} is universal (this assertion should be appropriately modified in the case of a multiple zero), while the value of g_{0c} itself depends on the regularization scheme. Therefore, in practice, it is better to perform easier calculations using for instance dimensional regularization. As it turns out, zero coupling is such a critical point with a triple zero in $\beta(g)$, and the two universal coefficients in the expansion

$$\beta(g) = -b_0 g^3 - b_1 g^5 + \mathcal{O}(g^7) \tag{22a}$$

are given, in the case of the $SU(n)$ gauge group, by

$$b_0 = \frac{11}{3} \frac{n}{(4\pi)^2}, \qquad b_1 = \frac{34}{3} \frac{n^2}{(4\pi)^4} \tag{22b}$$

While the principle of the computation is simple and follows the discussion of chapter 5, nontrivial technical details make it rather complex in the present case. This is due to an apparent conflict between the desire to maintain local invariance, and the necessity of introducing a gauge fixing term to avoid integration over a continuum of nondynamical gauge degrees of freedom. The way to overcome this difficulty is the subject of an elaborate theory presented in detail in textbooks on the quantization of continuous gauge fields. We only sketch the main steps. Using a basis t_a of the Lie algebra of G, with totally antisymmetric structure constants

f_{abc} such that

$$i[t_a, t_b] = f_{abc}t_c \qquad f_{acd}f_{bcd} = C_f^{(2)}\delta_{ab} \qquad \mathrm{Tr}\, t_a t_b = \tfrac{1}{2}\delta_{ab}$$

$$(23)$$

one writes $A_\mu = g\mathcal{A}_\mu^a t_a$. A term $(\partial_\mu \mathcal{A}_\mu^a)^2/2\alpha$ is added to the action in order to fix the gauge. Thus the operator in the quadratic part of the Lagrangian has no more zero modes and becomes invertible. This term is compensated by the introduction of anticommuting fields Φ_a (Faddeev–Popov ghost fields), the role of which under integration is to give a Jacobian necessary to maintain gauge invariance. Writing the action in the form

$$S = -\int \mathrm{d}^d\mathbf{x}\,\mathcal{L} \qquad (24a)$$

the resulting Lagrangian, written in terms of renormalized quantities, reads

$$
\begin{aligned}
\mathcal{L} = &\tfrac{1}{4}Z_3(\partial_\mu \mathcal{A}_\nu^a - \partial_\nu \mathcal{A}_\mu^a)^2 + \frac{1}{2\alpha}(\partial_\mu \mathcal{A}_\mu^a)^2 \\
&+ Z_1 g f_{abc}\partial_\mu \mathcal{A}_\nu^a \mathcal{A}_\mu^b \mathcal{A}_\nu^c + \tfrac{1}{4}g^2 \frac{Z_1^2}{Z_3} f_{abc}f_{ade}\mathcal{A}_\mu^b \mathcal{A}_\nu^c \mathcal{A}_\mu^d \mathcal{A}_\nu^e \quad (24b) \\
&+ Z_3^\Phi \partial_\mu \Phi_a^* \partial_\mu \Phi_a + Z_1^\Phi g f_{abc}\partial_\mu \Phi_a^* \mathcal{A}_\mu^b \Phi_c
\end{aligned}
$$

That the coefficient of the quartic term in $\mathcal{A}$ has the indicated form, is a consequence of Ward–Slavnov identities, expressing gauge invariance, which also implies the relation $Z_1/Z_3 = Z_1^\Phi/Z_3^\Phi$. The longitudinal part of the propagator is not renormalized. The calculation is performed in $d = 4 - \varepsilon$ dimensions, with a dimensionless coupling constant $u = g\mu^{-\varepsilon/2}$. The mass scale is therefore defined by μ. The subtraction terms $(Z_i - 1)$ are functions of u, α and ε, and are determined order by order in u in such a way that the limit $\varepsilon \to 0$ of physical quantities is finite. We adopt the minimal subtraction scheme, involving only divergent terms behaving as inverse powers of ε, without additional finite renormalization. Feynman rules are summarized in figure 2, where a heavy line indicates a gauge potential propagator, while a broken one is associated to a ghost propagator. We adopt the Feynman gauge $\alpha = 0$, which leads to more tractable calculations.

Subtraction terms are determined from the divergent two- and three-point functions. The corresponding one-loop diagrams are displayed in figure 3. Integration techniques for the dimensional continuation have been described in section 5.2.3. For instance, the *unsubtracted* gauge field propagator at the one-loop level (figure 3a)

a,μ ———— b,ν	$\dfrac{1}{(2\pi)^4}\dfrac{\delta_{ab}}{p^2}\left(\delta_{\mu\nu}-\alpha\dfrac{p_\mu p_\nu}{p^2}\right)$
 a - - -▶- - b	$\dfrac{1}{(2\pi)^4}\dfrac{\delta_{ab}}{p^2}$
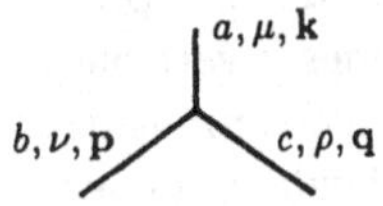 $a,\mu,\mathbf{k}$ $b,\nu,\mathbf{p}$ $c,\rho,\mathbf{q}$	$(2\pi)^4 ig f_{abc}[\delta_{\nu\rho}(q-p)_\mu + \delta_{\rho\mu}(k-q)_\nu + \delta_{\mu\nu}(p-k)_\rho]$
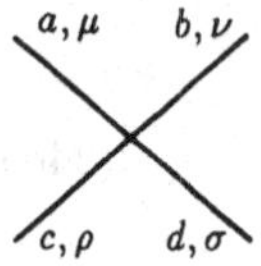 a,μ b,ν c,ρ d,σ	$-(2\pi)^4 g^2 f_{ace}f_{bde}(\delta_{\mu\nu}\delta_{\rho\sigma}-\delta_{\mu\sigma}\delta_{\nu\rho})-(2\ \text{perm.})$
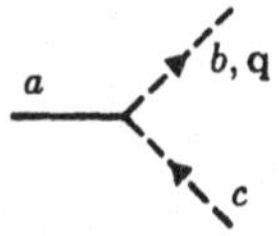 a $b,\mathbf{q}$ c	$(2\pi)^4 ig f_{abc}q_\alpha$

Fig. 2   Feynman rules for the continuous Yang–Mills Lagrangian.

reads

$$\Pi^{(1)}_{\mu a;\nu b}(p) = \tfrac{1}{2}g^2\pi^{d/2}C_f^{(2)}\delta_{ab}\Gamma(2-\tfrac{1}{2}d)[(4d-8)B(\tfrac{1}{2}d,\tfrac{1}{2}d)$$
$$+ (6-d)B(\tfrac{1}{2}d-1,\tfrac{1}{2}d-1)](p^2)^{d/2-2}(p^2\delta_{\mu\nu}-p_\mu p_\nu)$$

$$(25)$$

with

$$B(a,b) = \Gamma(a)\Gamma(b)/\Gamma(a+b)$$

Expressing this result in terms of the dimensionless variable u and expanding in ε, one derives at one-loop order the necessary counterterm leading to a finite result when $\varepsilon \to 0$

$$(Z_3 - 1)^{(1)} = \frac{10}{3}\frac{u^2}{16\pi^2}C_f^{(2)}\frac{1}{\varepsilon} \qquad (26)$$

The other three Green functions described in figure 3 can be similarly computed. The choice of zero external momenta in the 3-point functions leads to simpler calculations. One obtains the

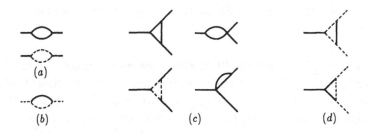

Fig. 3   One loop diagrams for 2- and 3-point functions. (*a*) Gauge potential propagator. (*b*) Ghost propagator. (*c*) Gauge vertex. (*d*) Mixed gauge and ghost vertex.

following results

$$
\begin{cases}
(Z_3^\Phi - 1)^{(1)} = \dfrac{u^2}{16\pi^2} C_f^{(2)} \dfrac{1}{\varepsilon} \\[2ex]
(Z_1 - 1)^{(1)} = \dfrac{4}{3} \dfrac{u^2}{16\pi^2} C_f^{(2)} \dfrac{1}{\varepsilon} \\[2ex]
(Z_1^\Phi - 1)^{(1)} = -\dfrac{u^2}{16\pi^2} C_f^{(2)} \dfrac{1}{\varepsilon}
\end{cases}
\tag{27}
$$

The function $\beta(u)$ is defined as

$$
\beta(u) = \mu \left. \frac{\partial u}{\partial \mu} \right|_{g_0, \alpha_0}
\tag{28}
$$

where in the general case the unrenormalized constants g_0 and α_0 are expressed in terms of the renormalized ones through

$$
g_0 = \frac{Z_1(u,\alpha)}{Z_3^{3/2}(u,\alpha)} g, \qquad \alpha_0 = Z_3(u,\alpha)\alpha
\tag{29}
$$

Hence,

$$
\beta(u) = \frac{-\frac{1}{2}\varepsilon u}{1 + u \frac{d}{du}[\ln Z_1 Z_3^{-3/2}]} = -\frac{1}{2}\varepsilon u - \frac{11}{3} u^3 \frac{C_f^{(2)}}{16\pi^2} + \cdots
\tag{30}
$$

which agrees, in the limit $\varepsilon \to 0$, with the leading term in equations (22*a*)–(22*b*) (since, for $SU(n)$, the quadratic Casimir invariant is given by $C_f^{(2)} = n$).

The two-loop calculations are performed in a similar way. The diagrams are much more numerous and include the one-loop counterterms. The Slavnov identity $Z_1/Z_3 = Z_1^\Phi/Z_3^\Phi$ which expresses gauge invariance, is imposed term by term at any order.

The subtraction constant Z_1 can then be determined from the other ones.

According to equation (22), the pure gauge theory is asymptotically free in the ultraviolet regime ($g_{0c} = 0$). Starting from a lattice version, the physical continuous region corresponds to large β_J according to equation (18). We introduce a standard mass scale for the lattice regularization (identified by the index L) through

$$\xi_L^{-1} = \Lambda_L \equiv \frac{1}{a}(b_0 g_0^2)^{-b_1/2b_0^2} \exp\left(\frac{-1}{2b_0 g_0^2}\right) \qquad (31)$$

This relation is obtained from the integral in equation (21) in the vicinity of $g_0 \sim 0$ using the two leading terms in the β-function. Any quantity expressed in this scale should have a finite limit when letting the lattice spacing a tend to 0. In other words, any physical length ξ will be computed using its finite limit $\xi \Lambda_L$ when β_J grows indefinitely.

It is important in practical applications of the lattice technique to be able to express the relation between Λ_L and similar quantities corresponding to other regularization schemes. Let us consider for instance another regularization characterized by a mass regulator m which will be sent to infinity. In the two schemes, the same correlation lengths $\xi_i \Lambda_L$ and $\xi_i \Lambda_m$ (expressed in the convenient scale) have finite limits. As the two models describe the same physics, the ratio of these limits is independent of the quantity ξ_i considered. The knowledge of the ratio Λ_L/Λ_m is not of purely academic interest. Indeed, one wants to relate perturbative predictions – achieved using the most convenient regularization, such as the dimensional regularization – to nonperturbative quantities, such as particle masses, computed numerically on a lattice.

This ratio, which involves a comparison of the ultraviolet cutoffs, would be known if the β function was explicitly computed in the lattice regularization. However, we did not perform the very complex two-loop calculations. There exists a trick which yields the required ratio by computing only one-loop diagrams. Fields are expanded in the neighborhood of a configuration $\{U_0\}$, which satisfies the classical motion equations. This is called the *background gauge*. Performing this expansion, the action is written as $S_J = S_0 + S_2 + \cdots$, S_2 being quadratic in the fluctuations, and $S_0 = S_J(U_0) = (-1/2g_0^2) \int d^4 x \, \text{Tr}(F_{\mu\nu}^0)^2$. Further terms will not be needed. The partition function reads

$$Z_{\mathrm{L}}(a) = e^{-S_0} \int \mathrm{d}U e^{-S_2}(1 + \mathcal{O}(g_0^2(a))) \tag{32}$$

Comparing with the similar formula obtained with the other regularization, one obtains

$$-\ln \frac{Z_{\mathrm{L}}(a)}{Z(m)} = \left[\frac{1}{2g_0^2(a)} - \frac{1}{2g_0^2(m)} + f(ma) \right] \int \mathrm{d}^4\mathbf{x}\, \mathrm{Tr}(F_{\mu\nu}^0)^2 \tag{33}$$

The function $f(ma)$, resulting from the calculation of the two path integrals (32), corresponds to a finite renormalization. Its explicit form can be predicted from the (logarithmic) divergences of the one-loop diagrams

$$f(ma) = \frac{11n}{48\pi^2}[\ln ma + C] \tag{34}$$

Formula (31) relates Λ_{L} to a and a similar relation can be written between Λ_m and m. Equation (33) is thus rewritten as

$$\ln \frac{Z_{\mathrm{L}}(a)}{Z(m)} = \frac{11n}{48\pi^2}\left(\ln \frac{\Lambda_{\mathrm{L}}}{\Lambda_m} - C \right) \int \mathrm{d}^4\mathbf{x}\, \mathrm{Tr}(F_{\mu\nu}^0)^2$$

As the two regularizations lead to the same physical results, the finite renormalization has to disappear, requiring that

$$\Lambda_{\mathrm{L}} = \Lambda_m e^C \tag{35}$$

A detailed and rather lengthy calculation enables one to extract the value of the constant C from the comparison of the two-point function for the field $F_{\mu\nu}^0$ in the two regularizations. A choice of gauge is necessary, introducing Faddeev–Popov ghosts. We only indicate here one final result, comparing the Wilson action to the minimal dimensional regularization (index MS) for the gauge group $SU(n)$

$$\ln \frac{\Lambda_{\mathrm{MS}}}{\Lambda_{\mathrm{L}}} = \tfrac{1}{2}(\ln 4\pi - \gamma) - 0.0224780n \tag{36}$$

The numerical constant is obtained from estimates of lattice integrals, and $\gamma \simeq 0.57721$ is the Euler's constant. Other formulae, corresponding to more general lattice actions, can be found in the references.

6.1.3 *Order parameter and Elitzur's theorem*

In the various types of models presented in previous chapters, finitely many local order parameters discriminate among the possible phases of a statistical system. For instance, the existence of a nonvanishing magnetization signals a spontaneous breaking of global symmetry in a spin model. One might imagine a similar mechanism in gauge theories, and study for instance the mean value $\langle U_{ij} \rangle$ of the field. This simple idea does not work. Elitzur has proved very generally that any *local* quantity (i.e. using only a finite number of fields) which is not gauge invariant (as $\langle U_{ij} \rangle$) has a vanishing mean value *at any temperature*, and therefore cannot be used as an order parameter. In other words, *local* observables cannot exhibit spontaneous breaking of *local* gauge symmetry.

Before proceeding to the proof of this theorem, let us discuss once more the significance of a nonvanishing magnetization. In the absence of an external field, such a phenomenon might at first seem strange, since the two opposite configurations $\{\sigma_i\}$ and $\{-\sigma_i\}$ contribute with the same Boltzmann weight. However, at low temperature, the two regions in configuration space representing opposite magnetizations can only be connected by a dynamical path involving the creation of an infinite interface at the price of an infinite energy cost. Hence the process cannot occur spontaneously. Another description uses a small external field h. The two fundamental configurations $\{s\sigma_i = +1\}$ and $\{\sigma_i = -1\}$ are now energetically separated by a gap $2Nh$, with N the total number of sites. One of these is thus eliminated in the thermodynamic limit $N \to \infty$, and when the symmetry breaking field is then set equal to zero, one is left with a nonvanishing magnetization. Very generally, the mean value of a quantity $f(\varphi)$ should be computed *in the presence of a vanishingly small external field J* and the thermodynamic limit $N \to \infty$ should be taken *before* taking the limit $J \to 0$; the two limits *do not commute*. We denote generically by φ the gauge field taking its values in a compact group manifold and define

$$\langle f(\varphi) \rangle = \lim_{J \to 0} \lim_{N \to \infty} \langle f(\varphi) \rangle_{N,J} \qquad (37)$$

with

$$\langle f(\varphi) \rangle_{N,J} = Z_{N,J}^{-1} \int \exp(S(\varphi) + J.\varphi)\, f(\varphi)\, \mathrm{d}\mu(\varphi) \qquad (38)$$

In the context of a local symmetry, the proof of Elitzur's theorem applies to a quantity $f(\varphi)$, not invariant under the gauge group $\varphi \to {}^g\varphi$ which leaves both the action and the field measure invariant. Non invariant means here that this quantity has no invariant component, i.e.

$$\int f(^g\varphi)\mathrm{d}g = 0 \qquad (39)$$

The gauge group is local and the above integral is understood on the lattice as performed over each $g(\mathbf{x})$ independently. Thus it is possible to consider the subgroup of those transformations which act only on the finite set $\{\varphi'\}$ on which the local function $f(\varphi)$ depends and which leaves invariant the complementary subset $\{\varphi''\} : \varphi'' = {}^g\varphi''$. Performing the change of variable $\varphi \to {}^g\varphi$ in the integral (38) and taking into account the invariance of the action and of the measure, the mean value reads (after averaging over this gauge subgroup)

$$\langle f(\varphi)\rangle_{N,J} = Z_{N,J}^{-1}\int \exp(S(\varphi) + J'\cdot{}^g\varphi' + J''\cdot\varphi'')f(^g\varphi)\mathrm{d}\mu(\varphi)\mathrm{d}g \qquad (40)$$

For a source term J bounded by a small enough quantity ε, the crucial point is the inequality

$$|\exp(J'\cdot\varphi') - 1| \le \eta(\varepsilon) \qquad (41)$$

with $\eta(\varepsilon)$ vanishing uniformly with ε, whatever the field configuration φ' and the total number of sites N. This inequality holds for gauge systems, because the subset $\{\varphi'\}$ is finite and independent of N, and because the fields φ are assumed to take their values in a compact domain. It does not hold for spin systems, because the partition $\{\varphi\} = \{\varphi'\} \cup \{\varphi''\}$ cannot be performed in the case of a global symmetry, so that $J\cdot\varphi$ is an extensive quantity proportional to N. The integral (40) is split into two terms, writing $\exp J'\cdot\varphi' = 1 + (\exp J'\cdot\varphi' - 1)$. The first one vanishes identically when integrating over g, due to condition (39). The second one is bounded using the inequality (41) and one gets

$$\left|\langle f(\varphi)\rangle_{N,J}\right| \le 2\eta(\varepsilon)\,\mathrm{Sup}f$$

Now the two limits are taken in the correct order and this leads to the expected assertion

$$\langle f(\varphi)\rangle \equiv 0 \qquad (42)$$

when the source term vanishes. As a result, any order parameter should be a nonlocal quantity. The physical interpretation of the gauge model has led Wilson to introduce such a quantity, henceforth called a *Wilson loop*. Let us consider a closed loop C and define

$$W(C) = \langle \operatorname{Tr} U_C \rangle \tag{43}$$

The trace refers here to the fundamental representation (although other choices might also be interesting). With the following assumptions

- β_J small (strong coupling region),
- pure gauge system, with *no coupled matter fields*,
- compact gauge group with a nontrivial center,

we will prove in section 3 that $-\ln W(C)$ is asymptotically proportional to the minimal area $\mathcal{A}(C)$ enclosed by the curve C, when C grows indefinitely (area law). The proof is based on the convergence of the strong coupling expansion for this quantity and uses the fact that the leading contribution of this series arises from diagrams involving plaquettes covering the minimal surface bounded by C.

In the weak coupling region, this assertion might be false. For discrete abelian groups such as Z_2, the area law fails in this regime, as will be proved below using duality arguments. We give here a heuristic perturbative argument, valid for a continuous symmetry group (it becomes, however, inexact for nonabelian, four-dimensional systems which are asymptotically free and where the strong coupling behaviour is expected to extend qualitatively down to small coupling). The parametrization (13) of the fields leads to the representation

$$W(C) = \left\langle \operatorname{P} \exp i \oint_C A_\mu dx^\mu \right\rangle = \exp\left(-\tfrac{1}{2} \oint \oint_C \Delta(x - y) dx dy + \cdots \right) \tag{44}$$

The propagator $\Delta(x - y)$ of the gauge field A_μ favours small separations because of its short-distance singularity. Hence the exponential argument in (44) is proportional to the length of the curve C.

The *string tension* is defined as the limit

$$K = \lim_{\substack{C \ large}} \left(-\frac{\ln W(C)}{\mathcal{A}(C)} \right) \tag{45}$$

It enables one to distinguish between a *confined* phase ($K \neq 0$) and a nonconfined one ($K \equiv 0$). The quantity K plays the role of an order parameter, albeit through a very nonlocal procedure, and, using this criterion, two phases are always seen to coexist in sufficiently high dimension. The above argument suggests that the confined phase is out of reach of continuous perturbative theory (in spite of many ingenuous attempts). This provides strong motivations for the introduction of lattice gauge theory.

What is the physical interpretation of K? Let us imagine the following process. At some time, a quark–antiquark pair is produced and separated up to a distance R. Its potential energy is then $V(R)$. The two quarks propagate during a time T, at a fixed separation R, and then annihilate. The corresponding probability amplitude is $\exp(-V(R)T)$ (since we are in Euclidean space, the usual expression $\exp(iEt)$ is replaced by $\exp(-Et)$). This amplitude is precisely the mean value of the Wilson loop operator, for the rectangular curve with area $R \times T$ bounded by the trajectory of the pair. At large distance and strong coupling, the potential is asymptotically linear $V(R) = KR$ and prevents any separation of the two particles which remain in a bound state. In the unconfined phase, where a perimeter law holds, $V(R)$ is bounded and it is possible to separate the two particles with only a finite energy. This interpretation explains why such an order parameter cannot be used in the presence of dynamical matter fields. Indeed, if the separation increases, an additional pair can be created and tends to restore a perimeter law for the Wilson loop parameter, at any coupling.

Compute the string tension K for a two-dimensional lattice gauge system.

6.1.4 Duality

We have already encountered duality when considering the two-dimensional Ising model. It enabled one to obtain relations which determined the critical point and played an important role in the exact solution. Duality can be generalized to arbitrary *abelian* gauge systems defined on an abstract *complex* of dimension d, with the associated notions of neighborhoods and boundaries. In particular, a lattice is a complex, the *cells* of which are (in increasing dimension) the sites, the links, the plaquettes, the cubes, etc. In general, a complex is a set of cells $c_k^{(i)}$ with the following properties.

1) With each cell $c_k^{(i)}$ is associated a nonnegative integral *dimension* k, and the upper bound of cell dimensions is the *dimension* d of the complex.

2) With each cell $c_k^{(i)}$ is associated a cell of *opposite orientation* denoted $-c_k^{(i)}$. This mapping is involutive: $-(-c_k^{(i)}) = c_k^{(i)}$.

3) With each pair of cells, the dimension of which differs by unity, is associated an integer $(c_k^{(i)} : c_{k-1}^{(j)})$ called the *incidence number*. If it does not vanish, $c_{k-1}^{(j)}$ is called a *face* of $c_k^{(i)}$. The incidence number changes sign whenever the orientation of either cell is reversed

$$(c_k^{(i)} : -c_{k-1}^{(j)}) = (-c_k^{(i)} : c_{k-1}^{(j)}) = -(c_k^{(i)} : c_{k-1}^{(j)}).$$

4) For any pair of cells c_k and c_{k-2}, one requires that

$$\sum_i (c_k : c_{k-1}^{(i)})(c_{k-1}^{(i)} : c_{k-2}) = 0$$

This relation is the basis of homology and cohomology theory.

Show that any lattice is a complex. We shall return to such constructions in chapter 11 (volume 2) when studying random geometry.

A *k-chain* with values in an abelian group G (denoted additively in this section, in contradistinction with our usual notation) is an odd function defined on the cells of dimension k in G

$$c_k^{(i)} \to \varphi_i \in G \qquad - c_k^{(i)} \to -\varphi_i$$

These functions have themselves the structure of an abelian group, and we denote a k-chain by $\sum_i \varphi_i c_k^{(i)}$. The *null chain*, denoted by 0, is the chain which assigns the zero in G to all cells. Finally, the boundary and coboundary operators denoted $\triangle$ and ∇ (this notation should not be confused with Laplacian and gradient) map the k-chains into the sets of $(k-1)$-chains and $(k+1)$-chains respectively, according to

$$\triangle \sum_i \varphi_i c_k^{(i)} = \sum_{i,j} \varphi_i (c_k^{(i)} : c_{k-1}^{(j)}) c_{k-1}^{(j)} \qquad (46)$$

$$\nabla \sum_i \varphi_i c_k^{(i)} = \sum_{i,j} \varphi_i (c_{k+1}^{(j)} : c_k^{(i)}) c_{k+1}^{(j)} \qquad (47)$$

The coefficients φ_i of a $(k-1)$-chain are now considered as the fields of a statistical model. A configuration of this system is thus characterized by a $(k-1)$-chain. We study the action

$$S\left(\sum_i \varphi_i c_{k-1}^{(i)}\right) = \sum_i \chi\left(\sum_j \varphi_j (c_k^{(i)} : c_{k-1}^{(j)})\right) + \sum_i V(\varphi_i) \quad (48)$$

The real functions V and χ on the group G describe respectively potential and interaction terms.

The case $k = 1$ and $G = Z_2 = \{0,1\}$ (*additive* group of integers modulo 2) corresponds to the Ising model. In this unusual notation the interaction term in equation (48) is the sum over all links ij of $\chi(\sigma_i + \sigma_j)$ where the function χ satisfies $\chi(0) = \beta$, $\chi(1) = -\beta$. Verify that gauge models correspond to $k = 2$.

The exponentiated action is expanded on group characters, which form a basis for the functions defined on an abelian group. The partition function reads

$$Z = \|G\|^{-n_{k-1}} \sum_{(k-1)-\text{chains}} \left\{ \prod_i \sum_r \beta(r) \chi_r \left(\sum_j \varphi_j (c_k^{(i)} : c_{k-1}^{(j)}) \right) \right\}$$

$$\left\{ \prod_i \sum_r \gamma(r) \chi_r(\varphi_i) \right\}$$

$$(49)$$

where the Fourier coefficients are

$$\begin{cases} \beta(r) = \|G\|^{-1} \sum_{\varphi \in G} \chi_r(\varphi) \exp \chi(\varphi) \\ \gamma(r) = \|G\|^{-1} \sum_{\varphi \in G} \chi_r(\varphi) \exp V(\varphi) \end{cases} \qquad (50)$$

In these formulae, the number of $(k-1)$-cells is denoted n_{k-1} and $\|G\|$ is the number of group elements. An arbitrary term in the expansion of the products enclosed within braces in (49) can be obtained by choosing a representation r for any index i. Hence this term can be considered as a chain with values in the *dual group* G^*, i.e. the set of irreducible representations of G. The dual G^* of an abelian group G has also the structure of an abelian group, with its group law defined as the product of representations of G. Such a group structure does not exist in the nonabelian case and this prevents a straightforward generalization of duality. The terms enclosed by the first braces in (49) are k-chains $g = \sum r_i c_k^{(i)}$ with values in G^*, and those enclosed by the second braces are $(k-1)$-chains $h = \sum -s_j c_{k-1}^{(j)}$. If we group all terms depending on a given field φ_j, we obtain

$$\left[\prod_i \chi_{r_i} \left(\varphi_j(c_k^{(i)} : c_{k-1}^{(j)}) \right) \right] \chi_{-s_j}(\varphi_j) = \chi_{-s_j + \sum_i r_i(c_k^{(i)} : c_{k-1}^{(j)})}(\varphi_j)$$

The summation over φ_j (i.e. over $(k-1)$-chains in equation (49)) can now be performed. The result vanishes unless the character index is 0 (0 denoting the trivial representation), and gives a factor $\|G\|$ in this case. This representation index is nothing but the component corresponding to the cell $c_{k-1}^{(j)}$ in the $(k-1)$-chain $-h + \triangle g$ with values in G^* (also written additively). Hence the partition function (49) takes the form

$$Z = \sum_{\substack{k-\text{chains} \\ \text{on } G^*}} \exp \left\{ \sum_i \ln \beta(r_i) + \sum_i \ln \gamma \left(\sum_j r_j(c_k^{(j)} : c_{k-1}^{(i)}) \right) \right\}$$

(51)

Comparing the argument of the exponential to the original action (48), one observes a close analogy. The fields r_i are associated to cells with dimension k (instead of $k-1$). They interact if their supporting cells have a common face (boundary interaction), while the fields in (48) interact if they bound the same cell (coboundary interaction). This difference is inessential and disappears if formula (51) is interpreted on the *dual complex*.

To define this new complex, take the same cells and just change their dimensions from k to $d-k$. The incidence numbers are unchanged. Let us denote by c_{d-k}^* the cell dual to c_k obtained using this substitution. As $(c_k : c_{k-1}) = (c_{d-k+1}^* : c_{d-k}^*)$, the characteristic properties for a *dual* complex are fulfilled.

   Geometrical interpretation. On a hypercubic lattice, the dual sites are identified with the centers of the hypercubes of the original lattice. A link joining two neighbouring dual sites is dual to the face separating the two corresponding hypercubes, and so on. Up to boundary effects,the dual lattice is also a hypercubical lattice, shifted with respect to the original one. Analogously the two-dimensional triangular and hexagonal lattices are dual to each other. Similarly, the face centered cubic and body centered cubic lattices are also dual in three dimensions.

Using this interpretation, the partition function (51) is rewritten on the dual lattice with an action

$$S^* \left(\sum_i r_i c_{d-k}^{(i)*} \right) = \sum_i \ln \gamma \left(\sum_j r_j (c_{d-k+1}^{(i)*} : c_{d-k}^{(j)*}) \right) + \sum_i \ln \beta(r_i)$$
$$+ n_{d-k}^* \ln \|G^*\| \tag{52}$$

Up to an inessential additive term, this action is exactly of the original type (48).

Let us proceed further in the case of a vanishing external field ($V \equiv 0$). This means that $\gamma(r)$ is a δ-function on the dual group, forcing its argument r to be the trivial representation 0 of the group G. Hence, the chain h is 0, which means that the chain $g = \sum_i r_i c_{d-k}^{(i)*}$, interpreted on the dual complex, has a zero coboundary $\nabla g = 0$. Any coboundary of a $(d - k - 1)$-chain is a solution to this equation, since the operators $\nabla\nabla$ and $\Delta\Delta$ identically vanish. Other solutions might, however, exist depending on the topology of the complex. For instance, a circle drawn on a torus is not necessarily the boundary of a surface on this torus. The homology (cohomology) group of order k is defined as the quotient of the image by Δ (∇) of the $(k + 1)$- $((k - 1)$-) chains by the corresponding kernel. Saying that the cohomology group of order $(d - k)$ is trivial means that $g = \nabla g'$ is the general solution of $\nabla g = 0$.

Show that this is the case for the complex formed by a hypercube and its faces. Verify that the (co)homology groups are stable for any subdivision of the cells and deduce that any lattice *with free boundary conditions* has trivial (co)homology groups.

The chain g' is defined up to the coboundary of a $(d - k - 2)$-chain. From these considerations, the expression of the dual action (in the absence of external fields) can be recast in the form

$$S^* \left(\sum_i r_i c_{d-k-1}^{(i)*} \right) = \sum_i \ln \beta \left(\sum_j r_j (c_{d-k}^{(i)*} : c_{d-k-1}^{(j)*}) \right) \quad (53)$$

$$+ n_{k-1} \ln \|G\| - \ln x_{k+1}$$

where x_{k+1} is the total number of $(k+1)$-chains without boundary (or, equivalently, the number of dual $(d - k - 1)$-chains without coboundary). This number can be extracted from the following recurrence relations

$$x_k = \|G\|^{n_{k+1}} / x_{k+1}, \qquad x_d = \|G\| \quad (54)$$

The dual action (53) is of a form similar to the initial one.

Check that in the case $k = 1$, $d = 2$, $G = Z_2$, one recovers the duality relations for the two-dimensional Ising model.

As an example, a model without an external field defined on cells of dimension $k = d$ is dual to a model with variables defined on sites, interacting with a constant external field. There are no collective effects, and we recover here the triviality of one-dimensional spin models and two-dimensional gauge models.

The four-dimensional Z_2 gauge theory is self-dual. It corresponds to the case $d = 4$, $k = 2$ without an external field, with the further property that Z_2 is self-dual. Hence the free energy per site satisfies the relation

$$F(\beta) = F(\beta^* = -\tfrac{1}{2} \ln \tanh \beta) + 3 \ln \sinh 2\beta \quad (55)$$

High and low temperature regions are related by duality, and formula (55) is similar to the corresponding one for the two-dimensional Ising model. If there is one, and only one, transition, it should occur at the fixed point of the transformation $\beta \to \beta^*$, i.e. at

$$\beta_c = \tfrac{1}{2} \ln(1 + \sqrt{2}) \quad (56)$$

Numerical simulations of this system confirm this result, the observed transition being here of first order.

Finally, let us consider the three-dimensional Z_2 gauge system with bosonic matter (Higgs) fields, the action of which reads

$(U_{ij} = \pm 1, \sigma_i = \pm 1)$

$$S = \beta \sum_p U_{ij}U_{jk}U_{kl}U_{li} + \gamma \sum_{(i,j)} \sigma_i U_{ij}\sigma_j \qquad (57)$$

For a given matter field configuration, we choose the *unitary* gauge corresponding to the transformation $U'_{ij} = \sigma_i U_{ij}\sigma_j$. With this choice, all reference to the matter fields disappears and the model described by equation (57) can be identified to the case $k = 2$ of the general expression (48). In three dimensions, the dual action reads

$$S^* = -\tfrac{1}{2}\ln\tanh\gamma \sum_p U^*_{ij}U^*_{jk}U^*_{kl}U^*_{li} - \tfrac{1}{2}\ln\tanh\beta \sum_{(i,j)} \sigma^*_i U^*_{ij}\sigma^*_j \qquad (58)$$

and the system is again self-dual. The exchange of gauge and matter couplings should be noticed, as well as the interchange of high and low temperature regions. In the convenient variables $\tilde{\beta} = \ln\sinh 2\beta$ and $\tilde{\gamma} = \ln\sinh 2\gamma$, the phase diagram is symmetric with respect to the line $\tilde{\beta} = \tilde{\gamma}$. In the particular case $\gamma = 0$, the three-dimensional pure gauge model is dual to the Ising model. It therefore admits a *second order* transition, and the numerical results for the Ising model allow to locate it at $\beta_c \approx 0.7613$.

(1) One can establish directly this equivalence between the two systems using a reasoning similar to the one presented in chapter 2. One starts from the high temperature diagrammatic expansion; graphs have the topology of closed surfaces made with p plaquettes and yield a contribution $\tanh^p \beta$. These surfaces can be characterized by the set of cubes filling their interior with variables σ_i assigned to the center of each cube: $\sigma_i = -1$ inside the surface, $+1$ outside. A plaquette separating two adjacent cubes i and j does or does not belong to the surface according to the value of $\sigma_i\sigma_j$. Hence it is possible to re-express the sum over all high temperature diagrams as the partition function of an Ising model.

(2) Apply duality to a cyclic group Z_n. This model depends on $\lfloor \tfrac{1}{2}n \rfloor$ parameters and duality yields a relation between these variables, leading to a symmetry of the phase diagram. In particular, deduce the position of the transition $\beta_c = \tfrac{2}{3}\ln(1+\sqrt{3})$ of self-dual models with Z_3 symmetry. Three phases arise in systems with $n \geq 5$, and it is no more possible to find the exact location of the transitions. Duality only relates their respective positions.

(3) Consider the $U(1)$ case. The dual group G^* is the additive group of integers. The corresponding duality has been exploited in chapter 4 for the XY-model.

(4) What is the observable dual to the Wilson loop, and, more generally, to any other observable in an abelian gauge system? Hint: the main difference in the diagrammatic expansion arising from the expression (51) consists in changing the representations r_i of the plaquettes on a given arbitrary surface with boundary C to $r_i \otimes f$ (f being the fundamental representation used to define the Wilson operator). Verify that the dual observable appears as the ratio of a frustrated partition function where some coupling constants have been modified, to the usual partition function. For instance, in three dimensions, for the Z_2 gauge model, the coupling of all links crossing a surface bounded by C is reversed. The high temperature diagrams of the dual Ising model should interlace the curve C in order to have different contributions in the two partition functions. Deduce from this fact that the Wilson loop parameter has a perimeter law at low temperature of the gauge model.

't Hooft has introduced an interesting gauge invariant observable for an arbitrary gauge group provided it has a nontrivial center. This includes of course nonabelian systems. Let us describe this observable in a four dimensional model (the case of interest) leaving the reader to give its generalization. Take a closed loop C^* on the dual lattice bounding a surface Σ^*. The 't Hooft observable is defined as the ratio of a frustrated partition function to an unfrustrated one. The frustration amounts to multiply all plaquette variables dual to those of Σ^* by the same element of the center. In general, this loop observable has a behaviour dual to the Wilson loop (i.e. a perimeter law at large coupling and an area law at small coupling). There are interesting cases, such as the Z_n models mentioned above, where both observables have a perimeter law in an intermediate massless phase. Wilson and 't Hooft loops have nontrivial relative topological properties, analogous to order and disorder variables.

6.2   Structure of the phase diagram

6.2.1   Mean field approximation

In spite of several *a priori* difficulties, such as Elitzur's theorem, the mean field method, as described in chapter 3, allows a first exploration of the phase diagram for a pure gauge system. A trial action is written as

$$S_H = \frac{H}{n} \sum_{(ij)} \text{Re Tr} \, U_{ij} \tag{59}$$

The corresponding partition function per link is

$$z(H) = \exp u(H) = \int \exp \left(\frac{H}{n} \text{Re Tr} \, U \right) dU \tag{60}$$

and hence each field has a nonvanishing average. In matrix notation, the mean value is proportional to the unit matrix

$$\langle U \rangle_H = z(H)^{-1} \int U \exp \left(\frac{H}{n} \text{Re Tr} \, U \right) dU = u'(H) \tag{61}$$

With such a gauge noninvariant trial action, it is clear that the conditions for the validity of Elitzur's theorem are not fulfilled. This defect will be cured later. It is nevertheless interesting to use this trial action to derive inequalities, since the results obtained in this way for invariant quantities are correct, as we shall see. The calculation scheme is the same as in chapter 3. The inequality (3.14) leads to

$$\frac{\ln Z}{Nd} \geq \text{Sup}_H \left\{ u(H) - H u'(H) + \tfrac{1}{2} \beta_J (d-1) u'(H)^4 \right\} \tag{62}$$

The right-hand side is the *mean field approximation* for the free energy. The function $u(H)$, defined through the integral (60) of a convex function (an exponential) over a compact domain, is also a convex function, which is minimal at $H = 0$ where it behaves as H^2. Hence the maximization of the r.h.s. of (62) leads to the solution $H = 0$, which is unique at $\beta_J = 0$ and in the vicinity of this point. In contradistinction with spin systems, this maximum always remains a local maximum, whatever the temperature. Indeed the term $u'(H)^4$ behaves as H^4, and therefore cannot change the curvature at $H = 0$. Nevertheless, this maximum does not remain the highest one as β_J increases. The behaviour of the function to be maximized is displayed in figure 4 for several values of $\beta^* = 2\beta_J(d-1)$. For a critical value $\beta^* = \beta_c^*$, the highest maximum suddenly changes its location. Hence we observe a first order transition, with discontinuities in the values of some physical observables. The corresponding characteristic shape of the free energy is displayed in figure 5. The dotted curve corresponds to the (unphysical) position of the minimum. Dashed parts display

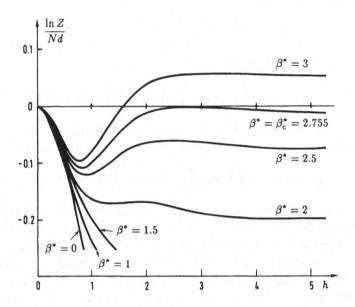

Fig. 4 Trial free energies for a Z_2 gauge group, to be maximized in H for different values of $\beta^* = 2\beta_J(d-1)$.

the position of secondary maxima, which are physically observable as metastable phases. Correlation lengths remain finite at the critical point T. Physical quantities admit analytic continuation at this point which represents metastable states. Therefore, there is *no continuous limit* in the present approximation. The end points A (located at infinity) and B of the metastable regions correspond to singularities of the observables in the variable β_J. The transition at T is a deconfining one. The Wilson loop observable, corresponding to a curve with a perimeter P, has a mean value

$$W(C) = \left\langle \mathrm{Tr} \prod_C U_{ij} \right\rangle = n[u'(H)]^P \qquad (63)$$

In the low temperature region ($H \neq 0$), it decreases according to a perimeter law, which shows the deconfined character of this phase. The area law of the confined high temperature region cannot

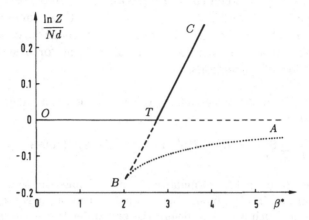

Fig. 5   The free energy in the mean field approximation (Z_2 group).

be obtained in this approximation without further corrections ($W(C) = 0$ in this region in the present context).

The simple pattern given by the mean field approximation is confirmed by numerical simulations. Let us discuss first the Z_2 case. The observed transition is indeed a first order one as soon as the dimension is greater or equal to four. Its location is reasonably predicted. In dimension four, mean field yields $\beta_c = 0.459$ to be compared with the position $\beta_c = 0.4407$, exactly known from duality. In dimension three, although the predicted order of the transition is incorrect – it should be a second order one, as the system is dual to the Ising model –, there is still some similarity between the position $\beta_c = 0.7613$ and the mean field prediction $\beta_c = 0.689$.

The upper critical dimension for nonabelian continuous gauge systems is expected to be $d = 4$, where the theory becomes asymptotically free. In higher dimensions, mean field predictions are confirmed by numerical simulations. For instance, the first order transition of the $SU(2)$ system for $d = 5$ is predicted to be at $\beta_c = 2.12$ to leading order and has been observed at $\beta_c = 1.64$. At $d = 4$, there is no evidence for any transition in the numerical simulations. However, observable quantities show rapid changes in a small region. If one explains this fact by a remembrance of the transition occuring in larger dimensions, one observes again some similarity between the observed and predicted positions (for $SU(2)$ e.g., 2.826 and 2.2 respectively). All these facts seem to indicate that this approximation provides a reasonable description of the system and is therefore a starting point for a systematic computation of corrections.

(1) Perform the same calculations using the axial gauge. The result reads

$$\frac{\ln Z}{N(d-1)} \geq \operatorname*{Sup}_{H} \left\{ u(H) - Hu'(H) + \tfrac{1}{2}\beta_J (d-2)u'(H)^4 + \beta_J u'(H)^2 \right\}$$

(64)

The last term in this formula might now change the sign of the curvature at $H = 0$. In sufficiently high dimension, it is too weak to produce such a change before the first order transition occurs. Its effect is just to bring back the end point A of the metastable strong coupling phase at a finite distance

$$\beta_A = [2u''(0)]^{-1}$$

(65)

As the dimension decreases, the swallowtail shape $TABT$ shrinks and disappears at a critical dimension

$$d_c = 2 - \frac{1}{6}\frac{u''''(0)}{u''(0)^3}$$

(66)

This choice of gauge seems therefore to be of interest in the implementation of the mean field approximation. Moreover, having made such a choice of gauge, the use of a noninvariant trial action (59) no longer presents theoretical problems. We will also see that, for continuous groups, one cannot escape an explicit choice of gauge for a determination of the corrections. Nevertheless, other problems appear. For instance, let us consider a Wilson loop of dimensions $R \times T$ using the time direction along which the gauge field has been set equal to unity. The approximation leads to a behaviour in R rather than $R + T$ since $W(C) = n[u'(H)]^{2R}$. This result prevents in particular an exponential decrease for fixed R as $T \to \infty$ which can be predicted from rigorous bounds.

This difficulty disappears when corrections are added in the case of a continuous gauge group, but remains when dealing with discrete groups. The mean field approximation appears in this case as a dilute gas approximation, each gauge field fluctuating independently in the mean field without interacting with its neighbours. This technique can be seen as a partial resummation of the low temperature expansion. The flip of a gauge field in a temporal direction is replaced, in the axial gauge, by the flips of an infinity of spatial fields (figure 6), but nevertheless leads to a finite energy change. The approximation does not take into account this phenomenon, and finite order corrections will not change this feature. In particular, the reader can check that the low temperature behaviour is incorrect. Using the above observation, it is now possible to correct this omission by adding the contribution of a dilute gas of such defects. The correct low temperature behaviour and the perimeter law for temporal Wilson loops are restored. However, the computations of corrections is much more complex, and one generally prefers to use the method which is described below and which does not require the gauge to be fixed.

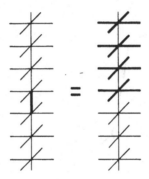

Fig. 6   The equivalent of flipping a temporal gauge field in the axial gauge.

(2) Using the mean field approximation, study the generalized action for a $SU(n)$ gauge group

$$S = \sum_{p} \left(\beta_{\mathrm{J}} \frac{1}{n} \mathrm{Re\, Tr}\, U_{\mathrm{p}} + \beta_{\mathrm{A}} \frac{1}{n^2 - 1} \left| \mathrm{Tr}\, U_{\mathrm{p}} \right|^2 \right) \qquad (67)$$

This extension is interesting, because it allows to probe the role of the center of the group. The additional term is invariant under transformations by an element belonging to the center Z_n: $U_{ij} \rightarrow \exp(2ik\pi/n)U_{ij}$ and is the fundamental Wilson action for the group $SU(n)/Z_n$. The two-dimensional phase diagram provides an interpolation between several limits

$\beta_{\mathrm{A}} = 0$ Wilson action for $SU(n)$.

$\beta_{\mathrm{J}} = 0$ Wilson action for $SU(n)/Z_n$.

$\beta_A = \infty$ action for the discrete gauge group Z_n, since this limit selects the configurations in which all plaquette variables U_p belong to the center. Hence, using a gauge transformation, all gauge fields U_{ij} may be chosen in the center Z_n.

$\beta_J = \infty$ forces all plaquette variables to be equal to unity and yields a trivial model, with only pure gauge configurations and without any transition.

Some care in the elaboration of the mean field approximation is needed in order to treat central elements correctly. We write $U_{ij} = V_{ij}\varepsilon_{ij}$, and treat independently V_{ij} and ε_{ij} with two associated mean fields H and h. One obtains a phase diagram which is diplayed in figure 7, with three different regions

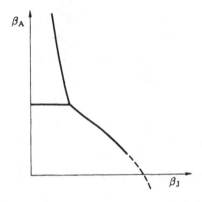

Fig. 7   Phase diagram predicted by the mean field approximation for extended theories with the action (67).

a) strong coupling phase for β_J, β_A small ($H = 0$, $h = 0$),
b) weak coupling phase of $SU(n)/Z_n$ ($H = 0$, $h \neq 0$) obtained as β_A is large and β_J is small (fields fluctuate around central values),
c) weak coupling phase ($H \neq 0$, $h \neq 0$).

All transition lines are of first order. Numerical simulations confirm this prediction. However, for $d = 4$, a part of the line (dashed in the figure) disappears. Phases *a* and *c* are continuously related, while the weak coupling region $SU(n)/Z_n$ remains isolated.

The end point is not very far from the axis $\beta_A = 0$. The crossover zone observed in $SU(n)$ systems ($n = 2, 3$) in which the properties of the system vary rapidly, can be explained by the presence of this neighbouring singular point. Note also that the model with $SU(n)/Z_n$ symmetry has no order parameter of the Wilson type (since this simple group is centerless). Nevertheless it exhibits a first order transition, even at $d = 4$.

6.2.2   Corrections to mean field and restoration of gauge invariance

We study now the corrections due to the fluctuations around mean field. This is done using the technique developed in chapter 3, which uses a Laplace transform of the measure on the fields. However, in the present case, it is not possible to use the identity (3.142). The calculation can nevertheless be performed, using a double steepest descent method that we now describe. The exponential representation of the δ-function

$$\delta(U_{\alpha\beta} - V_{\alpha\beta}) = \frac{1}{2i\pi} \int_{-i\infty}^{i\infty} \exp H^*_{\beta\alpha}(U_{\alpha\beta} - V_{\alpha\beta}) \, dH_{\beta\alpha} \qquad (68)$$

is inserted in the definition of the partition function for every matrix element $(U_{ij})_{\alpha\beta}$ of the gauge fields. Here V_{ij} and H_{ij} are $n \times n$ complex matrices on which we integrate, using flat measures

$$[dV] = \prod_{(ij)\alpha\beta} d(V_{ij})_{\alpha\beta} \qquad \text{and} \qquad [dH] = \prod_{(ij)\alpha\beta} d(H_{ij})_{\alpha\beta}$$

not to be confused with the Haar measure dU on the group (which appears as a submanifold of these $n \times n$ matrices). For each link (ij), only one pair (H, V) of these matrices is introduced, regardless of the link orientation, according to $V_{ji} \equiv V_{ij}^\dagger$, $H_{ji} \equiv H_{ij}^\dagger$. The partition function reads

$$Z(J) = \int \exp \left\{ S(V_{ij}) + \sum_{(ij)} \mathrm{Re\, Tr}\, J_{ji}^\dagger U_{ij} \right\}$$
$$\prod_{ij} \delta^{(n^2)}(U_{ij} - V_{ij})[dV] \prod_{ij} dU_{ij}$$

$$= \int [\mathrm{d}H][\mathrm{d}V] \prod_{ij} \mathrm{d}U_{ij}$$

$$\exp\left\{ S(V_{ij}) + \sum_{ij} \left(\mathrm{Re}\,\mathrm{Tr}(J_{ji} + H_{ji})^\dagger U_{ij} - \mathrm{Re}\,\mathrm{Tr}\,H_{ji}^\dagger V_{ij} \right) \right\}$$

$$(69)$$

Generalizing the function (60) to a matrix external field

$$u(H) = \int \exp\left(\frac{1}{n} \mathrm{Re}\,\mathrm{Tr}\,H^\dagger U \right) \mathrm{d}U \qquad (70)$$

the integration over gauge fields can be performed, and one gets

$$Z(J) = \int \exp\left\{ S(V_{ij}) + \sum_{ij} u(H_{ij} + J_{ij}) - \mathrm{Re}\,\mathrm{Tr}\,H_{ij}^\dagger V_{ji} \right\} [\mathrm{d}V][\mathrm{d}H]$$

$$(71)$$

This integral is now computed using the steepest descent method. For spin systems, the integration over V is Gaussian and can be done explicitly, without further corrections. This is no longer true in the present case. To leading order, we find

$$\ln Z = \sum_{ij} \left[u(\mathcal{H}_{ij} + J_{ij}) - \mathrm{Re}\,\mathrm{Tr}\,\mathcal{H}_{ij}^\dagger \mathcal{V}_{ij} \right] + S(\mathcal{V}_{ij}) \qquad (72)$$

This expression is stationary for $H = \mathcal{H}$, $V = \mathcal{V}$ given by

$$\mathcal{V}_{ij} = \frac{\partial u(\mathcal{H}_{ij} + J_{ij})}{\partial H_{ij}^\dagger} \qquad (73)$$

$$\mathcal{H}_{ij} = \frac{\partial S(\mathcal{V}_{ij})}{\partial V_{ji}} \qquad (74)$$

In the particular case of a uniform solution with constant matrices proportional to unity, these expressions can be identified with those in equation (62), so that

$$\mathcal{H}_{ij} = \frac{H}{n} I \qquad (75)$$

We are now in a position to compute systematic corrections to this approximation. But first, let us solve the paradoxes concerning gauge invariance. Saddle point equations (73)–(74), although gauge invariant, may have noninvariant solutions $\{\mathcal{V}_{ij}, \mathcal{H}_{ij}\}$; but then any gauge transformed configuration $\{g_i \mathcal{V}_{ij} g_j^{-1}, g_i \mathcal{H}_{ij} g_j^{-1}\}$ is also a solution. When computing the integral (71), one should

separate the degrees of freedom corresponding to gauge transformations from other "transverse"degrees of freedom. The for-
mer yield a multiplicative factor (to a power equal to the number
of sites) since the integrand is independent from these transformations. The steepest descent method can only be applied to
the latter. Note the transformation law of the sources $J_{ij} \rightarrow$
$g_i J_{ij} g_j^{-1}$, which induces corresponding transformations on the observables. Starting from the solution (75), a set of other saddle points can thus be constructed by gauge transformations.
For instance, the value $\langle U_{ij} \rangle = u'(H)$ for the solution (75) becomes $g_i \langle U_{ij} \rangle g_j^{-1}$. Integrating now on gauge degrees of freedom
gives

$$\langle U_{ij} \rangle = \int g_i u'(H) g_j^{-1} \prod \mathrm{d}g_i \equiv 0 \qquad (76)$$

As a consequence, Elitzur's theorem is satisfied. We also see
that the mean field approximation as presented in the previous
subsection yields correct results for gauge invariant quantities.
The virtue of the present formalism is to allow the systematic
computation of corrections.

In the case of a discrete group, the saddle points generated by
gauge transformations are well separated, and there are no zero
modes. The problem is to know whether the contributions of
the saddle points should be *summed* or *averaged*. The question
makes sense when the number of these saddle points can vary. The
answer is that they should be summed over. Let us prove this
for the example of an integral in one variable $\int \exp f(x) dx$, where
$f(x)$ is analytic with several maxima (figure 8). It is convenient
to introduce the variable $y = -f(x)$. The integral becomes
$\oint_C e^{-y} f'(x)^{-1} dy$ along a curve C in the complex y plane displayed
in figure 8b. It surrounds the singularities of the multivalued
function $x(y)$. Since we deal with analytic functions, it is possible
to distort the contour into two independent contours, as shown in
figure 8c. The contribution of each one is evaluated using a Taylor
expansion around the singular end point, and is found to coincide
with the saddle point contribution. This proves the summation
prescription over all equivalent saddle point contributions, as long
as the analyticity domain of the function is sufficient to allow the
distortion of contours. If the two maxima have different heights, the
higher one dominates exponentially over the lower. The main effect
of the minimum is to limit the radius of convergence of the Taylor

series of $1/f'(x(y))$. Hence, after integration, the series obtained by the steepest descent method are generally not convergent, but only asymptotic.

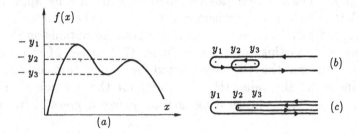

Fig. 8   Various integration contours (*b*) and their deformations (*c*) when using the steepest descent method for a function with several extrema (*a*).

In the application of this method to gauge systems, one deals with integrals in several variables. The analyticity properties required to distort the contours are harder to establish. Nevertheless, by analogy, we presume that the summation prescription is the correct one.

In particular, let us discuss the validity of the method near the critical dimension where a first order transition turns into a second order one. The transition point and the two end points of the metastable zones coalesce. Near the end point of the strong coupling region, the unique saddle point splits into gauge equivalent saddle points (2^N for Z_2). In this case, the minimum separating two neighbouring saddle points is very close, and the convergence of the method will be particularly poor. Note the additive factor $N \ln 2$ which appears in the expression for the free energy in the weak coupling phase, which is the effect of the summation prescription and is required to obtain the zero coupling entropy.

6.2.3   Discrete groups: $1/d$ expansion

It is expected that the mean field approximation gives good results in high dimension. For spin models, we have seen that corrections can be rearranged as a $1/d$ expansion. Let us examine now the case of gauge systems.

We will show, in chapter 7 (volume 2), that diagrammatic series can be ordered into successive corrections to the mean field

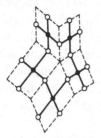

Fig. 9   A typical diagram occuring in the computation of corrections to the mean field approximation.

approximation, and we will describe the diagrams which are to be computed. The reader will find there all the technical details, and we only give here a quick description of the method. Looking at formula (71), one remarks that each term generated in the expansion of $\exp S(V_{ij})$ can be described as a set of plaquettes. The term $\operatorname{Re} \operatorname{Tr} H_{ij}^{\dagger} V_{ji}$ couples the fields V lying on the boundaries of these plaquettes to $u(H_{ij})$. Hence, if k plaquettes meet along a link ij, one gets a contribution $\partial^{k} u(H_{ij})/\partial H_{ij}^{k}$. The reader can also verify – this will be proved in chapter 7 – that *reducible* diagrams, i.e. those which can be disconnected by the omission of one link, should not be considered. Indeed, the self-consistent field H_{ij} is such that reducible contributions are automatically generated by formula (74).

In figure 9 we show a typical diagram in which each plaquette is represented by a black point located at its center and joined to the four midpoints of adjacent links (open dots). The irreducibility condition forbids the possibility of arborescences, the resulting diagram being strongly connected. This is the case displayed in figure 9, showing the general aspect of a one-loop diagram.

Let us construct such a loop with n plaquettes, that is with $2n$ segments of length $a/2$. As this loop should be closed, these segments can be associated in parallel pairs of opposite direction. Therefore, the loop uses at most n different directions chosen among the d possible ones. The choice of these directions yields a factor $d^{n}(1 + \mathcal{O}(1/d))$ in the contribution of the diagram. Moreover, each plaquette comes with a factor β_{J}, which is of order $1/d$ (because it appears only in a combination $2\beta_{\mathrm{J}}(d - 1)$ in the mean field approximation, so that this expression is kept fixed in the limit $d \to \infty$). The loop yields therefore at most a contribution of order $1 + \mathcal{O}(1/d)$, which is subdominant with respect to the leading term (62), of order d. Any additional loop in a more complex diagram will provide additional $1/d$ factors. Indeed, due to the irreducibility condition, it should share at least one link with previously constructed loops, the orientations of which have already been chosen.

Thus the diagrams contributing to the first $1/d$ correction have been characterized. They have only one loop and are all similar to the one drawn in figure 9. Their number is infinite, in contradistinction with spin models. In the latter case, a loop with n links is of order $\beta^n d^{n/2}$, behaving as $(1/d)^{n/2}$ as a function of its size n (for fixed βd). In the gauge case, despite the infinite number of diagrams, it is possible to re-sum their contributions. The diagram of figure 9 with n plaquettes contributes a term

$$\frac{d![2\beta_J u'(H)u''(H)]^n}{2n(d-n)!} \tag{77}$$

Prove this result.

For large d, these terms have to be summed for $n \geq 3$. The two plaquette term should also be added and the final correction reads

$$\frac{\ln Z}{Nd} = \text{(mean field contribution)}$$

$$+ \left[-\tfrac{1}{2}\ln(1-x) - \tfrac{1}{2}x + \tfrac{1}{8}x^2 + \tfrac{1}{4}xy + \tfrac{1}{16}y^2 \right]\frac{1}{d} + \mathcal{O}\left(\frac{1}{d^2}\right) \tag{78}$$

with

$$x = 2\beta_J(d-1)u'^2(H)u''(H) \qquad y = 2\beta_J(d-1)u''^2(H) \tag{79}$$

An additional contribution $d^{-1}\ln\|G\|$ which takes into account the multiplicity of the saddle points, as discussed in the previous subsection, should be added when H does not vanish.

Compute the next correction.

Figure 10 displays the free energy per link for the Z_2 gauge group, taking into account the first two corrections. The low

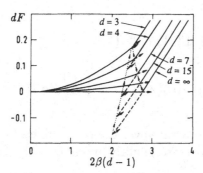

Fig. 10   Free energy of the Z_2 gauge system computed using the first two corrections in $1/d$.

temperature metastable region shrinks as the dimension decreases and disappears for $d_c \simeq 2.9$, very close to $d_c = 3$ which is expected to be the critical dimension for this model. However, the second order transition occuring in dimension three cannot be seriously analyzed by this method. Note also that the end point of the high temperature metastable region remains at infinity. We shall see later that its position behaves as $d^{3/4}$, and is thus out of reach of this $1/d$ expansion technique. From our previous discussion, these features are expected.

6.2.4   Continuous groups: computation of corrections

We have already outlined the necessity of gauge fixing for continuous gauge groups. The axial gauge choice leads to simpler calculations.   We treat in detail the Wilson action for a d-dimensional model with $SU(n)$ or $U(n)$ gauge group. We use the same method as above and write the partition function as

$$Z = \int \prod_{\substack{\text{spatial} \\ \text{links}}} dU_l \, \exp\left\{ \frac{\beta_J}{n} \mathrm{Re} \left[\sum_{\substack{\text{spatial} \\ \text{plaquettes}}} \mathrm{Tr}\, U_p + \sum_{\substack{\text{timelike} \\ \text{plaquettes}}} \mathrm{Tr}\, U_l U_{l'}^\dagger \right] \right\}$$

$$(80)$$

We insert

$$1 = \prod_{\substack{\text{spatial} \\ \text{links}}} \int \mathrm{d}^{2n^2} V_l \, \delta^{(2n^2)}(V_l - U_l)$$

$$= \prod_l \int \mathrm{d}V_l \mathrm{d}H_l (2\mathrm{i}\pi)^{-2n^2} \exp\left\{\frac{1}{n}\mathrm{Re}\,\mathrm{Tr}(U - V)\right\} \tag{81}$$

It is possible to integrate over the U variables. In the axial gauge, formulae (72)–(74) are slightly modified. To lowest order, the result becomes

$$\frac{\ln Z}{N} = (d-1)[u(\mathcal{H}) - \mathcal{H}\mathcal{V} + \tfrac{1}{2}\beta_{\mathrm{J}}(d-2)\mathcal{V}^4 + \beta_{\mathrm{J}}\mathcal{V}^2] \tag{82}$$

with the following saddle point conditions

$$\begin{cases} \mathcal{V} = u'(\mathcal{H}) \\ \mathcal{H} = \beta_{\mathrm{J}}\mathcal{V}[2 + 2(d-2)\mathcal{V}^2] \end{cases} \tag{83}$$

The action is then expanded up to quadratic terms around the saddle point. In order to simplify the calculations of the second derivatives needed for the determination of the first correction, we introduce projectors on the trivial and adjoint representations of $SU(n)$

$$\begin{cases} P^{(0)}_{\alpha\beta;\gamma\delta} = \dfrac{1}{n}\delta_{\alpha\beta}\delta_{\gamma\delta} \\[2mm] P^{(1)}_{\alpha\beta;\gamma\delta} = \delta_{\alpha\delta}\delta_{\beta\gamma} - \dfrac{1}{n}\delta_{\alpha\beta}\delta_{\gamma\delta} \end{cases} \tag{84}$$

The required derivatives can be parametrized as

$$\begin{cases} \left.\dfrac{\partial^2 u(H)}{\partial H_{\beta\alpha}\partial H_{\delta\gamma}}\right|_{H=\mathcal{H}1} = \left.\dfrac{\partial^2 u(H)}{\partial H^*_{\beta\alpha}\partial H^*_{\delta\gamma}}\right|_{H=\mathcal{H}1} = A P^{(0)}_{\alpha\beta;\gamma\delta} + B P^{(1)}_{\alpha\beta;\gamma\delta} \\[4mm] \left.\dfrac{\partial^2 u(H)}{\partial H_{\beta\alpha}\partial H^*_{\delta\gamma}}\right|_{H=\mathcal{H}1} = A' P^{(0)}_{\alpha\beta;\gamma\delta} + B' P^{(1)}_{\alpha\beta;\gamma\delta} \end{cases} \tag{85}$$

We split the operators according to their action on the antihermitian and hermitian parts of V. After a Fourier transformation, one finds respectively the following $(d-1) \times (d-1)$-matrices

$$\Delta_{\mu\nu}(k) = \left[2\cos k_d + 2\mathcal{V}^2 \sum_{\rho \neq \mu, d} \cos k_\rho\right]\delta_{\mu\nu} +$$

$$+ 4\mathcal{V}^2(1 - \delta_{\mu\nu})\exp(\tfrac{1}{2}\mathrm{i}(k_\mu - k_\nu))\sin\tfrac{1}{2}k_\mu\sin\tfrac{1}{2}k_\nu$$

$$\bar{\Delta}_{\mu\nu}(k) = \left[2\cos k_d + 2\mathcal{V}^2 \sum_{\rho \neq \mu, d} \cos k_\rho\right]\delta_{\mu\nu} +$$

$$+ 4\mathcal{V}^2(1 - \delta_{\mu\nu})\exp(\tfrac{1}{2}\mathrm{i}(k_\mu - k_\nu))\cos\tfrac{1}{2}k_\mu\cos\tfrac{1}{2}k_\nu \tag{86}$$

Functional integration yields to order one loop

$$Z = Z_{\text{mean field}}$$

$$\times \left[\det^{-\frac{1}{2}}(1 - 2\beta_{\mathrm{J}}n(A' - A)\Delta)\det^{-\frac{1}{2}(n^2-1)}(1 - 2\beta_{\mathrm{J}}n(B' - B)\Delta)\right]^N$$

$$\times \left[\det^{-\frac{1}{2}}(1 - 2\beta_{\mathrm{J}}n(A' + A)\bar{\Delta})\det^{-\frac{1}{2}(n^2-1)}(1 - 2\beta_{\mathrm{J}}n(B' + B)\bar{\Delta})\right]^N \tag{87}$$

Establish this formula and determine the next correction.

Without gauge fixing, the propagator $\Delta_{\mu\nu}$ takes a slightly different form. In both brackets in equation (86), the $\cos k_d$ term is multiplied by $\mathcal{V}^2$, and the indices μ and ν can take the value d. Hence the matrix Δ has a zero eigenvalue, which corresponds to the eigenvector $\sin\tfrac{1}{2}k_\mu\exp\tfrac{1}{2}\mathrm{i}k_\mu$ and which reflects the existence of local gauge transformations. The mass spectrum can be read on formula (87) as corresponding to the zero eigenvalues of the operators appearing in the determinants.

It is instructive to look at the weak coupling limit (H large). Taking into account the invariance of $u(H)$ with respect to unitary transformations, one finds

$$\begin{cases} B' - B = \dfrac{u'(\mathcal{H})}{2n\mathcal{H}} = \dfrac{\mathcal{V}}{2n\mathcal{H}} \\[2mm] A' - A = \begin{cases} B' - B & \text{for } U(n) \\ \mathcal{O}(1/H^2) & \text{for } SU(n) \end{cases} \end{cases} \tag{88}$$

Using the asymptotic forms of $u(H)$ for H large

$$u(H) = \begin{cases} H + \left(\ln \dfrac{\prod_{k=1}^{n-1} k!}{(2\pi)^{\frac{1}{2}n}} + \frac{1}{2}n^2 \ln n \right) - \frac{1}{2}n^2 \ln H \\[2mm] \quad + \dfrac{n^2}{8H} + \dfrac{n^4}{16H^2} + \cdots \qquad \text{for } U(n) \\[4mm] H + \left(\ln \dfrac{\prod_{k=1}^{n-1} k!}{(2\pi)^{\frac{1}{2}(n-1)}} + \frac{1}{2}(n^2 - 2) \ln n \right) - \frac{1}{2}(n^2 - 1)\ln H \\[2mm] \quad - \dfrac{n^2 - 1}{8H} - \dfrac{(2n^2 - 5)(n^2 - 1)}{48H^2} + \cdots \qquad \text{for } SU(n) \end{cases}$$

$$(89)$$

one finds

$$\mathcal{H} \sim 2\beta_J(d-1) \qquad \mathcal{V} \sim 1 - \frac{A_0}{4\beta_J(d-1)} \qquad (90)$$

A_0 being the coefficient of $-\ln H$ in the expansions (89). Similarly

$$\begin{cases} B \sim -B' \sim -\dfrac{1}{4nH} \\[3mm] A \sim -A' \sim \begin{cases} -\dfrac{1}{4nH} & \text{for } U(n) \\[2mm] \mathcal{O}(1/H^2) & \text{for } SU(n) \end{cases} \end{cases} \qquad (91)$$

Hence

$$\frac{\ln Z}{N} = \frac{\ln Z_{\text{mean field}}}{N} - \frac{1}{2}\left[\begin{matrix} n^2 \\ n^2 - 1 \end{matrix} \right] \operatorname{Tr} \ln \left(1 - \frac{\beta_J \mathcal{V}}{\mathcal{H}} \Delta \right) + \mathcal{O}\left(\frac{1}{\beta_J} \right)$$

$$(92)$$

Between the square brackets, the upper value n^2 refers to the $U(n)$ gauge group, and the lower one $n^2 - 1$ to $SU(n)$. Only one operator remains to be studied. At fixed momentum, this $(d-1) \times (d-1)$ matrix has two distinct eigenvalues. The first one

$$\lambda_1 = 4\beta_J(\mathcal{V}/\mathcal{H}) \sin^2 \frac{1}{2}k_d$$

vanishes at $k_d = 0$, and corresponds to the massless propagation of the gauge degrees of freedom which have not been fixed by the axial gauge choice (these are the time independent gauge transformations). The second one

$$\lambda_2 = 4\beta_J(\mathcal{V}/\mathcal{H}) \sin^2 \frac{1}{2}k_d + \mathcal{V}^2 \sum_{\mu=1}^{d-1} \sin^2 \frac{1}{2}k_\mu$$

is $d - 2$ times degenerate and vanishes only at $\mathbf{k} = 0$. Other hermitian channels are related to the operator $\bar{\Delta}$ and have no massless excitations. Indeed, it was expected that in this large β_J limit, the antihermitian channels represent the most important contribution to the corrections, since they generate the gauge potential in the continuous limit.

For large β_J, the energy per plaquette behaves as

$$E = \frac{2}{d(d-1)} \frac{\partial}{\partial \beta_J} \frac{\ln Z}{N} = \frac{1}{d}[(d-2)\mathcal{V}^4 + 2\mathcal{V}^2] \sim 1 - \frac{A_0}{d\beta_J} + \cdots \quad (93)$$

which is the expected result. Without gauge fixing, we would have obtained $1 - A_0/\beta_J(d-1)$ and this incorrect result would then have to be corrected to all orders. This phenomenon is general with continuous groups. At any order in $1/\beta_J$, only a finite number of terms contributes to the corrections to the mean field approximation when the gauge has been fixed. This was not the case for the discrete gauge groups, and we have seen that it was important not to fix the gauge in order to avoid defects of infinite order.

Treat the case of $SU(2)$. Some differences appear, because the representations of this group are real (equivalent to their complex conjugate representations). Using the parametrization $H = h_0 - i\mathbf{h}.\sigma$ in terms of Pauli matrices, one sees in particular that Imh plays no role. The integration over this variable yields a δ function for ImV which decouples to all orders.

Let us now look at the phase diagram, taking into account the corrections. The free energy of the strong coupling solution ($\mathcal{H} = \mathcal{V} = 0$) is compared to the weak coupling one (approximation (93) is often sufficient for this purpose). Predictions of first order transitions are very close to results from numerical simulations, as can be seen on table I. The most striking results concern $SU(2)$ and $SU(3)$ groups in four dimensions. As shown on figure 11, weak and strong coupling arcs do not meet, but remain parallel and very close to each other in a large domain. It is of course not possible to say that the corrected mean field predicts the absence of a transition, as is suggested in numerical simulations, but this is very encouraging for this technique.

Table I. Corrected mean field predictions versus numerical results for the location of the transition

d	group	mean field	Monte-Carlo
4	$U(1)$	1.03	1.01
	$U(2)$	3.45	3.30 ± 0.05
	$U(3)$	7.30	6.88 ± 0.05
	$U(4)$	12.5	12.14 ± 0.7
	$U(5)$	19.5	18.8 ± 1.1
	$U(6)$	28.0	27.0 ± 2.6
	$U(\infty)$	$0.38 \times 2n^2$	$0.375 \times 2n^{2*}$
	$SU(2)$	†	†
	$SU(3)$	†	†
	$SU(4)$	11.6	10.2
	$SU(5)$	18.0	16.4 ± 0.2
	$SU(6)$	26.5	24.0 ± 1.0
5	$SU(2)$	1.77	1.64

*extrapolation
†unobserved transition.

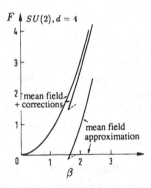

Fig. 11   Corrections to the mean field approximation for $SU(2)$ in four dimensions, according to H. Flyvbjerg, B. Lautrup and J.-B. Zuber, *Phys. Lett.* **110B**, 179 (1982).

6.3 Strong coupling expansions

6.3.1 Convergence

In the strong coupling region, it is natural to expand the Boltzmann weight as a power series in the small parameter β_J

$$\exp S = \prod_p \exp \beta_J \chi(U_p) = \prod_p (1 + \beta_J \chi(U_p) + \tfrac{1}{2}\beta_J^2 \chi^2(U_p) + \cdots) \quad (94)$$

One then further expands the product over the plaquettes, and each term is integrated over gauge fields. It is useful to associate a geometrical representation to each term – which we call a *diagram* or *graph* – as the set of plaquettes used in its contribution. A given plaquette may appear several times, according to the power of $\chi(U_p)$ selected in the expansion of the r.h.s. of (94). In the next subsection, we give a technique for resumming the contributions of plaquettes lying at the same location. This method makes full use of the structure of the gauge group and leads to easier integrations over gauge fields.

A given link cannot bound one and only one plaquette of the diagram since the integration over the corresponding gauge field leads to a zero contribution. Hence links are shared by at least two plaquettes. In particular, if one only considers diagrams such that any link bounds exactly two plaquettes, such graphs have the topology of simple closed surfaces made with plaquettes. A more complex diagram has also singular lines along which several open simple surfaces share their boundaries.

The general techniques needed for determining these diagrammatic series are described in detail in chapter 7 (volume 2). We only mention in this section the particular features of gauge theories and analyze the corresponding results. First, we want to examine the convergence of the expansions. The method consists of establishing an upper bound for the term of order n in the strong coupling expansion for the average $\langle X \rangle$ of a local observable X. Local means here that only a finite number of gauge fields appear in the definition of X. Let $\mathcal{L}$ be the finite, non-empty set of corresponding links. The convergence theorem will also be valid for the free energy per link, since its first derivative is the internal energy $E = \langle \chi(U_p) \rangle$ which is local.

We assume that the plaquette action $\chi(U_p)$ is bounded; this is the case for discrete or compact Lie groups, on which we focus our attention. We also assume that it is positive. This is not really a restriction, since this condition can be fulfilled by adding an inessential constant to the action. Hence, in some strong coupling region, the following inequality holds

$$0 \le \Omega(U_p) \equiv \exp \beta_J \chi(U_p) - 1 \le C_1 \beta_J \tag{95}$$

The lattice Λ is finite, but arbitrarily large. We take the thermodynamic limit at the end of the reasoning, after obtaining a radius of convergence independent of the lattice size. Using these conventions, the average of the observable reads

$$\langle X \rangle = Z_\Lambda^{-1} \int \prod_l dU_l \prod_{p \subset \Lambda} [1 + \Omega(U_p)] X \tag{96}$$

with

$$Z_\Lambda = \int \prod_l dU_l \prod_{p \subset \Lambda} [1 + \Omega(U_p)] \tag{97}$$

Let us expand the product in the integrand (96). Each term is interpreted as a diagram (set of plaquettes) $\mathcal{D}$ which can be split into two parts. The first, $\mathcal{D}_1$, is the union of all connected parts of $\mathcal{D}$ such that at least one link of $\mathcal{L}$ is adjacent to one plaquette of $\mathcal{D}_1$. The second is the complementary set $\mathcal{D}_2 = \mathcal{D} \backslash \mathcal{D}_1$. The integral (96) is factorized into the product of the two corresponding contributions. If $\mathcal{D}_1$ is fixed, all corresponding diagrams $\mathcal{D}$ are obtained by adding any diagram of the partition function of the lattice $\Lambda \backslash \overline{\mathcal{D}_1 \cup \mathcal{L}}$. This lattice is obtained from Λ by suppressing the set of plaquettes, (abusively) denoted as $\overline{\mathcal{D}_1 \cup \mathcal{L}}$, which are adjacent to a link of $\mathcal{L}$ or to a link bounding a plaquette of $\mathcal{D}_1$. Hence

$$\langle X \rangle = \sum_{\mathcal{D}_1} \left\{ \int \prod_{l \subset \mathcal{D}_1 \cup \mathcal{L}} dU_l \prod_{p \in \mathcal{D}_1} \Omega(U_p) X \right\} Z_{\Lambda \backslash \overline{\mathcal{D}_1 \cup \mathcal{L}}} Z_\Lambda^{-1} \tag{98}$$

It is now possible to find an upper bound for the various terms involved in this formula. As Ω is positive and since there are more plaquettes in Λ than in $\Lambda \backslash \overline{\mathcal{D}_1 \cup \mathcal{L}}$, the ratio $Z_{\Lambda \backslash \overline{\mathcal{D}_1 \cup \mathcal{L}}} Z_\Lambda^{-1}$ is less than unity. The function X is local and depends on a finite number of gauge fields varying in a compact or discrete group; its absolute value is therefore bounded by a constant C_2. Finally, the inequality (95) can be used to majorize the products of Ω over the plaquettes of $\mathcal{D}_1$, and one gets

$$|\langle X \rangle| \leq \sum_k n(k) C_2 (C_1 \beta_J)^k \qquad (99)$$

where k is the number of plaquettes of $\mathcal{D}_1$ and $n(k)$ is the number of diagrams $\mathcal{D}_1$ with k plaquettes. We only need now an estimate of $n(k)$. Let us define an iterative process, adding one plaquette at each step. At a given stage, we have an incomplete diagram with a set of unsaturated links onto which new plaquettes will be tied in the next steps. At the beginning, the set of unsaturated links is $\mathcal{L}$. Let us number the links of Λ arbitrarily; at each step, we decide to add the plaquette along the unsaturated link of lowest rank. There are at most $2(d-1)$ possibilities to choose this plaquette. Next, we have to see whether the four links bounding the added plaquette are unsaturated or not. This yields at most $2^4 = 16$ possibilities. To order k, the diagrams are those for which the set of unsaturated links is empty. Hence we get a generous upper bound

$$n(k) \leq [32(d-1)]^k \qquad (100)$$

This bound ensures that the r.h.s. of (99) is a convergent series provided $\beta_J < [32 C_1 (d-1)]^{-1}$. This completes the proof that the strong coupling expansion for $\langle X \rangle$ has a nonvanishing radius of convergence.

This upper bound is very crude, although it is sufficient for our purpose. Establish more restrictive bounds.

This convergence theorem has two important consequences. First, correlation functions behave exponentially. Let $X^{(\mathbf{x})}$ be the observable X translated by a vector $\mathbf{x}$ on the lattice. One gets the following bound on the correlation function

$$\left| \left\langle X X^{(\mathbf{x})} \right\rangle_c \right| = \left| \left\langle X X^{(\mathbf{x})} \right\rangle - \langle X \rangle \left\langle X^{(\mathbf{x})} \right\rangle \right| \leq \text{Cst } e^{-m|\mathbf{x}|} \qquad (101)$$

with

$$am \geq -4 \ln \beta_J + \text{Cst}$$

Indeed, only diagrams connecting the links of X and $X^{(\mathbf{x})}$ contribute to this quantity. They have at least $4|\mathbf{x}|/a$ plaquettes. This number corresponds to a minimal tube made of plaquettes and joining the two distant sets of links X and $X^{(\mathbf{x})}$. In the r.h.s.

series of (99), the summation starts at this value, and the result follows.

The second important consequence is the proof of the exponential decrease of the Wilson loop in terms of the minimal area $\mathcal{A}(C)$ bounded by the loop. Under hypotheses which have been given in section 1.3 and which will be justified below, the minimal diagram is made of the plaquettes covering the surface of minimal area bounded by the curve and contains $\mathcal{A}(C)$ plaquettes. The upper bound (99) yields a lower bound for the string tension (45)

$$K \geq -\ln \beta_{\mathrm{J}} + \mathrm{Cst} \tag{102}$$

6.3.2   *Character expansions*

Our goal here is to perform more effectively some of the group integrations. The diagrammatic interpretation of the expansion should however be modified in order to use the corresponding techniques. This modification yields an expansion analogous to the $\tanh \beta$ series (rather than β) for the Ising model. We recall first some useful results of the theory of compact groups. Let us consider the irreducible (unitary) representations r, of dimension d_r, of the group. Their matrix elements are a basis for all functions defined on the group. Using the invariant measure dU normalized to unity ($\int dU = 1$), they satisfy the following orthogonality and completeness relations

$$\int dU \mathcal{D}^r_{\alpha\beta}(U)\mathcal{D}^{s*}_{\gamma\delta}(U) = \frac{\delta_{rs}\delta_{\alpha\gamma}\delta_{\beta\delta}}{d_r} \tag{103}$$

$$\sum_{r,\alpha,\beta} d_r \mathcal{D}^r_{\alpha\beta}(U)\mathcal{D}^{s*}_{\alpha\beta}(V) = \delta(U,V) \tag{104}$$

On the r.h.s. of equation (104), the δ-function has the appropriate definition on the group manifold. The characters $\chi_r(U) = \sum_\alpha \mathcal{D}^r_{\alpha\alpha}(U)$, defined as the traces of irreducible representations, are thus an orthonormal basis for the *class* functions on the group. This statement is expressed by the relations

$$\int dU \chi_r(U)\chi_s^*(U) = \delta_{rs} \tag{105}$$

$$\sum_r d_r \chi_r(UV^{-1}) = \delta(U,V) \tag{106}$$

In particular, we have the following useful consequence

$$d_r \int dU \chi_r(U) \chi_s(U^{-1}V) = \delta_{rs} \chi_r(V) \qquad (107)$$

The decomposition in terms of characters generalizes Fourier series. Let us apply it to the exponentiated action

$$\exp \beta_J \chi(U_{\rm p}) = \sum_r \tilde{\beta}_r \chi_r(U_{\rm p}) \qquad (108)$$

with

$$\tilde{\beta}_r = \int dU \chi_r^*(U) \exp \beta_J \chi(U) \qquad (109)$$

We define

$$\tilde{\beta}_r = \tilde{\beta}_0 d_r \beta_r \qquad \text{for } r \neq 0 \qquad (110)$$

so that the partition function may be written as

$$Z = \tilde{\beta}_0^{n_{\rm p}} \int \prod_{\rm p} \left[1 + \sum_{r \neq 0} d_r \beta_r \chi_r(U_{\rm p}) \right] \prod_{ij} dU_{ij} \qquad (111)$$

As usual, each term of the expansion of the product in the integrand is interpreted as giving rise to a diagram. Each one is thus a set of *distinct* plaquettes, to each of which a nontrivial representation has been assigned. We emphasize that, in contradistinction with the expansion (94), two plaquettes should not share the same location on the lattice. Note that the plaquette action χ is real, but that this need not be the case for χ_r. It is necessary to choose for each plaquette a conventional orientation. Changing this convention replaces $U_{\rm p}$ by $U_{\rm p}^{-1}$, that is replaces the representation r by its conjugate $\bar{r}$, $(\mathcal{D}_r(U^{-1}) = \mathcal{D}_r^\dagger(U) \sim \mathcal{D}_{\bar{r}}(U))$, which is generally not equivalent to r. Nevertheless, as the original χ is real, one has $\beta_r = \beta_{\bar{r}}$ and the choice of a conventional orientation does not affect the results (as was of course to be expected).

(1) Give the explicit form for the Fourier components β_r for the two groups $SU(2)$ and $SO(3)$. Matrices U can be parametrized as

$$U = \cos \tfrac{1}{2}\theta + i\boldsymbol{\sigma}.\hat{\mathbf{n}} \sin \tfrac{1}{2}\theta, \qquad 0 \le \theta \le 4\pi \qquad (112)$$

in terms of Pauli matrices. The Haar measure is normalized to
unity and reads

$$dU = \sin^2 \tfrac{1}{2}\theta \frac{d\theta}{2\pi} \frac{d^2\hat{n}}{4\pi} \tag{113}$$

The corresponding irreducible representations are classified by a
non-negative integer or half-integer spin j and the characters are
given by

$$\chi_j(U) = \frac{\sin(j + \tfrac{1}{2})\theta}{\sin \tfrac{1}{2}\theta} \tag{114}$$

A direct application of (109) yields the coefficients $\tilde{\beta}_j$ in terms of
modified Bessel functions, and therefore

$$\exp(\tfrac{1}{2}\beta_J \operatorname{Tr} U) = \exp \beta_J \cos \tfrac{1}{2}\theta = \sum_j 2(2j + 1)\frac{I_{2j+1}(\beta_J)}{\beta_J}\chi_j(U) \tag{115}$$

The group $SO(3)$ leads to analogous expressions, discarding the
representations with noninteger spin j. Table II displays the values
of these coefficients for some usual gauge groups, computed for
Wilson's action. Finally, the geometrical action derived from the
heat kernel (9) is directly written in the Fourier representation.
For example, the values of the quadratic Casimir operator $C_j^{(2)} = j(j + 1)$ for the $SU(2)$ group lead to formula (10).

(2) Derive the asymptotic form of the coefficients β_r for the
Wilson action in the two limiting cases, β_J small or large.

In the expansion (109) of the exponentiated action, the orthogonality relation (105) gives a vanishing result as long as the representation r does not occur in the expansion of the corresponding
power of χ, the character for the sum of the fundamental representation and of its conjugate. Therefore β_r vanishes as an integral
power of β_J. This yields the small β_J behaviour. For instance,
for the $SU(2)$ group, at least $2j$ spins $\tfrac{1}{2}$ are needed to construct a
spin j, and hence $\beta_j \sim \beta_J^{2j}$. Note also that β_r increases monotonically from 0 to 1 as β_J varies from 0 to infinity. For a continuous
gauge group, the large β_J behaviour is obtained through a steepest
descent method around $U = I$

$$\tilde{\beta}_r \sim \operatorname{Cst} d_r \frac{\exp(\beta_J\chi(1))}{\beta_J^{X/2}} (1 + \mathcal{O}(\beta_J^{-1})) \tag{116}$$

where X denotes the dimension of the Lie algebra (i.e. the number
of infinitesimal independent generators of the group). Up to a shift
in the action, amounting to subtracting the value of the character

Table II. Coefficients in the character expansion of the Wilson action

group	$\bar{\beta}_0$	τ and $\chi_\tau(U)$	d_τ	$\bar{\beta}_\tau$
Z_2 $U(1)$	$\cosh\beta_J$ $I_0(\beta_J)$	$\sim Z_2 U$ $ne^{in\varphi}$	1 1	$\sinh\beta_J$ $I_n(\beta_J)$
$SU(2)$	$2I_1(\beta_J)/\beta_J$	$j\,\tfrac{1}{2}$ integer $\dfrac{\sin(j+\frac{1}{2})\theta}{\sin\frac{1}{2}\theta}$	$2j+1$	$I_{2j}(\beta_J) - I_{2j+2}(\beta_J)$
$SO(3)$	$e^{\beta_J/3}\left(I_0(\tfrac{2}{3}\beta_J) - I_1(\tfrac{2}{3}\beta_J)\right)$	j integer $\dfrac{\sin(j+\frac{1}{2})\theta}{\sin\frac{1}{2}\theta}$	$2j+1$	$e^{\beta_J/3}(I_j(\tfrac{2}{3}\beta_J) - I_{j+1}(\tfrac{2}{3}\beta_J))$
$U(n)$	$\det I_{i-j}(\beta_J/n)$ $(1\le i,j\le n)$	$l_1 \ge \cdots \ge l_n\quad U = \text{Diag}(e^{i\alpha_j})$ $\chi = \dfrac{\det(\exp i(l_j+n-j)\alpha_i)}{\det(\exp i(n-j)\alpha_i)}$	$\displaystyle\prod_{i<j}\frac{l_i - l_j + j - i}{j - i}$	$\det I_{l_i - j + i}(\beta_J/n)$
$SU(n)$	$\displaystyle\sum_{k=-\infty}^{\infty} \det I_{i-j+k}\left(\frac{\beta_J}{n}\right)$	as $U(n)$ with $l_n \equiv 0$	as $U(n)$	$\displaystyle\sum_{k=-\infty}^{\infty} \det I_{l_j -j+i+k}\left(\frac{\beta_J}{n}\right)$
$U(\infty)$ or $SU(\infty)$	$\exp\left(\dfrac{\beta_J}{2n}\right)^2$ in the limit $n \to \infty$ with fixed $\beta_J n^2$			

at the identity, the coefficient $\tilde{\beta}_r$ behaves according to a power of β_J for a continuous group. On the other hand, for a discrete group, this coefficient decreases exponentially.

The diagrammatic strong coupling expansion may be analyzed from a topological point of view. A given subset of plaquettes is homeomorphic to a simple surface if any link bounds at most two plaquettes of this subset. The boundary of this surface is made of the links bounding exactly one plaquette, and is homeomorphic to a set of simple closed curves. Any diagram can be decomposed as a set of maximal simple surfaces by cutting it along the links bounding more than two plaquettes. Each of the boundary curves is

- either a true free boundary, bounding only one simple surface,
- or a singular branch line along which n ($n \geq 3$) simple surfaces meet.

The classification of simple surfaces is made according to the following criteria.

a) Orientability. Two neighbouring oriented plaquettes are said to be *coherently oriented* if they induce opposite orientations on the common adjacent link. A surface is *orientable* if there exists an orientation convention for all its plaquettes such that any pair of neighbouring plaquettes is coherently oriented.

b) The number b of holes, i.e. the number of simple closed curves bounding the surface.

c) The *genus* $g = 2-(n_2-n_1+n_0+b) \geq 0$, where n_2 is the number of plaquettes, n_1 the number of distinct links and n_0 the number of sites located on the surface. Any orientable surface of genus g (which is always even in this case) is homeomorphic to a sphere with $\frac{1}{2}g$ handles and b holes. The torus corresponds to the case $g = 2$, $b = 0$, and the disk to $g = 0$, $b = 1$. Conversely, if the surface is not orientable, it is homeomorphic to a sphere drilled with $b + g$ holes. Among them, g (≥ 1) are closed by a Möbius band. For example, the Möbius band corresponds to $b = 1$, $g = 1$; the projective plane to $g = 1$, $b = 0$; and the Klein bottle to $g = 2$, $b = 0$. In later chapters, when dealing exclusively with orientable surfaces, we shall denote by g, and still call genus, half of the above (even) quantity.

Now, let us perform the integration over the gauge fields. Relation (103) forbids the existence of free boundaries for diagrams pertaining to the partition function. For a given observable, the free boundary coincides with the set of links occuring in the definition of this observable. Relation (107) allows the integration over the internal links of the simple surface components corresponding to a given diagram. In particular, the plaquettes of a simple surface component should carry the same representation when they are coherently oriented. As a consequence, non orientable surfaces carrying complex representations (representations which are not equivalent to their conjugate) should be excluded. The integration over inner links of a simple component yields

$$\beta_r^{n_2} d_r^{2-g-b} \prod_{\text{boundaries}} \chi_r(U_{\text{boundary}}) \tag{117}$$

Check this result.

The final step consists of integrating gauge fields along the singular branch lines. This is the most delicate part, and it depends on the Clebsch–Gordan coefficients coupling the representations of the group.

Show that, if there is only one singular closed line which is the common boundary shared by p simple surfaces with representations $r_1, r_2, \ldots, r_p$, the integration over gauge fields yields an integral factor $N_{r_1,\ldots,r_p}$, which counts the number of times the trivial representation is contained in the product $r_1 \otimes r_2 \otimes \cdots \otimes r_p$.

Let us consider in more detail the possible topologies of diagrams contributing to the Wilson loop. The simplest topology is a disk, the boundary of which is the loop. As we have already seen, this kind of diagram leads to an exponential decrease of the Wilson loop parameter with respect to the minimal area enclosed by the loop. However, other topologies may occur, e.g. the toric diagram displayed in figure 12. This kind of diagram obviously leads to a perimeter law, unless its contribution vanishes identically. It describes a process in which the charges are screened and can thus be separated, without confinement. The leading behaviour of the terms associated with the disk and the toric topologies are respectively $\beta_J^{L^2}$ and $\beta_J^{16(L-1)}$, so that the transition between an area law and a perimeter law is expected to occur

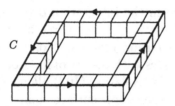

Fig. 12   A diagram leading to a perimeter law for the Wilson loop.

for a loop side $L \simeq 15$. This size exceeds the possibilities of present computers and, from numerical simulations, one may hastily conclude that the asymptotic behaviour follows an area law. Fortunately, the toric diagrams vanish when the center of the group is nontrivial. Indeed, using the result from the last exercise, one sees that the contribution vanishes when $r \otimes \bar{r}$ does not contain the representation used in the definition of the Wilson loop. Since on the one hand $r \otimes \bar{r}$ induces a trivial representation on the center (which is an invariant subgroup), while on the other the fundamental representation is non trivial on this subgroup, the afore mentioned contribution has to vanish. For instance, with the $SU(2)$ group, one cannot screen a spin $\frac{1}{2}$ with $j \otimes j$ combinations, which are always of integral spin. With the $SO(3)$ gauge group, the fundamental representation of which is of spin 1, this screening does occur and the Wilson loop operator cannot be used to distinguish the phases of the system.

6.3.3   Free energy

We have now all the prerequisites to derive various expansions. The problem of enumerating and counting all graphs is treated in chapter 7 (volume 2). We consider first the series for the free energy per site, for a d-dimensional hypercubic lattice. The results given below are valid for any gauge group. Let us introduce first some coefficients corresponding to the various possible topologies. These are

• The sphere

$$S_n = \sum_r \beta_r^n d_r^2$$

- The torus

$$T_n = \sum_r \beta_r^n$$

- p disks sharing the same closed curve as boundary

$$\Theta_{n_1\ldots n_p} = \sum_{r_1\cdots r_p} \beta_{r_1}^{n_1} \cdots \beta_{r_p}^{n_p} d_{r_1} \cdots d_{r_p}$$

- A cylinder closed at both extremities by duplicated disks

$$\zeta_{n_1\cdots n_5} = \sum_{r_1\cdots r_5} \beta_{r_1}^{n_1} \cdots \beta_{r_5}^{n_5} d_{r_1} d_{r_2} d_{r_4} d_{r_5} N_{r_1 r_2 r_3} N_{r_3 r_4 r_5}$$

- Four simple arcs sharing the same endpoints; taken by pairs, they bound six different disks

$$\Phi_{n_{12}n_{13}n_{14}}^{n_{34}n_{24}n_{23}} = \sum \prod_{i<j} d_{r_{ij}} \beta_{r_{ij}}^{n_{ij}} \int dU_i \prod_{i<j} \chi_{r_{ij}}(U_i U_j^{-1})$$

The free energy, up to and including diagrams with 16 plaquettes, is then given by

$$\frac{2\ln Z}{d(d-1)} = \ln \tilde{\beta}_0 + (d-2)[\tfrac{1}{3}S_6 + (2d-5)S_{10} + \tfrac{10}{3}(d-3)S_{12}$$
$$+ (20d^2 - 106d + 143)S_{14} + (84d^2 - 504d + 757)S_{16}$$
$$+ \tfrac{1}{4}(d-3)T_{16} - (2d - \tfrac{29}{6})S_6^2 - (44d^2 - 226d + 294)S_6 S_{10}$$
$$+ (2d-5)\Theta_{551} + \tfrac{2}{3}(d-3)(2d-5)(\Theta_{555} + \Theta_{5551})$$
$$+ \tfrac{8}{3}(d-3)(3\Theta_{842} + 3\Theta_{10,4,2} + 2\Theta_{933} + \tfrac{3}{2}\Theta_{664} + \Phi_{444}^{111})$$
$$+ (20d^2 - 108d + 149)(2\Theta_{951} + \zeta_{51415}) + o(\beta_j^{16})]$$

$$(118)$$

This general formula can now be applied to any given gauge group or any variant one-plaquette action. This is just a straightforward, although tedious, exercise in algebra. At this level of complexity, the help of a computer algebra system is useful. Once the series with numerical coefficients has been obtained, it can be analyzed using the various general techniques reported in chapter 7 (volume 2).

We recall that pure gauge theories have a first order transition above some critical dimension. The free energy is not singular at this point, as was suggested by the mean field approximation. In this case, the two branches corresponding to the weak and strong coupling phases are two determinations of the same multivalued

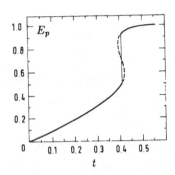

Fig. 13   The plaquette energy E as a function of $t = \tanh \beta_J$ for the Z_2 gauge system.

analytic function. It is *theoretically* possible, from the series, to perform an analytic continuation of the function and, following the adequate path on Riemann sheets, to compute the weak coupling determination and therefore to have access to the transition point. *In practice*, available series are much too short to allow for this process.

Let us analyze in more detail the four-dimensional Z_2 model. A direct analysis of the series in β_J shows evidence for a singularity located around $\tanh \beta_J = 0.48 - 0.49$, which is beyond the transition point $\tanh \beta_c = \sqrt{2} - 1$, exactly known by duality. We interpret this singularity as the end point of the metastable strong coupling phase. As suggested by the mean field approximation, the coupling constant might be a regular, uniform function of the internal energy per plaquette E. This can be checked by inverting the series $E(\beta_J)$ into a new series $\beta_J(E)$. The analysis confirms the expectation; no evidence for a singularity is found, confirming the idea of a first order transition which admits an analytic unfolding. Figure 13 displays the resummed series and allows the transition to be located, compatible within errors with the duality prediction.

Four-dimensional nonabelian gauge systems are more difficult to analyze. Numerical simulations seem to imply no transition. Nevertheless one finds a peak in the specific heat (figure 14). Various resummation techniques of the series diverge in this region, and it is rather difficult to reproduce the structure seen in figure 14. Additional hypotheses are necessary to improve the fit. For instance, one may conjecture the existence of a pair of

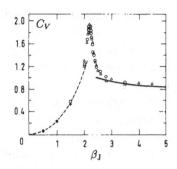

Fig. 14 The specific heat of the four-dimensional $SU(2)$ system (from B. Lautrup and M. Nauenberg, *Phys. Rev. Lett.* **45** (1980) 1755).

conjugate complex poles in this region. For practical purposes, the best technique consists in matching the series into the weak coupling region by using results from asymptotic freedom. An example will be given for the string tension.

In the limit of a large dimension, strong coupling expansions can be reordered in a way different from the $1/d$ series of successive corrections to mean field. Looking at formula (118), one sees that the coefficient of each term is a polynomial in d with a degree increasing by unity every fourth term. It is thus suggested to keep the quantity $\beta_J^4 d$ constant as d tends to infinity and to reorder the series as an expansion in $d^{-\frac{1}{4}}$. Let us consider the following reduction process on a connected diagram. Suppress a slab of the lattice limited by two consecutive hyperplanes $k < x_\mu < k + 1$ and containing at least one plaquette of the diagram. As the diagram is closed, at least four plaquettes disappear. Moreover, other plaquettes may be duplicated when the two half spaces are stuck together and must be replaced (combining their group representations) either by a single plaquette, or no plaquette at all. In any case, this operation decreases the number of plaquettes by at least four, while the *dimensionality* of the diagram (i.e. the number of different directions effectively used by the plaquettes of the diagram, or equivalently the degree of the polynomial in d entering in its contribution) decreases at most by one unit. It is also clear that the dominant terms in the above limit are obtained whenever exactly four plaquettes disappear at each step in the reduction, until one is left with a simple diagram having the shape of a cube. This characterizes the dominant diagrams as the boundary of connected trees made with three-dimensional cubes, with all plaquettes sharing the same group representation.

Since each cube has six faces, it has at most six neighbours in the tree diagram. In a step-by-step reconstruction, the addition of a new cube on a given plaquette of a cube leads to $2d-5$ ($\approx 2d$) possibilities. Some of these may be forbidden by excluded volume effects, but their number is negligible with respect to $2d$. The expectation value $p_r = \langle \chi_r(U_p) \rangle / d_r$ of a single plaquette in the representation r, satisfies in this limit the self-consistency equation

$$p_r = \beta_r + 2d p_r^5 \tag{119}$$

This is parametrized as

$$\begin{cases} f_r \equiv \dfrac{p_r}{\beta_r} = (1 - u_r)^{-1} \\ x_r \equiv 2d\beta_r^4 = u_r(1 - u_r)^4 \end{cases} \tag{120}$$

The free energy is recovered by integration of the internal energy p_r per plaquette

$$\begin{aligned} \tilde{F} &\equiv \frac{\ln Z}{N} - \frac{d(d-1)}{2} \ln \tilde{\beta}_0 \\ &= \frac{d^{3/2}}{12\sqrt{2}} \sum_{r \neq 0} d_r^2 x_r^{1/2} \frac{u_r(1 - 3u_r)}{(1 - u_r)^2} (1 + \mathcal{O}(d^{-1/4})) \end{aligned} \tag{121}$$

With the Wilson action, only the fundamental representation and its conjugate give a dominant contribution to (121), and the substracted free energy $\tilde{F}$ takes a universal form displayed on figure 15. It presents a cusp at $u = 1/5$, i.e. for $2d\beta_J^4 = 4^4/5^5$. As the rearrangement of the series is only possible within the radius of convergence, only the arc OA is physically acceptable. This is the whole strong coupling phase, including the metastable region. The second derivative of the free energy, which could be interpreted as a plaquette–plaquette susceptibility, behaves as $(\beta_A - \beta_J)^{-1/2}$ in the vicinity of the point A, as d goes to infinity. This is reminiscent of a continuous phase transition with a corresponding correlation length diverging as $(\beta_A - \beta_J)^{-\nu}$ for an exponent $\nu = \frac{1}{4}$ (see chapter 11 in volume 2). The weak coupling branch is not seen here. Indeed, from the analysis done in the mean field framework, this phase lies along the vertical axis, using the present variables.

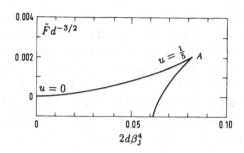

Fig. 15   The leading order in a $d^{-\frac{1}{4}}$ expansion for the rescaled (and subtracted) free energy in large dimension plotted as a function of $2d\beta_j^4$.

Compute the next order correction to this approximation.

6.3.4   *String tension and roughening transition*

In nonabelian gauge theories without transitions in four dimensions, one wants to extrapolate strong coupling series for the string tension to the weak coupling regime. This would provide a link with the scaling behaviour and allow useful information on confinement in the continuous limit to be obtained. However, two difficulties arise. The first one has already been discussed and is related to the delicate resummation in the intermediate region. The second one is specific to this nonlocal observable and corresponds to the *roughening transition* which does not affect bulk quantities. Its name originates from crystal growth theory where a similar phenomenon occurs.

As a simple example, let us consider a three-dimensional Ising model, dual to a Z_2 gauge theory, with boundary conditions such that there exists at low temperature an interface separating two regions of opposite orientation. Boundary spins are constrained to point up above a given plane, and down under this plane. The surface tension is related to the difference in free energy as compared with a homogeneous system (chapter 2). If A denotes the area of the minimal interface, and ΔF the difference in free energy,

$$\Delta F = -KA \qquad (122)$$

with K the string tension of the corresponding Z_2 gauge model in its strong coupling phase.

At zero temperature, the interface is a crystallographic plane. As T increases, it fluctuates more and more. When a temperature T_R, generally smaller than T_c, is reached, these properties change drastically. The fluctuations are so important that it is no longer possible to say that the interface is localized. Let us be more precise and analyze quantitatively the phenomenon. Let h_i be the height of the interface above its equilibrium position $\mathbf{r}_i$ in the reference plane, and consider the quantity

$$p(h, \mathbf{r}_i - \mathbf{r}_j) = \left\langle \delta(h_i - h_j - h) \right\rangle \qquad (123)$$

which expresses the probability that two points separated by a distance $r_{ij} = \left| \mathbf{r}_i - \mathbf{r}_j \right|$ have a height difference of h. The *width* of the interfacial region can be defined as the second momentum Δh of this distribution in the large distance limit

$$\sum h^2 p(h, r_{ij}) = \left\langle (h_i - h_j)^2 \right\rangle \quad \rightarrow \quad (\Delta h)^2 \text{ as } r_{ij} \rightarrow \infty \quad (124)$$

We say that the system undergoes a *roughening transition* if this quantity diverges as the temperature reaches (from below) a *roughening temperature* T_R. An equivalent description is to consider the free energy fL needed for the construction of a step of length L on the interface. The quantity f vanishes for $T \geq T_R$, so that the large-scale deformations of the surface are energetically costless. The surface tension K has no reason to vanish at T_R, but is singular at this temperature, as will be seen below.

So far this description has been qualitative. It is based on converging evidence coming from the study of exact models, Monte Carlo simulations and experiments on crystal growth. A convenient description is the *solid on solid (SOS)* model where one considers only a restricted class of possible deformations, excluding overhangs and disconnected inclusions of one phase into the other. It leads to an action of the type $\beta \mathcal{H} = \beta \sum_{(i,j)} \left| h_i - h_j \right|$, which can be also approximated by $\beta \mathcal{H} = \beta \sum_{(i,j)} \left(h_i - h_j \right)^2$, a model belonging to the same universality class. One recognizes the discrete Gaussian model, dual to the XY-model which has been studied in chapter 4. Its transition is the roughening transition. The surface tension, which is the free energy of this system, admits an essential singularity of the form

$$K = A(\beta) + B(\beta) \exp\left(-\frac{C}{\sqrt{T_R - T}}\right) \qquad (T < T_R) \quad (125)$$

The step free energy f is identified with the inverse correlation length, hence

$$f \sim \xi_{XY}^{-1} \sim \exp(-\frac{C}{2\sqrt{T_{\mathrm{R}} - T}}) \tag{126}$$

The scale governing the phenomena is ξ_{XY} when $T < T_{\mathrm{R}}$ (this corresponds to the high temperature region of the XY-model). For $T > T_{\mathrm{R}}$, ξ_{XY}^{-1} vanishes and the scale reduces to the size L of the sample. Therefore, we introduce the quantity λ defined by $\lambda = \ln(\xi_{XY}/a)$ for $T < T_{\mathrm{R}}$, $\lambda = \ln(L/a)$ for $T > T_{\mathrm{R}}$. At large distances, the probability $p(h, r)$ which has been introduced in (123) reads

$$p(h, L) \sim \frac{1}{\sqrt{4\pi\lambda}} \exp(-\frac{h^2}{4\lambda}) \tag{127}$$

Hence the interface width is

$$\Delta h = \sqrt{2\lambda} \tag{128}$$

while the probability for a point of the interface to be at its equilibrium position is

$$p(0) = \frac{1}{\sqrt{4\pi\lambda}} \sim \frac{1}{\Delta h} \sim \begin{cases} (T_{\mathrm{R}} - T)^{1/4} & \text{for } T < T_{\mathrm{R}} \\ \dfrac{1}{\sqrt{\ln L}} & \text{for } T > T_{\mathrm{R}} \end{cases} \tag{129}$$

Such behaviour is presumably universal, and should also apply to the interface in the original Ising model. According to the standard lore, the roughening transition does not affect bulk quantities like the free energy or its derivatives, nor the spin–spin correlation functions.

From duality, these conclusions remain valid for the Z_2 gauge model in three dimensions, where the interface free energy per unit area is interpreted as the string tension. The discussion can also be extended to higher dimensions. A rigorous proof of the existence of this transition is still lacking. To get clear evidence, we need a better indicator than the Wilson loop which only presents a very weak singularity of the type (125). Other quantities diverge at T_{R} and yield a sharper signal. For instance,

$$\eta^2 = \frac{\sum_{\mathbf{x}_\perp} E(\mathbf{x})\mathbf{x}_\perp^2}{\sum_{\mathbf{x}_\perp} E(\mathbf{x})} \tag{130}$$

with

$$E(\mathbf{x}) = \frac{\langle W(C)\chi(U_\mathrm{p})\rangle - \langle W(C)\rangle \langle \chi(U_\mathrm{p})\rangle}{\langle W(C)\rangle} \qquad (131)$$

Intuitively, $E(\mathbf{x})$ measures, using a test plaquette p parallel to the Wilson loop plane, the density of chromoelectric energy at point $\mathbf{x}$, and $\mathbf{x}_\perp$ is the distance from p to the minimal surface bounded by the Wilson loop. As a consequence, η^2 generalizes the expression (124), and one expects a divergence according to a power law. Corresponding series can be computed and analyzed, and one gets a clear indication for the expected singularity. Remarkably, the position of the roughening transition, expressed in terms of the gauge coupling β_f relative to the fundamental character, seems to be largely independent of the gauge group (up to uncertainties in the numerical analysis), which confirms its geometrical nature.

$$\begin{cases} d = 3 & \beta_{f\mathrm{R}} \simeq 0.46 \\ d = 4 & \beta_{f\mathrm{R}} \simeq 0.40 \\ d = 5 & \beta_{f\mathrm{R}} \simeq 0.37 \end{cases} \qquad (132)$$

The respective positions of the roughening and of the bulk transition are not fixed. A heuristic argument claims that the deconfining phase transition will be of first order if the roughening transition occurs at a lower temperature. Indeed, if the deconfining transition is of second order, the interquark potential behaves as $1/R$ at the critical point, whereas it is natural to expect an exponential fall-off of the corrections to the leading behaviour up to the roughening point.

One may wonder what is the critical theory at T_R. In three dimensions, we saw that it belongs to the universality class of the 2-dimensional XY-model. In d dimensions, the same reasoning suggests a set of $(d-2)$ two-dimensional decoupled XY-models. The indicator η would thus have a critical exponent $\frac{1}{4}$, while the step free energy vanishes with an essential singularity. This vanishing plays an important role; the surface can then freely move in the directions orthogonal to its minimal plane and behaves thus as a continuous surface. Translation and rotation symmetries are then restored. This has been checked both in strong coupling calculations of the interquark potential, and in numerical simulations measuring the potential.

The analysis of strong coupling series for the string tension is of course complicated by the presence of the roughening transition. Generally, the effect of the nonanalyticity at β_R is to give the fallacious impression that the string tension vanishes prematurely. Figure 16 displays a comparison of strong coupling expansions to numerical simulations. In the transition region between strong and weak coupling regimes, the curve can be tangentially connected to the predictions of the renormalization group and asymptotic freedom. The string tension has a dimension of $length^{-2}$. Expressed in physical units, the parameter $\sigma = Ka^2$ can be interpreted in the framework of the old relativistic string model and is related to the Regge slope by $\sigma = (2\pi\alpha')^{-1}$. The Regge relation of particle physics is the almost linear expression of spin versus mass square for low lying particles and resonances. Experimentally, $\alpha' \simeq 0.90\text{GeV}^{-2}$ so that $\sqrt{\sigma} \simeq 420\text{MeV}$. Formula (31) provides the weak coupling behaviour

$$K \sim \sigma\Lambda_{\text{L}}^{-2}(b_0 g_0^2)^{-b_1/b_0^2}\exp(-1/b_0 g_0^2) \qquad (133)$$

The corresponding curve looks almost like a straight line in the logarithmic scales used in figure 16. Relating this curve to the series extrapolations in the transition region allows the parameter Λ_{L} to be determined. It is in fair agreement with the values obtained by other techniques.

The string tension gives only the coefficient of the linearly rising part of the quark–antiquark static potential. Much effort has been devoted to obtaining its overall shape for a description of heavy quark–antiquark meson bound states, accessible experimentally. It is not our purpose here to review this important aspect of quantum chromodynamics, directly tied to particle phenomenology. Nor will we discuss a possible finite temperature deconfining transition of importance in astrophysics and heavy ion collisions. We will rather limit ourselves with a short description of the excitation spectrum of pure gauge theories.

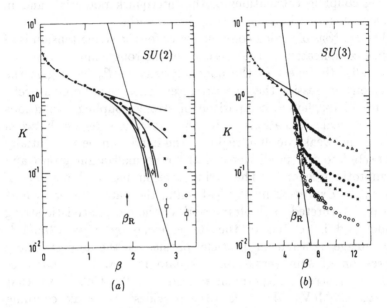

Fig. 16   The string tension for $SU(2)$ (a) and $SU(3)$ (b) systems. Data
are taken from M. Creutz, *Phys. Rev.* **D21** 2308 (1980) and M. Creutz and
K.J.M. Moriarty, *Phys. Rev.* **D26** 2166 (1982).

6.3.5   Mass spectrum

The mass spectrum of pure gauge theories seems to be the best
candidate for applications of strong coupling series. Effects such
as the roughening transition will not be present, and it is therefore
expected that the predictions of strong coupling and asymptotic
freedom can be easily matched in the transition region.   The
corresponding states have no quark components and are searched
for experimentally under the name of *glueballs*.

The derivation of strong coupling series for the quantities of
interest is, however, somewhat delicate. Correlation functions are
expected to behave exponentially at large distance according to

$$G(\mathbf{r}) = \langle \mathcal{O}_1(0)\mathcal{O}_2(\mathbf{r})\rangle_c \sim A_1(\mathbf{r})e^{-m_1(\hat{\mathbf{r}})r} + A_2(\mathbf{r})e^{-m_2(\hat{\mathbf{r}})r} + \cdots \quad (134)$$

This formula involves several masses $m_1 \leq m_2 \leq \cdots$ which depend
on the direction $\hat{\mathbf{r}}$ of observation. The simplest idea is to compute
the correlation function between two distant plaquettes, using
diagrams connecting their boundaries, and to derive $m_1(\hat{\mathbf{r}}) =$

$\lim_{r\to\infty} \ln G(\mathbf{r})/r$. However, a degeneracy problem appears and prevents a direct exponentiation of the series.

In order to explain this phenomenon, let us compute the correlation length of the Ising model in a direction parallel to a principal axis.

Let n be the distance (in lattice spacing units) of the two spins. The lowest order diagram is drawn as the straight line joining the two locations. The next correction adds opposite kinks located at sites i and j $(0 \leq i < j \leq n)$. Therefore

$$\langle \sigma_0 \sigma_n \rangle = (\tanh \beta)^n + (d-1)(\tanh \beta)^{n+2} n(n+1) + \cdots \quad (135)$$

The correlation length is obtained by taking the logarithm, which is linear in n to lowest order

$$\xi = -\frac{1}{\ln \tanh \beta}(1 + \mathcal{O}(\tanh \beta)) \quad (136)$$

The subsequent correction to the logarithm of the correlation function is unfortunately quadratic in n and the series for the correlation length seems to have infinite terms. This paradox is due to the fact that the transfer matrix have several equal eigenvalues. Taking a direction parallel to a principal axis amounts to combining the corresponding eigenvectors in an unfortunate way which leads to this disaster. From translational invariance, a correct diagonalization can however be carried out using a Fourier transformation

$$G(\mathbf{p}_\perp, x_0) = \sum_{\mathbf{x}_\perp} \langle \sigma_0 \sigma_{x_0, \mathbf{x}_\perp} \rangle e^{-i\mathbf{p}_\perp \mathbf{x}_\perp} \sim e^{-x_0/\xi} + \cdots \quad (137)$$

Now, the first correction to the lowest order diagram is just a single kink (spins are shifted by one unit in perpendicular directions)

$$G(\mathbf{p}, x_0) = (\tanh \beta)^n + (\tanh \beta)^{n+1} 2d(n+1) \sum_\mu \cos p_\mu + \cdots \quad (138)$$

There are no more difficulties in computing the logarithm, even including further corrections.

The previous exercise shows that it is necessary to diagonalize the transfer matrix, taking into account the various symmetries. This can be partly done by going over to momentum space. One has also to take into account the discrete rotational invariance of the lattice and "C-parity", related to charge conjugation. We

Fig. 17   Random walk of a plaquette.

introduce the following Green function

$$G^{\pm}_{\mu\nu,\rho\sigma}(\mathbf{k}) = \sum_{\mathbf{x}} \left\langle \mathrm{Tr}\, \tfrac{1}{2} \left(U_{\mu\nu}(0) \pm U^{\dagger}_{\mu\nu}(0) \right) \right.$$

$$\left. \times \mathrm{Tr}\, \tfrac{1}{2} \left(U_{\rho\sigma}(\mathbf{x}) \pm U^{\dagger}_{\rho\sigma}(\mathbf{x}) \right) \right\rangle_{\mathrm{c}} e^{i\mathbf{k}\cdot\mathbf{x}} \quad (139)$$

where $U_{\mu\nu}(\mathbf{x})$ is the plaquette variable U_{p} with directions μ and ν, the center of which is located at $\mathbf{x}$. Instead of computing G, let us, as usual, discuss its inverse. The lowest order contributions to this quantity are fourth order terms connecting two nearby plaquettes and completing the boundary of a cube. This is described in momentum space by a matrix $M_{\mu\nu,\rho\sigma}(\mathbf{k})$ obeying the symmetry relations

$$M_{\mu\nu,\rho\sigma} = M_{\rho\sigma,\mu\nu} = C M_{\nu\mu,\rho\sigma} \qquad (C = \pm 1) \qquad (140)$$

Explicitly,

$$\begin{cases} M_{\mu\nu,\mu\nu} = 2 \sum_{\rho\neq\mu,\nu} \cos k_\rho \\[2ex] M_{\mu\nu,\mu\rho} = \begin{cases} 4\cos\tfrac{1}{2}k_\nu \cos\tfrac{1}{2}k_\rho & \text{when } C = +1 \\ 4\sin\tfrac{1}{2}k_\nu \sin\tfrac{1}{2}k_\rho & \text{when } C = -1 \end{cases} \end{cases} \qquad (141)$$

Other components vanish. Returning to the propagator G expressed as

$$\left[G^{-1}(\mathbf{k}) \right]_{\mu\nu,\rho\sigma} = (1 - \beta_r^4 M_{\mu\nu,\rho\sigma}(\mathbf{k})) \qquad (142)$$

one obtains a plaquette random walk description depicted on figure 17. In configuration space, this amounts to approximate G by retaining only diagrams in the shape of tubes connecting an initial to a final plaquette locations, in complete analogy with Brownian motion. Further corrections are obtained by adding various "decorations" to the propagator tube.

We have not yet taken into account the discrete rotational symmetry of the lattice. In four dimensions and for a propagation along the fourth direction, we have to decompose the propagator according to the representations of the discrete cubic symmetry (rotations keeping invariant the fourth dimension). The following quantities are to be considered

- A singlet $\operatorname{Re}\operatorname{Tr}(U_{12} + U_{23} + U_{31})$
- A doublet $\{\operatorname{Re}\operatorname{Tr}(U_{13} - U_{23}); \operatorname{Re}\operatorname{Tr}(U_{13} + U_{23} - 2U_{12})\}$
- A triplet $\{\operatorname{Im}\operatorname{Tr}U_{12}; \operatorname{Im}\operatorname{Tr}U_{23}; \operatorname{Im}\operatorname{Tr}U_{31}\}$ which exists only for complex representations of gauge groups.

Check this decomposition.

We refer to the literature for details on the explicit computation of the glueball mass spectrum. Let us just quote a result for the fundamental singlet state 0^{++} for a $SU(3)$ colour gauge group. Series have been derived up to eighth order. To obtain stable extrapolations, it is better to re-express the scalar mass in terms of the plaquette energy before attempting to estimate the series. One finds

$$\frac{m_S}{\Lambda_{\rm L}} \approx 340 \pm 40 \qquad (143)$$

This value agrees with numerical simulations and would lead to a glueball mass of approximately $800\,\mathrm{MeV}$. This result should be taken with a grain of salt, due to several uncertainties, including the effects of quarks neglected in this calculation. There seems also to exist a 2^{++} state with a mass close to the fundamental one. Higher excited states might also be present. It is still an open problem to identify the glueballs experimentally. An extensive numerical effort is made to ascertain the hadronic mass spectrum involving quarks. There are however a number of technical difficulties, described below, in the case where the latter have a small (and in the limit vanishing) mass.

6.4   Lattice fermions

6.4.1   The doubling problem

The main goal of lattice gauge theories is to yield a regulated model for quantum chromodynamics. Hence, one is specially interested in fermionic fields. We have described in section 1 the technique which transforms a globally invariant action (as the bosonic action (1)) into a gauge invariant one. This method remains valid for fermions, but serious problems appear when one wants to write down a satisfactory global action for physical fermions.

Starting from the continuous action for free fermionic fields $\bar{\psi}$ and ψ of spin $\frac{1}{2}$

$$S_{\text{cont}} = \int \mathrm{d}^d x \left(\tfrac{1}{2} \mathrm{i} \bar{\psi} \gamma_\mu \partial_\mu \psi - m \bar{\psi} \psi \right) \tag{144}$$

the first idea is to replace derivatives by finite differences involving only nearest neighbours $\partial_\mu f \to (f_{i+a\hat{\mu}} - f_i)/a$. This leads to the so-called *naive fermionic action*

$$S_{\text{nf}} = \tfrac{1}{2} \mathrm{i} a^{d-2} \sum_{(ij)} \bar{\psi}_i \gamma_{ij} \psi_j - m a^d \sum_i \bar{\psi}_i \psi_i \tag{145}$$

with

$$\gamma_{ij} = \gamma_\mu (x_i - x_j)^\mu$$

For short, we do not write explicitly all the indices of the Grassmannian fields $\bar{\psi}_i$, ψ_i. Besides the spatial location i referring to a lattice site, there are a spinor index (saturated by the Dirac matrices γ), a colour index (related to the internal symmetry which will become a local one and on which the gauge fields will act), and possibly flavour indices (corresponding to other global internal symmetries which we ignore here; the mass m may depend on them). The Euclidean Dirac matrices obey the anticommutation rule

$$\left\{ \gamma_\mu, \gamma_\nu \right\} = -2\delta_{\mu\nu} \tag{146}$$

They will be written in some faithful matricial representation of minimal dimension $2^{\lfloor d/2 \rfloor}$. The corresponding matrices generate the algebra of all $\lfloor d/2 \rfloor \times \lfloor d/2 \rfloor$ matrices. In even dimension, the matrix

$$\hat{\gamma}_5 = \mathrm{i}^{d/2} \gamma_1 \gamma_2 \cdots \gamma_d \qquad \hat{\gamma}_5^2 = 1 \tag{147}$$

is linearly independent of, and anticommutes with, the γ_μ's. Added to the d previous matrices (and multiplied by i), it will be the $(d+1)$-th Dirac matrix for the odd dimension $d+1$. In the case where d is even, the γ_μ's can be chosen antihermitian and

$\hat{\gamma}_5$ hermitian. The notation $\hat{\gamma}_5$ is borrowed from four-dimensional space-time physics.

Let us compute the propagator corresponding to the free field action (145). In the momentum representation, it reads

$$\Delta(\mathbf{p}) = \left(\frac{1}{a}\gamma_\mu \sin p_\mu a + m\right)^{-1} \tag{148}$$

The masses of the particles are determined by the position of the poles. In the limit $a \to 0$, poles lie in the vicinity of points $\bar{\mathbf{p}}$ with components equal to either 0 or π/a (recall that the components of the momenta are defined on the lattice modulo $2\pi/a$). Therefore, there are 2^d such poles and not a unique one as expected. This species multiplication reflects a discrete symmetry of the naive fermionic action. Let us introduce in even dimension the d operators $T_\mu = \gamma_\mu \hat{\gamma}_5(-1)^{x_\mu/a}$, $\mu = 1,\dots,d$. They generate a discrete group of order 2^{d+1}, and it can be easily checked that the action (145) is invariant under their action ($\psi \to T_\mu \psi$, $\bar{\psi} \to \bar{\psi} T_\mu^\dagger$). Each of these transformations T translates the momentum by one of the quantities $\bar{\mathbf{p}}$ defined above and acts on spinor indices through a matrix $\mathcal{S}(\bar{\mathbf{p}})$. In the vicinity of each of these poles, the propagator (148) reads

$$\Delta(\mathbf{p} = \bar{\mathbf{p}} + \mathbf{k}) = \mathcal{S}(\bar{\mathbf{p}})\frac{m - \gamma_\mu k_\mu}{\mathbf{k}^2 + m^2}\mathcal{S}(\bar{\mathbf{p}})^{-1} + \mathcal{O}(a) \tag{149}$$

The effect of the transformations $\mathcal{S}$ is just to change the representation of the Dirac matrices into equivalent ones. This formula shows that the 2^d Dirac particles play exactly the same role. At this level of free fields, it is possible to introduce in the action (145) a projector on only one of these states and therefore to eliminate the unwanted partners. This is not possible any more, using local operators, when the fields interact, since the unwanted particles can be produced by pairs and contribute to intermediate states.

Study the degeneracy problem in odd dimensions.

Several solutions have been proposed to eliminate this problem. However, every one presents some disadvantages. We will show in the next subsection that it is in fact impossible to define a fermionic action on the lattice with all desired properties. A first attempt at a cure is to partly delocalize the mass term on neighbouring sites in an amount proportional to a parameter r. The continuous limit $a \to 0$ is formally not changed by this operation. One obtains in this way the fermionic *Wilson action*

$$S_{\text{Wf}} = \tfrac{1}{2} \sum_{(ij)} \bar{\psi}_i (i\gamma_{ij} a^{d-2} + r a^{d-1}) \psi_j - \left(m + \frac{dr}{a} \right) \sum_i \bar{\psi}_i \psi_i \quad (150)$$

The corresponding free propagator reads

$$\Delta(\mathbf{p}) = \left(\frac{1}{a} \gamma_\mu \sin p_\mu a + m + \frac{r}{a} \sum_\mu (1 - \cos p_\mu a) \right)^{-1} \quad (151)$$

In the neighborhood of momentum $\bar{\mathbf{p}}$ (with k components equal to π/a, and $d - k$ vanishing components), the particle has, in the limit $a \to 0$, a mass equivalent to $m + 2kr/a$. In the continuous limit, only one state near $\mathbf{p} = 0$ remains with a finite mass and other particles decouple, their mass becoming infinite.

The disadvantage of this formulation is to break explicitly the chiral invariance of the massless limit

$$\psi \to e^{\lambda \hat{\gamma}_5} \psi \quad , \qquad \bar{\psi} \to \bar{\psi} e^{\lambda \hat{\gamma}_5} \quad (152)$$

which was a (noncompact) symmetry of the continuous action and of the naive lattice fermionic action as the mass m vanishes. In particle physics, this symmetry plays a very important role in the theory of weak interactions. Indeed, it then follows that the two states with opposite chirality ± 1, obtained from the action of the projectors $\tfrac{1}{2}(1 \pm \hat{\gamma}_5)$, are decoupled in a massless theory and remain so, even when turning on the interaction. It is thus possible to write an action for particles with a definite chirality, without introducing partners with the opposite one (Weyl fermions). The standard model of Glashow–Weinberg–Salam for electroweak interactions uses this property in a crucial way. It would therefore be desirable to construct a chiral invariant regularized lattice model, without the unwanted feature of particle doubling.

6.4.2   The Nielsen–Ninomiya theorem

A theorem due to Nielsen and Ninomiya states that this lattice construction is impossible under some general hypotheses. It was first proved in a Hamiltonian formalism. We give here a proof adapted to the Euclidean formulation. The ingredients of this no-go theorem are the following. Consider a *discrete* fermionic system on a *regular* lattice, with a *local, hermitian* and *translational invariant* action. Then, there are as many states with chirality $+1$ than states with chirality -1. We will make more precise the hypotheses as we proceed with the proof.

Assume to the contrary that it is possible to write an action for fermions with chirality $+1$. The most general quadratic term is of the form

$$\sum_{i,j,\mu} \bar{\psi}_i \gamma_\mu \tfrac{1}{2}(1 + \hat{\gamma}_5) i \mathcal{F}_\mu(\mathbf{x}_i - \mathbf{x}_j) \psi_j \tag{153}$$

Translational invariance requires the function $\mathcal{F}_\mu$ to depend only on the difference of arguments, and hermiticity that $\mathcal{F}_\mu^*(\mathbf{x}) = \mathcal{F}_\mu(-\mathbf{x})$ (assuming that the lattice has a symmetry $\mathbf{x} \to -\mathbf{x}$). We shall now show that this action automatically contains opposite chirality partners. The Fourier transform

$$\tilde{F}_\mu(\mathbf{k}) = \sum_i \mathcal{F}_\mu(\mathbf{x}_i) e^{i\mathbf{k}\cdot\mathbf{x}_i} \tag{154}$$

is analytic in $\mathbf{k}$, periodic with the periodicity of the reciprocal lattice, and real. Rather than analyticity, we only need that $\tilde{F}_\mu(\mathbf{k})$ be smooth, and this is the case as soon as $|\mathbf{x}|^d \mathcal{F}_\mu(\mathbf{x})$ vanishes sufficiently fast at large distance. Hence the theorem applies for a very large class of models using interactions beyond nearest neighbours.

The quadratic action (153) is diagonal in the momentum representation, and each physical state corresponds to a simple isolated zero of $\tilde{F}_\mu(\mathbf{k})$ (these are usually called "critical points" for the vector field $\tilde{F}_\mu(\mathbf{k})$). In the neighborhood of such a point, for a once differentiable $\tilde{F}_\mu$, we have

$$\tilde{F}_\mu(\mathbf{k}) = \mathcal{M}_{\mu\nu}(\mathbf{k} - \bar{\mathbf{k}})_\nu + \mathcal{O}((\mathbf{k} - \bar{\mathbf{k}})^2) \tag{155}$$

We assume a generic situation with $\partial/\partial k_\nu \tilde{F}_\mu(\mathbf{k})\big|_{\mathbf{k}=\bar{\mathbf{k}}}$ nonsingular. The real matrix $\mathcal{M}$ with elements $\mathcal{M}_{\mu\nu}$ can be decomposed as the product $\mathcal{M} = \mathcal{O}\mathcal{S}$ of an orthogonal matrix by a symmetric positive definite one. The orthogonal matrix $\mathcal{O}$ can always be absorbed into a redefinition of the fermionic fields. Indeed there always exists a matrix $\mathcal{R}$ such that

$$\gamma_\mu(1 + \hat{\gamma}_5)\mathcal{O}_{\mu\nu} = \mathcal{R}^{-1}\gamma_\nu(1 + \varepsilon\hat{\gamma}_5)\mathcal{R} \tag{156}$$

The sign $\varepsilon = \det \mathcal{O} = \pm 1$ (which is also the sign of $\det \mathcal{M}$) is called the *index* of the vector field $\tilde{F}_\mu(\mathbf{k})$ at the critical point $\bar{\mathbf{k}}$. To prove this relation, we observe that (even) d-dimensional space can be considered as the subspace of a $d + 1$-dimensional one. Extend in this larger space the orthogonal transformation $\mathcal{O}_{\mu\nu}$ to an element $\tilde{\mathcal{O}}_{\mu\nu} \in SO(d + 1)$ by acting on the last coordinate axis through multiplication by $\varepsilon = \det \mathcal{O}_{\mu\nu}$. Since $SO(d + 1)$ admits a spinor representation, there exists a matrix $\mathcal{R}$ in the Clifford algebra such that

$$\mathcal{R}^{-1} \begin{pmatrix} \gamma_1 \\ \vdots \\ \gamma_d \\ \hat{\gamma}_5 \end{pmatrix} \mathcal{R} = \begin{pmatrix} \gamma_\mu \mathcal{O}_{\mu\nu} \\ \varepsilon \hat{\gamma}_5 \end{pmatrix}$$

thus justifying (156).

Prove equation (156) in details for $d = 2$.

The positive definite matrix S generates a transformation in momentum space around $\bar{\mathbf{k}}$. It amounts to a positive rescaling of each component of $\mathbf{k} - \bar{\mathbf{k}}$ in a rotated frame. This does not affect the momentum space orientation. As a consequence, the critical point (155) can be interpreted as giving rise to a physical state of chirality ε.

The real vector field $\tilde{F}_\mu(\mathbf{k})$ is defined on a compact manifold, here the torus T^d. Indeed, as the lattice is regular, the function $\tilde{F}_\mu(\mathbf{k})$ is periodic. A theorem due to Poincaré and Hopf states that the sum of indices ε of a real vector field defined on a compact manifold is equal to the Euler characteristic of this manifold, here 0 for the torus T^d. This theorem is a generalization of the following simple one-dimensional property: a continuous periodic function in one variable vanishes as many times with a positive derivative as with a negative one. This example suggests how one should deal with degenerate cases, when some critical points coincide. The conclusion is that there are as many physical states with chirality $+1$ as with chirality -1.

As a consequence of the Nielsen–Ninomiya theorem, it is necessary to violate one of the hypotheses if one wants to construct a chiral invariant lattice model. A possible way out might be the use of a random lattice (chapter 11, volume 2). Another one is the choice

$$\tilde{F}_\mu(\mathbf{k}) = k_\mu$$

for $-\pi/a < k_\mu < \pi/a$. This function is discontinuous at the limit of the first Brillouin zone $\left|k_\mu\right| < \pi/a$, and the corresponding action is not local. In particular, any pair of sites interacts with the same strength, whatever the distance is. This introduces serious difficulties in further developments, such as the introduction of gauge interactions.

The above discussion pertains only to free fermions. It is however presumably related to the anomalies found in the continuum theories of chiral fermions interacting with gauge fields. The latter have to be compensated for in realistic models of electroweak interactions.

6.4.3   Staggered fermions

An alternative technique has also been introduced to reduce the number of species in the naive fermionic action and to keep some discrete aspects of chiral symmetry (Kogut, Susskind). Unfortunately, it still does not allow a discretization of Weyl fermions. The idea is to use Grassmannian fields with only one component (i.e. without a spinor index) and to reconstruct the physical multicomponent spinor using the various poles of its propagator. In other words, the different components of a fermion are distributed on the sites of 2^d "superlattices" with a doubled spacing and interpreted as a unique one component Grassmannian field. A version of this model reads

$$S_{\mathrm{KS}} = -\tfrac{1}{2}a^{d-1}\sum_{i,\mu}(-)^{x_{i,1}+\cdots+x_{i,\mu-1}}(\bar{\chi}_i\chi_{i+\hat{\mu}}+\bar{\chi}_{i+\hat{\mu}}\chi_i)+ima^d\sum_i\bar{\chi}_i\chi_i$$

(157)

where the Grassmannian fields $\bar{\chi}$, χ have only one component. The technique of reconstruction of the physical spinor $\bar{\psi}$, ψ with $2^{\lfloor d/2\rfloor}$ components is suggested by the following exercise.

Diagonalize the kinetic part of the quadratic naive fermionic action (145) with respect to the spinorial indices and show then that the states can be decoupled. For that purpose, one defines unitary matrices A_i depending on the site i such that the change of fields

$$\psi_i = A_i \chi_i, \qquad\qquad \bar{\psi}_i = \bar{\chi}_i A_i^\dagger \qquad\qquad (158)$$

diagonalizes simultaneously the Dirac matrices

$$A_i^\dagger \gamma_\mu A_{i+\mu} = \Delta_{i\mu} \qquad\qquad (159)$$

It can be checked that

$$\begin{cases} A_i = \gamma_1^{x_{i,1}} \cdots \gamma_d^{x_{i,d}} \\ \Delta_{i,\mu} = -(-1)^{x_{i,1}+\cdots+x_{i,\mu-1}} \end{cases} \qquad\qquad (160)$$

is a possible solution. Using the transformation (158), the action takes the form (157), except that $\bar{\chi}$, χ have as many components as $\bar{\psi}$, ψ; however, they are decoupled, and will remain decoupled after the introduction of gauge fields. It is thus possible to keep only one of these components, and this leads to the action (157).

Physical fermions are reconstructed as follows. Each lattice site x_i is characterized by its position y in a lattice with doubled spacing and its location η in the corresponding cluster of 2^d original sites, according to

$$x_{i,\mu} = 2y_\mu + \eta_\mu, \qquad\qquad \text{with } \eta_\mu = 0 \text{ or } 1.$$

The field χ is considered as a field with 2^d components (indexed by η) on the doubled lattice

$$\begin{cases} \chi_i \equiv (-1)^{y_1+\cdots+y_d} \chi_\eta(y) \\ \bar{\chi}_i \equiv (-1)^{y_1+\cdots+y_d} \bar{\chi}_\eta(y) \end{cases} \qquad\qquad (161)$$

After some straightforward algebra, the action (157) takes the form

$$S = a^d \sum_{y,\mu,\eta,\eta'} \bar{\chi}_\eta(y) \Gamma^\mu_{\eta,\eta'} \Delta_\mu \chi_{\eta'}(y) + a\bar{\chi}_\eta(y) \tilde{\Gamma}^\mu_{\eta,\eta'} \delta_\mu \chi_{\eta'}(y) \qquad (162)$$

In this formula there appear two finite difference symmetric operators Δ_μ and δ_μ which tend respectively towards the first and second derivatives in the continuous limit of a vanishing lattice spacing a

$$\begin{cases} \Delta_\mu f(y) \equiv \dfrac{1}{4a}[f(y+\hat{\mu}) - f(y-\hat{\mu})] \xrightarrow[a\to0]{} \partial_\mu f(y) \\ \delta_\mu f(y) \equiv \dfrac{1}{4a^2}[f(y+\hat{\mu}) - 2f(y) + f(y-\hat{\mu})] \xrightarrow[a\to0]{} \partial_\mu^2 f(y) \end{cases} \qquad (163)$$

The matrices Γ and $\tilde{\Gamma}$ introduced above are defined through

$$
\begin{cases}
\Gamma^\mu_{\eta,\eta'} \equiv 2^{-\lfloor\frac{1}{2}d\rfloor} \operatorname{Tr}(\Gamma^\dagger_\eta \gamma_\mu \Gamma_{\eta'}) \\
\tilde{\Gamma}^\mu_{\eta,\eta'} \equiv 2^{-\lfloor\frac{1}{2}d\rfloor} \operatorname{Tr}(\Gamma^\dagger_\eta \gamma_\mu \Gamma_{\eta'})(\delta_{\eta',\eta-\hat\mu} - \delta_{\eta',\eta+\hat\mu}) \\
\Gamma_\eta \equiv \gamma_1^{\eta_1}\cdots\gamma_d^{\eta_d}
\end{cases}
\tag{164}
$$

starting from any given representation of the Dirac matrices. Finally, formula (162) can be rewritten in terms of physical fermions with a usual spinorial index α and an additional flavour index λ, both taking $2^{\lfloor\frac{1}{2}d\rfloor}$ different values

$$
\begin{cases}
\psi_y^{\alpha\lambda} \equiv 2^{-3\lfloor\frac{1}{2}d\rfloor/2} \sum_\eta (\Gamma_\eta)_{\alpha\lambda}\chi_\eta(y) \\
\bar\psi_y^{\alpha\lambda} \equiv 2^{-3\lfloor\frac{1}{2}d\rfloor/2} \sum_\eta (\Gamma_\eta)^*_{\alpha\lambda}\bar\chi_\eta(y)
\end{cases}
\tag{165}
$$

The final form of the action is

$$
\begin{aligned}
S_{\mathrm{KS}} = (2a)^d \sum_{y,\mu} &\left[\bar\psi_y(\gamma_\mu\otimes 1)\Delta_\mu\psi_y + a\bar\psi_y(\hat\gamma_5^\dagger\otimes\gamma_\mu^\dagger\hat\gamma_5^\dagger)\delta_\mu\psi_y\right] \\
&+ (2a)^d im \sum_y \bar\psi_y\psi_y
\end{aligned}
\tag{166}
$$

Its interpretation is in terms of $2^{\lfloor\frac{1}{2}d\rfloor}$ different species of physical fermions. These different flavours are coupled on the lattice by the second term of the first summation, even if this term is irrelevant in the continuous limit $a\to 0$.

In spite of all their shortcomings, these various fermionic discretizations have been used efficiently in numerically probing quantum chromodynamics in the confined regime, a problem that is still under active investigation.

Notes

The original idea of lattice gauge theory is due to K.G. Wilson, *Phys. Rev.* **D10** 2445 (1974). The corresponding action in the abelian case had been proposed by F.J. Wegner, *J. Math. Phys.* **12** 2259 (1971), as an example of a model with duality properties and no local order parameter. Some of the original papers, including the work of R. Balian and the authors, are reprinted in

C. Rebbi's book, *Lattice Gauge Theories and Monte-Carlo Simulations* (World Scientific, Singapore 1983). Numerous reviews are also available, such as those by J.B. Kogut, *Rev. Mod. Phys.* **51** 659 (1979), J.-M. Drouffe and J.-B. Zuber, *Phys. Reports* **102** 1 (1983), on theoretical aspects, M. Creutz, L. Jacobs and C. Rebbi, *Phys. Reports* **95** 201 (1983), on numerical simulations. They include a large bibliography. The book by M. Creutz, *Quarks, Gluons and Lattices* (Cambridge University Press, Cambridge 1983) offers a nice introduction to the subject. A number of rigorous properties are studied in E. Seiler's monograph *Gauge Theories as a Problem of Constructive Quantum Field Theory and Statistical Mechanics*, Lecture Notes in Physics **159**, Springer Verlag, Berlin (1982).

We have not discussed in the text the Hamiltonian formalism due to J. Kogut and L. Susskind, *Phys. Rev.* **D11** 395 (1975). Among variant actions, we quote the program for an improved lattice action initiated by K. Symanzik, in *Mathematical Problems in Theoretical Physics*, R. Schrader et al eds, Lecture Notes in Physics, **153**, Springer Verlag, Berlin (1982).

The quantization of Yang–Mills fields in the continuum is presented in the books of J.C. Taylor *Gauge Theories of Weak Interactions*, Cambridge University Press, (1976), and L.D. Faddeev and A.A Slavnov, *Gauge Fields — Introduction to Quantum Theory*, Benjamin Cummings (1980). Asymptotic freedom was discovered by D.J. Gross and F. Wilczek, *Phys. Rev. Lett.* **30** 1343 (1973), and H.D. Politzer, *Phys. Rev. Lett.* **30** 1346 (1973). For the determination of the Λ_L parameter, see A. Hasenfratz and P. Hasenfratz, *Phys. Lett.* **93B** 165 (1980). R. Dashen and D.J. Gross, *Phys. Rev.* **D23** 2340 (1981) have proposed to use the background gauge. See also H. Kawai, R. Nakayama and K. Seo, *Nucl. Phys.* **B189** 40 (1981); P. Weisz, *Phys. Lett.* **100B** 331 (1981); A. Hasenfratz and P. Hasenfratz, *Nucl. Phys.* **B193** 210 (1981); A. Gonzalez-Arroyo and C.P. Korthals Altes, *Nucl. Phys.* **B205 [FS5]** 46 (1982). A review on duality can be found in R. Savit, *Rev. Mod. Phys.* **52** 453 (1980). The work of S. Elitzur on the absence of order parameter is in *Phys. Rev.* **D12** 3978 (1975). Dual non local observables were introduced by G. 't Hooft, *Nucl. Phys.* **B138**, 1 (1978).

The steepest descent method is presented in the book of R.B. Dingle, *Asymptotic Expansions — Their Derivation and*

Interpretation, Academic Press, New York (1973). Systematic corrections to mean field are found in E. Brézin and J.-M. Drouffe, *Nucl. Phys.* **B200 [FS4]** 93 (1982). For corrections to mean field, see H. Flyvbjerg, B. Lautrup and J.-B. Zuber, *Phys. Lett.* **110B** 279 (1982).

The topological classification of surfaces can be found in P.S. Aleksandrov, *Combinatorial topology*, Graylock 1956.

Numerous strong coupling series are available in the literature. The series given for the free energy is taken from the work of one of the authors *Nucl. Phys.* **B170 [FS1]** 91 (1980). For the string tension, see G. Münster, *Nucl. Phys.* **B180 [FS2]** 23 (1981); G. Münster and P. Weisz, *Nucl. Phys.* **B180 [FS2]** 13 (1981); J.-M. Drouffe and J.-B. Zuber, *Nucl. Phys.* **B180 [FS2]** 253, 264 (1981). Series for the glueball mass are found in N. Kimura and A. Ukawa, *Nucl. Phys.* **B205 [FS5]** 637 (1982); K. Seo, *Nucl. Phys.* **B209** 200 (1982); G. Münster, *Phys. Lett.* **121B** 53 (1983). The resummation of strong coupling series in large dimension, including a discussion of a lattice model for random surfaces, is studied in J.-M. Drouffe, G. Parisi and N. Sourlas, *Nucl. Phys.* **B161** 397 (1979).

Staggered fermions are introduced in T. Banks, J. Kogut and L. Susskind, *Phys. Rev.* **D13** 1043 (1976), and their interpretation as flavoured fermions follows H. Kluberg-Stern, A. Morel, O. Napoly and B. Petersson, *Nucl. Phys.* **B220 [FS8]** 447 (1983). The original proof of the no-go theorem is due to H.B. Nielsen and M. Ninomiya, *Nucl. Phys.* **B185** 20 (1981). We have patterned our presentation on the work of L.H. Karsten, *Phys. Lett.* **104B** 315 (1981). The Poincaré-Hopf theorem is standard in differential geometry; see e.g. J.W. Milnor, *Topology from the Differentiable Viewpoint* (The University Press of Virginia, Charlottesville, 1965). For a review on anomalies in the context of continuous field theory, see S.B. Treiman, R. Jackiw, B. Zumino and E. Witten, *Current Algebra and Anomalies*, World Scientific, Singapore (1985).

INDEX

E

Effective potential, 156, 240, 249
Eguchi, Ooguri, 578
Elitzur's theorem, 341
Elliptic functions, 609
Energy–momentum tensor, 507
Enveloping algebra, 583
ε-expansion, 235, 290, 311
Equation of state, 166
Equations of motion, 145
Euclidean fields, 21
Euler characteristic, 740, 765, 793
Euler's formula, 705
Euler's pentagonal identity, 549, 610
Euler's relation, 409

F

Faddeev–Popov determinant, 785
Faddeev–Popov ghost fields, 337, 436
Feigin–Fuchs, 533
Ferdinand, Fisher, 574
Fermi levels, 714
Feynman integrals, 246
Feynman rules, 154, 244, 337
Finite size effects, 478
Fisher circles, 139
Fisher, 233, 235, 290, 480
Fisher–Gaunt, 157
Fock–Bargmann space, 50, 676
Fokker–Planck equation, 487, 649
Free fields, 22
Freudenthal formula, 626
Friedan, 780, 798
Friedan, Qiu, Shenker, 543, 592, 618, 638
Frustrated partition function, 72, 86
Frustration, 352, 588
Fujikawa, 776

G

Gaudin, 699
Gauge fields, 329
Gauge fixing, 435
Gauge invariance, 329
Gauss series, 561
Gauss sum, 612
Gauss–Bonnet formula, 766
Gaussian discrete model, 205
Gaussian integrals, 22
Gaussian model, 33, 550, 592
Gell-Mann–Low formula, 522
Generating functionals, 236
Genus, 705
Gepner, Qiu, 582
Gepner, Witten, 581
Glueball mass spectrum, 390
Goddard, Kent, Olive, 636, 642
Goldstone modes, 107, 118, 120, 127
Goldstone's theorem, 298, 300, 435, 495
Grading, 524
Graph, 405
Grassmannian integrals, 48
Green function, 8
Green functions, 238
Gross–Neveu model, 490, 720
Group characters, 331
Group theoretical factors, 259

H

Haar measure, 427
Hadron spectrum, 494
Haffnian, 95
Hall conductance, 729
Halperin, 647
Hard hexagon model, 558
Hardy–Ramanujan formula, 548
Harris criterion, 716, 720
Hausdorff dimension, 6, 802, 808
Heat bath algorithm, 460

M

Ma, 456
Majorana field, 93, 554, 721
Majority rule, 183
Marginal operators, 174, 218, 307
Markov process, 5, 458
McBryan–Spencer, 222
McCoy–Wu, 78
Mean field approximation, 108, 352
Meijering, 743
Mermin–Wagner theorem, 143, 193, 197, 219
Metropolis algorithm, 462
Microcanonical simulations, 465
Migdal–Kadanoff approximation, 178
Minimal coupling, 330
Minimal models, 546
Minimal subtraction scheme, 337
Modular group, 567
Modular invariance, 564
moduli, 786
Monodromy, 562
Monte Carlo renormalization group, 456, 481
Monte Carlo sweep, 460
Multicritical points, 317
Möbius transformation, 510

N

n-vector model, 25, 118, 259, 427, 436, 595
Nambu–Goto action, 781
Neveu–Schwarz and Ramond boundary conditions, 577, 616
Nickel, 311
Nielsen–Ninomiya theorem, 396
Niemeijer–Van Leeuwen, 183
Noether current, 508
Noether's theorem, 97

nonlinear σ-model, 26, 107, 436, 769
Number of loops, 247

O

One particle irreducible Green functions, 239
One-loop correction, 155
Onsager, 48, 58, 68, 75
Orbifold, 554, 594
Order–disorder variables, 87
Oscillator wavefunctions, 698

P

Padé approximants, 448
Painlevé equations, 98
Parisi, Sourlas, 686, 715
Parisi, Wu Yong Shi, 486
Pasquier, 579, 588, 595
Path integral, 12
Peierls' argument, 432
Perimeter law, 345
Perturbation theory, 242
Perturbative renormalization, 264
Peter–Weyl theorem, 427
Pfaffian, 57
Phenomenological renormalization, 224
Pippard–Ginsberg, 165
Planar approximation, 703
Planar limit, 805
Plaquette, 60, 331
Poisson distribution, 665
Poisson formula, 204, 580
Polyakov loops, 469
Polyakov, 738, 779, 780, 798
Potts model, 176, 557, 595
Power counting, 261
Primary divergent diagrams, 262
Primary field, 512
Pseudofermion method, 494

Printed in the United States
By Bookmasters